The Money Miners

By the same author:

Vintage Pierpont

Two Centuries of Panic

Operation Dynasty

The Bold Riders

The Money Miners

The Great Australian Mining Boom

Trevor Sykes

Allen & Unwin

TO DAYA, whose slumber was made hideous by the clatter of a typewriter.

First published in 1978 by
Wildcat Press
54 Park Street, Sydney, NSW 2000

Reissued in 1995 by
Allen & Unwin
83 Alexander Street
Crows Nest NSW 2065
Australia
Phone: (61 2) 8425 0100
Fax: (61 2) 9906 2218
E-mail: info@allenandunwin.com
Web: www.allenandunwin.com

National Library of Australia
Cataloguing-in-Publication entry:

Sykes, Trevor.
The money miners.

Includes index.
ISBN 1 86373 844 4

1. Mineral industries—Australia—History.
2. Speculation—Australia. I. Title.

332.67220994

Printed and bound in Australia by The SOS Print + Media Group.

10 9 8 7 6 5 4 3 2 1

Contents

I know more about finance and mining now than I did when it was written, but it captures the pace and excitement of those raging days. A more knowledgeable book would probably have been duller to read.

The Money Miners was originally published by Australian Consolidated Press under the imprint of Wildcat Press—a great name for the book publishing adjunct of *The Bulletin.* Indeed the then editor of *The Bulletin,* Trevor Kennedy, formed Wildcat Press to publish this as its first book. Alas, there was never enough money in publishing, so Wildcat Press ceased and *The Money Miners* went out of print. Every month or so I receive a request for the book, so this edition by Allen & Unwin will fill some public demand, however tiny. Meanwhile, I would like to thank Trevor Kennedy once again for launching me as an author. I have no doubt that if he had stuck to book publishing he would still be as poor as I am, so I commend his good sense in finding another career.

In February 1990, I attended a historic gathering at Hannan's Club in Kalgoorlie. Ron Manners of Croesus Mining had organised a dinner to celebrate the 20th anniversary of Poseidon hitting $280. Those present included Norm Shierlaw, Ken Shirley, Keith Biggs, Geoffrey Hynam and Basil Lewis. We took a bush lawyer's view that after 20 years the statute of limitations must have expired so everyone could tell the truth. A lot of truth was told, the main problem being to remember it next day (or even that night). The story of the discovery of Windarra told here is substantially true. Keith Biggs, Poseidon's very alert neighbour, bought heavily into the stock. Shierlaw, knowing Poseidon was onto nickel and determined to defend his company, bought strongly in defence and the two of them must have accounted for most of the rise from $1 to $5. After that momentum took over. One point that still bemused Shierlaw was how the drill supervised by geologist Geoff Burrill stopped overnight just a few feet from the ore. He was also bemused by the heavy buying in Poseidon the next day as soon as the ore was struck. I'd brought along one of my few remaining copies of *The Money Miners,* which was signed by all the principals present then auctioned on behalf of the Royal Flying Doctor Service. The top bid was $2000 by Keith Biggs. So whatever you pay for this copy, you're getting a bargain.

Trevor Sykes
February 1995

Introduction

Australians like to gamble. In the summer of 1969-70 they were gambling in millions of dollars in the wildest stock market boom the nation has ever seen. But many of the gamblers knew little about either shares or mining and they would have been no wiser if they had relied on the statements of some of the companies involved.

Some of Australia's greatest mines rose to fame in this era. Other companies were money miners, aiming for the speculators' pockets.

This book is a history of that boom, written in the hope that if the Australian public understands what happened in 1969-70 they will be better prepared for the next boom.

Because there will certainly be another.

CHAPTER ONE

Origins

Mining-share booms are no strangers to Australia. The Poseidon boom of 1969-70 was the biggest yet, but mining booms have played an important role in Australia's history.

In the early 19th century gold rushes, the share trading was largely confined to the actual fields, as miners traded shares in syndicates. Thus a royal commission which visited Ballarat in 1854 noted the remarkable rise of syndicates and companies among the diggers and said that "share dealing has become a multifarious business". Later booms were to see organised stock exchanges.

As the rich alluvial deposits, which could be worked by a few men, petered out the miners' efforts shifted to the quartz reefs. Extraction of the gold required machinery, which meant more capital, so larger companies began to be formed. They issued shares, and whenever mining prospered, there were bouts of speculation on the exchanges. In the 1870s gold fever was running so high in Victoria, following the discoveries at Bendigo, that a Dr I.L. Smith advertised in the newspapers warning that "Over-speculation, with its attendant anxieties, will produce frightful nervous effects ... THE REACTION IS SURE TO TELL".

There were booms in mining shares in every decade of the last half of the 19th century. The biggest, and the one most like the Poseidon boom, was the speculation in the 1890s, sparked by the gold discoveries at Coolgardie and Kalgoorlie.

(Professor Geoffrey Blainey had delineated the colorful history of Australian mining in *The Rush That Never Ended.* An excellent popular history of Coolgardie and Kalgoorlie is *The Mile That Midas Touched* by Ted Mayman and Gavin Casey.)

The excitement of these booms waned in the 20th century as mining became a wages industry. For much of the first half of the 20th century depressed or controlled metal prices took the incentive out of prospecting. The gold towns prospered again in the 1930s, but they were the only bright spot on the black face of the Depression. Mt Isa Mines, floated in Sydney in 1924, took 23 years to pay its first dividend.

The second half of the 20th century saw a rekindling of the old speculative fires. As the 1950s dawned, prospectors were active again, and before long they were making discoveries.

The first impetus came in wolfram and beach sands. Australia had always been wolfram country and the shortages of World War II began a fresh search for this ore. Wolfram and scheelite are the ores of tungsten, a metal used in batteries, steelmaking and radio valves. Two Commonwealth officials, inspecting a small, old mine on King Island in Bass Strait, discovered a huge orebody there and a company called King Island Scheelite was formed. The beach-sands industry had been operating since 1934, producing ilmenite and rutile. From these minerals was produced titanium, a valuable element in making paints and lightweight steels.

Rising rutile prices spurred fresh finds in Queensland and New South Wales, and the mineral-sands industry prospered.

One by-product of World War II was immediate interest in the awesome new element, uranium. In 1949 a veteran Northern Territory prospector named Jack White discovered uranium at Rum Jungle, only 50 miles south of Darwin. The deposit was mined for the Commonwealth Government by Consolidated Zinc, an old Broken Hill mining company now under British control. The £25,000 bounty paid to White by the Commonwealth stimulated further searching. In 1954 a syndicate headed by Mt Isa taxi driver Clem Walton struck more uranium halfway between Isa and Cloncurry. The deposit was named Mary Kathleen after a syndicate member's wife who had died only a few weeks before.

By that time the nation was already in the grip of an oil boom. Ampol Exploration Ltd, drilling at Rough Range in W.A., struck oil in December 1953. On December 4, Ampol's chairman Bill (later Sir William) Walkley stopped traffic in Pitt Street as he walked along it in a blazing red ten-gallon hat. Some pressmen had given the hat to Walkley during an earlier visit to Carnarvon and he had promised to wear it down Pitt Street if Ampol struck oil. Ampol Ex shares roared up from 18/- to £6/5/- in a single day, sparking speculation in other oil stocks such as Woodside and Santos. Rough Range was too small to be commercial, but Ampol was a partner in the consortium which later found oil offshore at Barrow Island. When Mary Kathleen was found eight months after Rough Range, Australia had a lively share boom.

An important factor of the 1953 and 1954 share booms was that they ushered in "new" minerals. Neither oil nor uranium had been the subject of previous rushes. Australians knew they had gold, copper, tin, wolfram and base metals, but these new minerals added an exotic touch to the mining scene. Nothing in the financial world ignites public imagination more fiercely than the sudden discovery of a precious new commodity. The cumulative effect of new discoveries multiplies the excitement. The discovery of one, and then another, new mineral that was previously unknown on the continent fanned speculation that there might be more rich, unknown deposits. This is the indispensable belief upon which booms are founded. As the 1960s unfolded there was little to dispel it.

The 1950s and 1960s saw an unbroken string of great mineral discoveries in Australia, several of them of world importance. When the Bell Bay aluminium smelter opened outside Launceston in 1955, Australia was importing bauxite from Malaysia. Within a few years this scene changed dramatically. The bauxite deposits of the Darling Ranges in Western Australia were developed from 1957 by Western Mining Corporation Ltd and other members of the Melbourne-based Collins House group. In 1961 They invited Alcoa of the US in as a majority partner and started Australia's first major bauxite mine. It was overshadowed by the parallel development of the gigantic Weipa deposit on Cape York by Comalco Ltd.

The red cliffs of Weipa had first been noted by Matthew Flinders in 1801-2. Harry Evans, the Consolidated Zinc geologist who first explored the deposit in 1955, found it so huge that it was hard to believe it was all bauxite. Its 2200 million tonnes contained one quarter of the world's known reserves at the time of the discovery.

The new discoveries were shattering old beliefs that Australia was deficient in some minerals. In 1938, Dr W.G. Woolnough, geological adviser to the Commonwealth Government, reported gloomily: "It is certain that if the known supplies of high grade ore are not conserved Australia will in little more than a generation become an importer rather than a producer of iron ore." The Commonwealth consequently embargoed the export of iron ore, which meant there was no incentive to search for it. One man who knew where iron was to be found was the Pilbara pastoralist Lang Hancock. In November 1952, Hancock and his wife had been flying across the Hamersley Ranges in a frail single-engined Auster when they were forced down low by a violent thunderstorm. Flying through a deep gorge, Hancock noticed the rain streaming down sheer, 60 metre walls that appeared to be iron. In 1959 he met representatives of Rio Tinto Zinc Corporation of the UK and after the Commonwealth lifted its iron ore export embargo in 1960, Hancock flew them to Hamersley and showed them the ore. It was low-grade limonite but the geologists were impressed by the bands of iron across the ranges, totalling several hundred kms in length. Two geologists from Conzinc Riotinto of Australia Ltd, a subsidiary of RTZ, searched the banded iron formations and discovered Mt Tom Price, containing 600 million tonnes of high-grade ore. The discovery of the Hamersley orebody led to rapid discovery and development of other deposits at Shay Gap, Mt Newman and eventually Robe River. Australia became a major force in the world iron-ore market.

Another long-held belief was that there was no oil south of the Equator, coupled with a belief that Australia's rocks were too old to contain oil. These beliefs were also proved wrong in the 1960s as the Barrow Island and Moonie fields opened. Then in 1966 Australia's biggest company, Broken Hill Proprietary Co Ltd, and the world's biggest oil company, Esso, struck oil in the Bass Strait. The size of this field was small by world standards, but Bass Strait dwarfed the previous oilfields found in Australia.

Meanwhile, trade had been flourishing in established minerals. The Australian mineral-sands industry was by 1969 producing about 95 percent of the world's rutile, 60 percent of its zircon and about 25 percent of its ilmenite and monazite. The British-controlled Associated Minerals Corporation Ltd was the largest beach sands miner in the world.

Coal was also prospering. Although this long-neglected mineral gained most of its glamor after the OPEC price increases of 1973, it was improving its position far earlier. Between 1955 and 1965, coal exports from Australia to Japan jumped from 200,000 to seven million tonnes, amounting to 45 percent of Japan's consumption. The big Blackwater mine outside Gladstone began developing in 1964 and in 1968 shipped to Japan its first coal of a 10-year contract for 31 million tonnes.

The mid-1960s saw gyrating copper prices, resulting in some

spectacularly good years for large established miners such as Peko-Wallsend, Mt Isa and Mt Lyell. Broken Hill South, another member of the Collins House group,[1] was encouraged to reopen the old Cobar copper mine in central New South Wales.

From Groote Eylandt, in the Gulf of Carpentaria, Commonwealth geologists reported promising outcrops of manganese, another mineral new to Australia. BHP proved up a huge manganese deposit on the island in 1962. Yet another new mineral was found in 1966 when Broken Hill South uncovered the large phosphate deposit at Duchess in western Queensland.

There were inevitably some disappointments during these years, but they were small alongside the visions of wealth being opened up by the vast new mineral finds. After witnessing an almost unbroken run of successes, Australians became convinced their country was uniquely rich in minerals. As one mine after another was opened — often of global significance — there was no reason to believe the run of luck would stop.

These discoveries and successes excited the share market, but did not always provide a convenient vehicle for speculation. Comalco and Hamersley did not list until well after their discoveries. Shares in successful well-established companies such as Peko and Mt Isa rose strongly in the late 1960s, but while they were glamorous they were not speculative counters. There would have been more excitement if an important find were made by a weak company with a low share price. This happened on January 28, 1966, when Western Mining Corporation struck nickel on the shores of Lake Lefroy in Western Australia. The nickel boom had begun.

---

[1] The public companies in the Collins House group are Western Mining, Broken Hill South, North Broken Hill, EZ Industries, APPM and Beach Petroleum.

CHAPTER TWO

The Devil's Metal

The nickel mine that was to be called Kambalda had an odd history. The orebody had been waiting like a wallflower, occasionally noticed but long ignored.

The region, 35 miles south of Kalgoorlie, is dominated geographically by the shimmering white saltbed of Lake Lefroy. The lake, which covers 123 square miles and is normally dry, is bordered by the red dirt and dusty salmon gums typical of the area.

Following the discovery of alluvial gold at Kalgoorlie in 1892, prospectors quickly fanned out across the scrub-covered countryside in search of further promising outcrops. Among countless other discoveries, they found the Butterfly leases on a low ridge called Red Hill on the shores of the lake. In 1897 the leases were being worked by the Red Hill (WA) Gold Syndicate Ltd and a small settlement had sprung up near the site. Government surveyors laid out a small townsite at the base of Red Hill and the Surveyor General, William Rowley, named it Kambalda.

Like many other lodes, the gold petered out at depth and by 1907 had run out. The township was deserted and the mine forgotten. Also forgotten was the fact that nickel traces had been detected in the Red Hill ore.

On the other side of the continent in March 1933, Western Mining was incorporated in Victoria. Its objectives were to search for gold in WA. In Australia's grim depression years, gold mining was almost the only expanding industry. The flight from paper money had increased the value of gold and many gold companies were floated in these otherwise hard times.

Starting with the thoroughly professional outlook which it has always maintained, WMC began with a team of geological consultants from Harvard University, who mapped and analysed structures under option in the Eastern Goldfields. (The Eastern Goldfields are around Kalgoorlie. To West Australians Kalgoorlie is east.)

In the years to World War II the company brought a number of gold mines into production, including those of Gold Mines of Kalgoorlie Ltd in 1934 and Central Norseman Gold Corporation N.L. in 1935. During the war some mines closed while others continued at lower rates of operation. After the war the mines reopened and operations at GMK were expanded by amalgamation with several other companies on Kalgoorlie's Golden Mile.

By the mid-1950s the industry was back in the doldrums and suffering from the pegged international price of gold at a time of rising costs. This prompted the company to diversify its exploration effort into other metals. The need for the diversification can be seen from the accompanying table which shows Western Mining's profit, dividend and share prices during the latter half of the 1950s. Net profits were around the £400,000 mark and showing no growth. On a constant dividend of 1/- per share, the share price moved sluggishly between 10/- and 15/-. It is a profile of a company going nowhere.

WESTERN MINING CORPORATION LTD

REVENUE, PROFIT & SHARE PRICE

Years to end March	Group Revenue	Net Profit
1956	£755,484	£387,092
1957	£769,973	£387,237
1958	£785,639	£403,015
1959	£892,909	£461,605
1960	£726,533	£352,597

SEVEN-YEAR SHARE PRICE RANGE

Calendar Year	1954	1955	1956	1957	1958	1959	1960
High	20/6	14/-	14/6	15/-	13/6	14/6	14/3
Low	13/6	10/10	11/6	11/-	9/3	12/6	10/3

Western Mining's survival was due to intelligent policy, not luck. The company's management realised it had to diversify and searched for opportunities. They were also willing to investigate offers. They acquired a reputation for being interested in odd minerals and it became a standard saying in Kalgoorlie whenever a prospector found a strange rock to tell him: "Take it to Western Mining". In this way, Western Mining had as potential prospectors almost the entire population of Kalgoorlie, because on the goldfields prospecting is a way of life. At weekends, or whenever he has a day off, a Kalgoorlie resident will wander off into the bush with a water-bag and a prospector's hammer.

As a result of this diversification program the company proved the existence of substantial bauxite deposits in the Darling Ranges near Perth, leading to the formation of Alcoa of Australia Ltd, in which WMC now holds 20 percent equity. The group was also an early iron-ore explorer and successfully developed the Koolanooka mine near Geraldton, although this project was soon eclipsed by the huge iron ore mines of the Pilbara.

During the early 1960s the company was concentrating on exploration of the Kalgoorlie region for gold and other metals. Extensive diamond drilling on the outskirts of Kalgoorlie proved the existence of a large low-grade gold deposit at Mt Charlotte. This was to be the last of the profitable large-scale gold mines in Kalgoorlie's long history.

In 1962 Western Mining was extending its detailed geological mapping south of Kalgoorlie and by 1964 it had almost reached Kambalda. This jogged the memories of two prospectors, John Morgan and George Cowcill. They had been prospecting the area intermittently for 20 years and in 1947 had noticed a long thin outcrop of dark gossan near Red Hill. (A gossan is an iron-oxide deposit resulting from the oxidisation and leaching of a sulphide mineral body containing such minerals as pyrite. A gossan can be thought of as an ironstone cap above a sulphide occurrence.)

The gossan contained no gold, but they revisited it in 1954 in the hope that it might contain uranium. The samples they had assayed contained no uranium either, but the Kalgoorlie School of Mines reported anomalous nickel values. The assayers' report said: "In view of the meagre Australian resources of nickel and the relatively high prices, it might be worthwhile to explore the extent of the occurrence and to submit a sample for nickel assay".

Morgan and Cowcill did not bother to follow the recommendation and their samples lay forgotten for another 10 years, when Morgan took some of the old samples to Roy Woodall, a Western Mining geologist. The samples assayed 0.7

percent nickel, sufficiently high for Western Mining to begin serious exploration in the area.

Following intensive surface exploration, the company decided in 1965 to drill several holes to investigate the nature of the sulphides which had caused the limonitic outcrops. On January 28, 1966, the first hole drilled on the shores of Lake Lefroy intersected almost three metres of massive sulphides containing 8.3 percent nickel at a depth of nearly 500 ft. The discovery was deep but unexpectedly rich.

Western Mining worked quickly. Subsequent drilling indicated the presence of a significant deposit of massive nickel sulphides and on April 4, 1966, the company announced the discovery and its development plans for Australia's first nickel mine. The mine was brought into production rapidly under a program directed by the present chairman, Sir Arvi Parbo, then the company's deputy general manager in WA. Western Mining were old mining hands and realised the importance of bringing in the mine while prices were high. When the price of a metal is high and demand is outstripping supply, a "new boy" has his best chance of entering the market as a producer. Mine development costs can be written off more quickly against the strong cash flow and it is easier to gain access to customers. It was imperative that WMC develop its mine with the utmost speed to catch the market. "We made some mistakes," Sir Arvi recalled[1] "but the mistakes could be corrected later and the important thing was to bring the mine in to catch the market".

One mistake was the new township of Kambalda. In its haste, WMC forgot one of the basic rules of mine development: never build the town until you've drilled the ground for ore. The township turned out to be almost on top of an orebody later called the Fisher Shoot. Construction of houses stopped and the rest of the town was built on a previously explored site a mile away. Newcomers are still puzzled by the spectacle, in barren scrubland miles from anywhere, of a small town divided into two widely separated suburbs.

No modern Australian mine was ever brought into production more quickly than Kambalda. Shaft-sinking began on the Lunnon Shoot in July 1966. Concentrator-construction began in October and in March 1967 the first ore was produced from the mines. By August 1967, the first consignment of concentrate was shipped overseas and in the following month the official opening ceremony was held at Kambalda.

Western Mining had won the race. Today it is an established

nickel producer with around 7 percent of world production. This figure is small compared to International Nickel (Inco) and Falconbridge, the Canadian giants of the nickel world, but it is comparable with the size of Societe Le Nickel of France. Western Mining ranks as the western world's fourth largest nickel producer and Sir Arvi Parbo can take a great deal of the credit for the achievement.

Since the heady days of 1966 the world nickel price has stagnated and costs of both operations and infrastructure have escalated wildly, particularly after 1973. At today's costs and prices, it would not be economically viable to bring Kambalda into production. It is worth devoting the rest of this chapter to explaining why this is so. The history of nickel sheds light not only not only on why Western Mining succeeded at Kambalda, but also on why Poseidon later failed at Windarra.

Nickel in the form of mineral compounds is relatively abundant, comprising about 0.009 percent of the earth's crust, compared to 5.8 percent for iron and 0.006 percent for copper[2]. Nickel occurs almost exclusively in melanocratic and ultramafic igneous rocks, or in altered versions of them. Ultramafic rocks are the chief host rocks for nickel. These rocks are also called basic or ultrabasic, according to slightly different definitions, because they have a low quartz content, in contrast to acidic rocks such as granite. Among ultrabasic rocks, important host types for nickel are peridotite, pyroxenite, gabbro and norite. Ultrabasic rocks are not common compared to granite. They appear to occur mainly as intrusions along zones of intense movement in the earth's crust, such as the earthquake belt around the Pacific Ocean.

Economic concentrations of nickel occur in two types of deposit — sulphide and lateritic. Nickel also occurs in significant concentrations in sea-bed nodule deposits which may prove economic in the distant future. Sulphide deposits of nickel generally occur in either noritic gabbro intrusions (notably the huge norite dike at Sudbury, Ontario) or in various intrusions of serpentinised peridotite, i.e. peridotite partially or wholly changed by metamorphosis to serpentinite. Just how the concentrations of sulphide nickel form within these two basic host rocks is a matter for dispute, but the favored theory is that the ultramafics are formed by masses of cooling magma (molten volcanic rock) and the nickel sulphides gradually concentrate near the base of the magma during the cooling process.

In these sulphide deposits, nickel occurs chiefly as the

mineral pentlandite, which is a ferro-nickel sulphide. Metals found in association with nickel can be iron, copper, cobalt, silver, gold and the platinum group of metals. Most of the world's nickel occurs in lateritic deposits, formed by the alteration through weathering of nickel-bearing igneous ultramafic rocks, particularly peridotite, serpentinite and pyroxenite. Nickel is generally more expensive to extract from the lateritic ores, which puts a premium on sulphide deposits.

Nickel was first encountered, if not identified, in the treatment of copper ores in Saxony, Bohemia and Sweden in the 17th century. Recovering copper from these ores proved troublesome due, as was later discovered, to contamination by nickel. The ore was called "kupfernickel", or "devil's copper".

In 1751 the Swedish chemist Axel Cronstedt identified nickel as a hard, lustrous, silvery-white metallic element, but it was not until the 1820s that an important use was found for the metal. As an alloy with copper and zinc, called German silver, it was used in tableware. At about the same time in England, Faraday's experiments with nickel-iron alloys led to the evolution of stronger armorplate. From the 1880s nickel steels were being produced in Britain and France mainly for armorplating. After the Russo-Japanese War in 1905 demonstrated the practical effectiveness of hardened armorplate, demand for nickel steels grew rapidly as the European powers armed before and during World War I.

In the late 1960s, demand for nickel was expanding rapidly in the booming western economies. Mainly it was being used in stainless steel, which accounted for roughly half of all production. The metal also had extensive industrial applications in alloys and nickel-plating. Smaller quantities were being used in aerospace and the munitions industry. Although the amounts used in the last two industries were small, these uses had an important impact on the market. Aerospace was then a glamor industry (Neil Armstrong landed on the moon on July 20, 1969) with a seemingly boundless future — and it used nickel. The Vietnam War was at its height and seemed endless — and used nickel. These visible and spectacular uses of nickel enhanced its value to investors.

During the second half of the 1960s, when demand was running high, supply was restricted by under-capacity of the existing mines, which was exacerbated by labor troubles. At that time International Nickel Corporation (Inco) of Canada was easily the world's largest producer, supplying half of the western world's metal. Another Canadian producer,

Falconbridge, was the second largest, and Societe Le Nickel, mining the huge lateritic deposits of New Caledonia, was third. These three were the majors and other producers were very small. The Communist bloc was roughly self-sufficient.

The supply of nickel was tightened drastically in these years by two big disputes between Inco and the Sudbury miners. The miners, said to be the highest paid in North America, struck in 1966 when negotiating a three-year contract, and supplies of nickel became scarce. Consumers had not long absorbed the impact of this strike when in 1969 there was an even bigger one, in which the Falconbridge miners eventually joined. The world was starved of nickel.

There were two prices for nickel — the producers' price and the free price. The producers' price was relatively low, but supplies at that price were rationed. Customers at one stage were receiving only 70 percent of their requirements. The free market was supplied by excess production from Communist countries and by western merchants who received allocations from the producers and resold them. During 1969, the free price of nickel was to rise spectacularly from £1500 a ton in January to nearly £7,000 in November.

All through the late 1960s, nickel miners could sell every pound of nickel they produced, and at rising prices. It was an ideal time for a big new mine to come on-stream and Western Mining's discovery in 1966 caught the market perfectly. But the soaring price of nickel in later years, up to the end of 1969, could not last. There were three reasons for this.

The first, and most obvious, was that sooner or later the Canadian miners would end their strike. When they returned, more nickel would be produced, although there would still be insufficient to meet demand. This was well recognised.

The second was that no one has yet managed to repeal the age-old laws of supply and demand. As the price of a metal rises, there is a stronger incentive to search for new orebodies. Also, known deposits which were previously marginal become economic. As the metal's price rises, supplies from these new sources will come on stream, bringing the price back down. This happens regularly in many metal markets, tin and copper being good examples. No metal is so rare that its price can keep rising indefinitely, and although nickel mining was restricted to three producers in the mid-1960s, it is not a rare metal. New mines were bound to open and they did.

Western Mining provided the world's major new source of supply in the late 1960s, but as the 1970s opened the nickel market — ironically, at the time the Australian share boom was at its height — was moving back into surplus as other suppliers began entering the market.

South Africa, Rhodesia, Cuba, the Dominican Republic, the Philippines, Greece and Indonesia have all become significant suppliers since 1970, at a time when Comecon producers and existing western mines have been developing acute over-capacity. On present projections there will be a world nickel glut until at least 1987, with even further capacity coming on-stream in Cuba, Indonesia, the Philippines, Greece, Yugoslavia and Guatemala.

Thirdly, another effect of the rising nickel price was to force consumers to look for substitutes. They discovered that a number of substances could be substituted for nickel in many of its uses, with the result that nickel demand began falling in the mid-1970s.

However, in 1966 all this was well into the future, and Western Mining had found Kambalda at the ideal time. The company's shares, in the doldrums only a year before, stood at $6.96 at the end of 1966. By January 1968, with production on-stream, the shares had raced to $43.50. The following month saw the 50c shares split into 10c units. Later that year the 10c shares reached $16 — the equivalent of $80 per share for the old capital. Net profit, which had slumped to $692,000 in 1965-66, soared to $14 million in 1969-70 — multiplying by 20 times in four years. Among those who made fortunes were clients of the Melbourne broking firm F.R. Morgan & Co, whose principal was closely associated with WMC. The breathtaking rise of WMC shares galvanised the whole mining board, starting rises in many other stocks.

In contrast with the tumult of the market, Kambalda is a neat little town, looking rather like a garden suburb set in silent scrubland. The green lawns of the houses come as a shock to a traveller from the waterless bush nearby.

A tourist can stand alongside the Silver Lake shaft above the Lunnon shoot and admire the scenery. On the dazzling white salt flats of Lake Lefroy the wheelmarks of exploration rigs seem like ski tracks through snow. From the shoreline of the lake the country undulates gently upward, covered with salmon gums. Two hundred yards to the rear of the tourist, in a shallow gully, is the small hole where Cowcill and Morgan dug their samples. Ahead, on a low ridge close to the lake, stands the long abandoned poppet head of the old Red Hill mine. There are few mine sites in Australia which blend present prosperity so well with a sense of history and the silent, timeless beauty of the Australian bush.

---

[1] Interview with author, February 1976.

[2] For most of the technical material in this chapter I am indebted to Australian Mineral Economics Pty Ltd of Sydney.

CHAPTER THREE

Showing Endurance

The great Poseidon boom from October 1969 to February 1970 stands out so strongly in the memory that it is often forgotten there was a very strong share boom just before, in 1968. Calendar 1968 was the best year for stockbrokers in living memory. Stock exchange indices were rising beautifully (in retrospect, pundits would say, too quickly) and many did not expect 1969 to be as good.

In 1968, the Sydney non-ferrous metals index gained a phenomenal two-thirds from 2216.58 to 3692.49. The all-ordinaries did only half as well but still gained a very satisfactory 34 percent during the 12 months. Traditionally, the Christmas-New Year break has been a quiet time when a broker returned to the bosom of his family, but the share traders were too active for any holidays in 1968. The market ended the year on a high note, as traders pushed mining share prices up on December 31 in a half-day session. The Australian reported: "National share markets yesterday farewelled 1968 with a healthy exuberance. The rising tone of the last two months was enhanced".

There were several reasons for this bubbling market. The nation had been through a long period of tranquillity and steadily rising prosperity, interrupted only by the short, sharp break of the 1961 credit squeeze. Disposable income was rising and unemployment was negligible. Economists were concerned by the rising rate of inflation, but it too was negligible by later standards. The Consumer Price Index rose by 2.6. percent in 1968-9. The only time the CPI rose by more than three percent in the whole decade of the 1960s was in the final quarter of 1969. Trade unions were going through a quiescent period, apart from protests over the Vietnam War. The conservationist movement, later to be the bane of the mining industry, was in its infancy.

Even the easy affluence of these years might not have lured Australians toward stock market speculation if it had not been for the uncovering of vast and previously unsuspected mineral wealth by Comalco, Hamersley, BHP, Western Mining and others. The growing belief that Australia was destined for mineral greatness was reinforced in 1968 by the publication of Sir Harold Raggatt's *Mountains of Ore.*

Sir Harold's qualifications were impressive. He was a former secretary of the Commonwealth Department of National Development, chairman of the Australian Atomic Energy Commission and director of the Bureau of Mineral Resources. Since retiring from the public service he had been a consulting geologist to BHP.

Sir Harold was an enthusiast for the mining industry, having predicted in 1965 that mineral exports would overtake wool within a decade as Australia's major foreign-exchange earner. In the preface of his book he wrote: "I hope this book helps many Australians to catch something of the pioneering spirit which drives the mining men who are changing the face and economy of our country. We cannot all be pioneers in the same way as the miners, but we can support them by investing in exploration and mining companies. Every Australian who has money to spare for investment should have a stake in our mineral industry."

The title of Sir Harold's book was taken from a quotation by the late Tom Price, vice-president of Kaiser Corporation, after returning from a visit to the Hamersley Ranges. Price said: "There are untold millions of iron ore in the Pilbara deposits. I think this is one of the most massive orebodies in the world. There are mountains of ore there ... it is just staggering. It is like trying to calculate how much air there is."

Mountains of Ore was one of those books which was not widely read, yet wielded considerable influence. It is doubtful whether one of the small investors who plunged into the stock exchange maelstrom of 1969 had ever read through the book, with its technical descriptions, tables, maps and graphs; but they had probably all read or heard a description of it at some time, because it was well publicised. They might not even have consciously remembered the book, but it was one more factor reinforcing their conviction that Australia was rich in minerals, as indeed it was.

The Australian public did not need much encouragement from Sir Harold to buy mining stocks. The indispensable element in any share boom is a mood of optimism in the investing public. This had been established by the repeated successes of the mining industry. The success rate had been so high it had become almost monotonous. Large and small investors began to divert more money into mineral stocks.

This resulted in 1968 being one of the best years ever for Australian stock exchanges. The strong bull market was led by BHP. After first finding rich natural gas at Barracouta, Esso and BHP had found oil and gas at Marlin in 1966. Next year they struck oil at Halibut and again at Kingfish, which is still the largest oilfield ever found in Australia. In 1968 the partners found more gas at Snapper. This early run of successes dazzled the market. There was a wave of buying, including some heavy US orders, for BHP. BHP shares had opened the year just over $15. In June they had peaked at $25.75 and they stayed high for several months.

In bull markets such as this, one useful measurement of a stock's prospects is its market capitalisation. In this simple calculation, the market price of the stock is multiplied by the number of shares on issue. This shows the total value which the market is putting on the company. In January, BHP had 129 million shares on issue and their mid-price for the month was $16.90. The market was therefore capitalising BHP in January at 129 million x $16.90 = $2,180 million. At the peak of $25.75 in June, on the same number of issued shares, BHP was capitalised at $3322 million. The company's total assets at the time totalled only $1200 million. Shareholders' funds, a better comparison with market capitalisation, amounted to only $800 million. Market capitalisation, used in this way, is a valuable aid in recognising when a company's share price has gone out of whack. It is not the only way of measuring the validity of a share price, but it is one worth remembering during a boom. BHP's price was unreal unless it was on the verge of a discovery of global significance. Nevertheless, sharebrokers were still urging clients to buy the stock. A young newspaperwoman I knew had been left a small legacy and was told by her broker to buy on the grounds that, "You can't go wrong in BHP". She bought at $24 and it slid to nearly $16 in four months. She was almost in tears every time she asked about it. She could hardly have done worse in any stock on the board.

On the way up, BHP's surge had carried the rest of the oil board with it. Magellan went from a year's low of $1.66 to $5.20, while Ampol Exploration rose from $1.71 to $4.70. Some spectacular gains were registered by the small explorers.

Kamilaroi (then on the oil boards) went from 24c to $3.70, Abrolhos leapt from 18c to $1.10 and Exoil shares ran from 21c to $1.30 while the options went from six cents to 90.

On the mining boards the pacesetter for much of the year was Great Boulder Gold Mines Ltd. In mid-1967 Great Boulder, in partnership with North Kalgurli (1912) Ltd, had made Australia's second commercial nickel discovery at Scotia, 40 miles north of Kalgoorlie. Like Western Mining, Great Boulder and North Kalgurli were old gold mining companies, long established in the region.

Great Boulder's shares, which had been as low as $4.90 in 1968, soared to $20 in August. One side-effect was a wave of support for other gold stocks. Central Norseman Gold NL hit $7.50 and Loloma $7.70. Lake View & Star went to $8.40 and Emperor, which had been 60c at one stage, reached $4.65.

Many of the older stocks registered their highest prices for the whole boom period in July-August 1968. Great Boulder, Western Mining, CRA, Ampol Exploration and Peko all hit their boom peaks or went close to them in these months. These were the golden months for the heavyweights of the industry. There were relatively few speculative miners listed in 1968, but the share performances of some of them gave an inkling of things to come. Barrier Exploration NL caught the imagination of speculators when it began drilling its promising Cowan West prospect on the Jimberlana Dyke in W.A. toward the end of the year. Barrier shares, which had been as low as 10c earlier in 1968, ran to $1.60. Even more exciting was Silver Valley Minerals NL, which sprang into prominence early in the year on hopes of a major silver deposit near Armidale in NSW. The shares, which had bottomed at 13 cents in 1967, ran from 60 cents to $4.60 in May 1968 then slumped when assays from the New England deposit proved disappointing.

But the real star of the 1968 'specs' was Endurance Mining Corporation NL, a sleepy Tasmanian tin miner which was suddenly catapulted from obscurity into national fame. It was taken over by Sydney entrepreneurs, the shares ran to $10.60 after the company claimed to have half a million tons of scheelite, then the shares began to slide, the entrepreneurs sold out and much later it was discovered than an unpublished geologist's report had greatly downgraded the deposit.

The NSW Government appointed two inspectors to make a special investigation into Endurance. The inspectors reported to the then Attorney-General, Sir Kenneth McCaw, in March

1971 but it was not until November 1974, after a delay of over three years, that the report was made public. Only then did the investing public learn what had happened inside Endurance. The inspectors, John Lockhart, QC, and Mark Mealey, a Companies Office inspector, condemned directors for not revealing the downgrading of the deposit and estimated that insiders had made nearly $5 million profit selling out while the report lay in Endurance's files. The following summary of the Endurance story is compiled from the inspectors' report.[1]

Endurance was incorporated in Tasmania in 1922 and for nearly half a century led a blameless existence. It was listed on the Hobart and Melbourne Stock Exchanges and until 1968 had been a small tin miner and explorer. It had a tiny capital and had last paid a dividend in 1966. In its past three years Endurance had made small losses of between $10,000 and $25,000. This may sound like a formula for a quiet life, but such companies are always liable to a sudden takeover. A public company is a valuable vehicle which can be put to a variety of uses. If its issued capital is small, a raider may not have to buy many shares to gain effective control. If it has been making losses, its shares will be cheap to buy.

In Endurance's case the capital was very small indeed, amounting to only 189,424 shares with a par value of 50 cents each. In early 1968 they had been holding a little above par, being seldom traded. The largest shareholder, who had the jaw-breaking name of Szulc-Szczgielski, held 47,000 shares, or around one-quarter of the issued capital. He had been trying to quit his holding for some time. In June 1968 Szulc-Szczgielski arranged to sell his block of shares for 61 cents each to a 10-man syndicate. The syndicate was split into five equal segments as follows: Paul Murray and John Maguire (corporate consultants), 20 percent; Peter Murray (financial adviser — no relation to Paul), 20 percent; Laurence O'Neill and Richard Baker (company directors), 20 percent; Leonard Hall and Richard Relph (consulting geologists), 20 percent; George Jones, Graham Hurst and Andrew Clinton (partners of the Sydney broking firm of Clinton & Co), 20 percent.

Having bought 25 percent of the company, the syndicate ensured control of Endurance by buying another 18 percent in a first-come-first-served bid. A first-come bid is one where a bidder enters the market with the intention of buying a set number of shares for a set price, in this case 18 percent . at 85 cents. Later amendments to the Companies Act attempted to ban this form of bid, but have proved unsuccessful. At the time it was a perfectly legal and frequently used device.[3] As a result of the first-come offer, the syndicate acquired a further 34,460 shares. They now had 43.5 percent of Endurance's capital, which represented effective control. It had cost them $59,828, or an average of 72.7 cents per share.

A layman might wonder why anyone would spend $60,000 acquiring control of a tiny company with few assets and a $25,000 loss in its last financial year. The answer is that the value of a public company's shares can be enhanced by selling it some exciting asset. So before continuing with the Endurance story it is necessary to explain the mechanics of a phenomenon which will crop up occasionally in this book, the back-door listing.

The beauty of a listed public company is that its shares are quoted daily on the stock exchange and are generally easy to buy and sell. If the price of the shares is rising, they can be sold at a profit. The normal way of listing a company is to issue a prospectus, sell shares to the public, then apply to the exchange for listing. This costs, effort, time and money. The prospectus is a legal document disclosing details of the company's assets, earnings, directors and intentions. It takes time to draw up and care to check, because any false statement in a prospectus is punishable by severe penalties, including jail. Selling shares to the public also takes time and may require an underwriter. (An underwriter guarantees that all the shares in the company will be placed and in return charges a commission for his risk and service. If there is a shortfall in the issue, the underwriter has to take it himself.) Finally, the home stock exchange might refuse listing, in which case all the effort would have been for nothing.

A faster and easier way of listing is to buy a company which is listed already and then have it take over another company or asset. The listed company, having little cash, pays for its new acquisition by issuing shares to the vendors (sellers). The shares so issued will amount to a controlling interest in the listed company. At the completion of the deal, the listed company has a new asset, and the vendors of the asset have control of the company. The listed company's former activities will shrink in importance and while its old shareholders will still have the same number of shares, they will own a smaller percentage of the total issued capital. Their consolation is that, optimistically, the newly-acquired asset will enhance the value of their holdings.

Endurance had an existing board of three directors. They were retained on the board but in practice their interests were confined to the company's traditional tin mining in Tasmania. They played no significant role in the events that followed and according to the inspectors' report were not even informed of major events. On September 11, Paul and Peter Murray joined the board and on December 12 O'Neill and Maguire also became directors.

The inspectors described the syndicate as "highly disciplined and skilled share traders". Their aims were set out in a memorandum dated July 17, 1968, prepared by Paul Murray and distributed to all members. Item 12 of this document said: "It is planned to raise additional capital more or less immediately to finance an exploration program on the company's areas and other prospects which will be offered to the company. Application moneys paid in respect of this issue will be fully deductible for tax purposes." Item 13 said: "As a result of these activities it is expected that keen market interest will be generated in the company's shares, and, being tightly held, a substantial improvement in price should result."

The syndicate's takeover saw the beginning of an explosion in the size of Endurance's capital. In August 1968, Endurance directors agreed to split the company's capital from 50c into 25c units, then have a three-for-one issue of shares paid to 10c each. This resulted in issued stock being increased to 1.5 million shares with a paid value of $210,798.

The inspectors' report concluded that the capital changes engineered by the syndicate had been planned well before it gained control of the company. "The syndicate members had little if any interest in the mining activities of Endurance," the inspectors concluded. "They were concerned to take over Endurance for the purpose of making as much money as possible by trading in its shares. They realised their ambition. The whole scheme is a typical example of what is sometimes called the market pool manipulation system which was fairly common in the United States some 40 years ago."

The report evokes the image of a detailed conspiracy by the syndicate members, but on closer examination their purposes seem less definite. There is no reason to presume they were not genuinely interested in developing Endurance's tin mines. Between August 1968 and October 1970 about $250,000 was spent on these properties, the money having been raised when the syndicate took over. Paul Murray and Maguire, while brokers with Mullens and Co, had tried to help Oil and

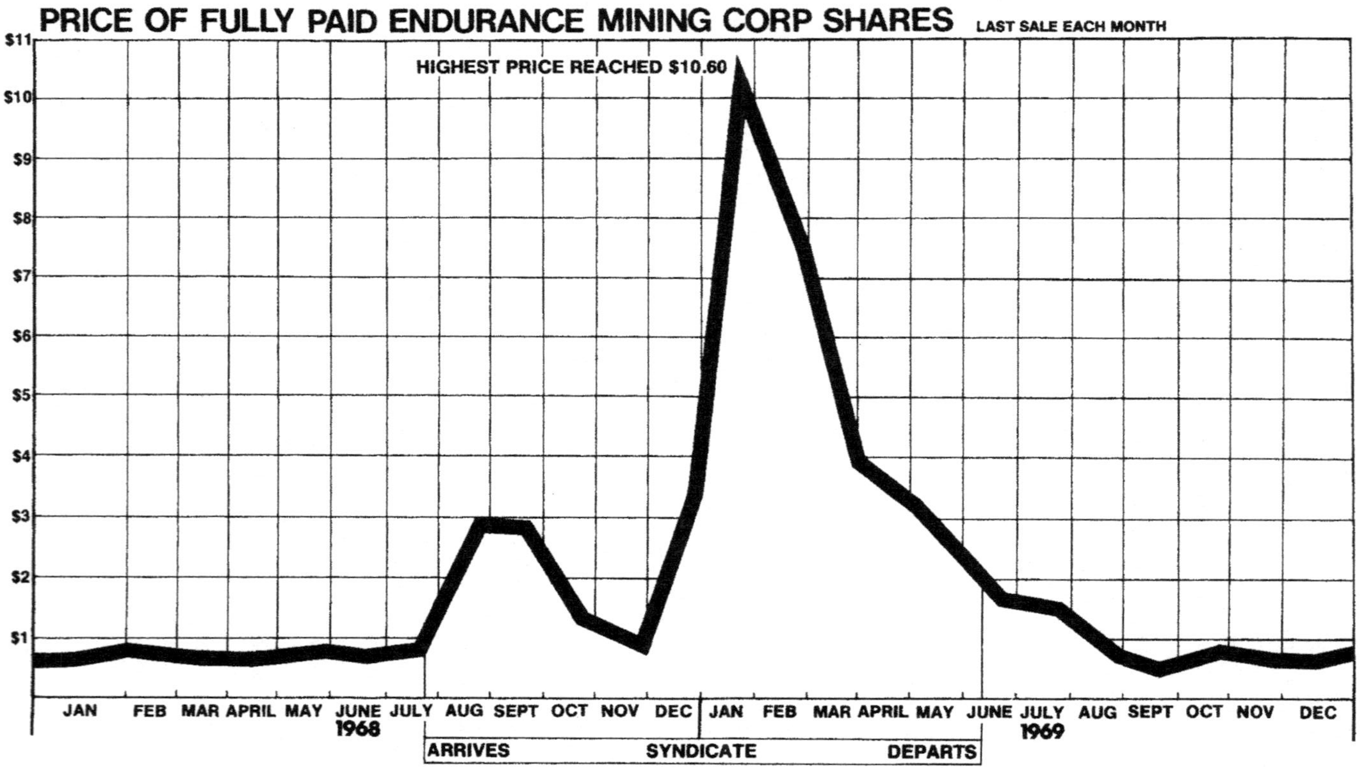

The graph shows the dramatic rise and fall of Endurance's share price, coupled with the arrival and departure of the syndicate. There was heavy selling by insiders immediately after their departure. *(Graph from the inspectors' report.)*

Minerals Quest NL raise money to buy tin leases in NSW. When BHP defined tin reserves at Monarch mine in Tasmania, the syndicate-controlled board acquired the deposit for Endurance (it was too small to interest BHP further).

Of course, the syndicate wanted to make profits, and as high as possible. So does everyone who buys a mining share, let alone buying control of a company. But on their actions immediately after gaining control, the syndicate members were initially interested in tin. Like any mining company on the board they were interested in buying a good mining prospect if they found one, but there is no evidence that they knew of Attunga's existence when they took over Endurance.

Attunga Mining Corporation Pty Ltd had been incorporated in November 1967 to mine copper at Attunga in northern NSW. This mine, last worked in World War II, now had high but patchy copper values and the consultant geologist thought the best way to exploit it would be on a small scale, hand picking the ore.

The first directors of Attunga were Kenneth Brudenell-Woods, Dr James Furber and Colin Graham. These three men, together with their families and companies, also controlled nearly all the issued capital of the company. Attunga had only 100 shares, divided between 15 shareholders.

In February 1968, scheelite, an ore of tungsten, had been found outcropping on the prospect. There was a chance that instead of owning an uneconomic copper mine, Attunga might have a tungsten mine. In August and September 1968 a total of 54 percussion drill-holes were put down to assess the deposit. Then two diamond drill-holes were sunk. The first hole produced a weighted average grade of 0.86 percent tungstic oxide (WO3) over 412 ft.

By this time the members of the Endurance syndicate had become aware of Attunga. Brudenell-Woods had been share trading in June and July 1968 through an account with Clinton & Co. Hall Relph and Associates Pty Ltd, whose two principals were members of the syndicate, were invited by Brudenell-Woods in mid-September to become exploration managers at Attunga. The supervising geologist on the site, employed by Attunga, was a young man named Lloyd Hamilton.

To understand what happened next it is necessary to have some knowledge of mining exploration procedure. A diamond drill produces a smooth core sample of the rock through which it passes, although when it passes through areas of poor ground some of the material can be lost and core recovery is low. As the core comes to the geologist he inspects it for mineralisation. Samples are then taken from the core and sent to a laboratory for assay. At the drilling site, the core is laid in trays and numbered carefully. This core, in the case of 100 percent recovery, is a smooth cylinder of rock two to three inches in diameter, depending on the size of the drill bit being used. To take a sample, the core is split lengthwise with a hammer and chisel. One half of the split core is then divided into samples — perhaps every 10 feet — and sent to the laboratory. The other half is kept in the field for possible future checking. The core can be split into several parts and samples sent to three or four laboratories if the company feels it necessary to have more than one assay. The normal procedure is for the field geologist to log all the cores. As each core section comes in, he numbers it. The core is placed in a tray with that number. When a sample is taken from the core it is put into a bag with the same number. The laboratory receives a bag full of rock with a number on it from a company and is told to assay the sample for nickel, tungsten, uranium or whatever. This procedure maintains a fair degree of security, although a shrewd employee of an assaying laboratory could probably make a good estimate of what a small company was doing. Having assayed the samples, the laboratory sends back to the company a list of the assays recorded by the various numbered samples. The geologist then matches this against his log book, which shows where the cores came from and this builds up his picture of the orebody.

This procedure appears to have worked satisfactorily with the first diamond drill hole at Attunga. The first 20 assay results from this hole reached Attunga on October 31. Further samples were sent to Geological and Mineral Laboratories Pty Ltd the next day. By this time, Attunga was trying to sell the deposit. At a meeting on October 30, Brudenell-Woods had reported to directors that "subject to confirmation of the tungsten field" he had received substantial offers. The board discussed the offers and authorised him to continue negotiating.

It was the second diamond drill hole which would cause all the trouble. Samples from DDH2 were sent for assay on November 18. On November 21, Brudenell-Woods told Paul Murray that the reserves of scheelite at Attunga were 200,000 tons of 0.64 percent ore. Neither Hamilton nor Hall had made such an estimate. On that same day, November 21, Paul

Murray proposed to the board of Endurance that it take an option over Attunga's scheelite deposit. On November 28, Hall Relph and Associates reported favorably to Endurance on the scheelite prospect. Further samples from DDH2 were sent for assay on November 30 and December 5.

On December 10, an option agreement was signed between Attunga and Endurance. On December 17, the assay results from DDH2 came to hand. They showed an average grade of 1.44 percent tungstic oxide over a total depth of 600 ft.

A grade of 1.44 percent may sound low to a layman, but it was an economic grade for a tungsten mine. The assays, which appeared in the press the following day, were good news. Endurance shares had enjoyed a healthy run up to $3 in July-August, when exploration was starting, had since drifted back to 80 cents. Now they shot up to $4, hitting $5 near Christmas. Time and a lot of burned fingers would later show these assays were badly overstated.

A factor in the bubbling share price had been the press coverage of Attunga's prospects. The first significant press release had been by A.G. Healing Ltd, whose subsidiary A.E. Goodwin Ltd had taken a 10 percent stake in Attunga in June 1968. Healings announced in July that Goodwins had undertaken to supply Attunga with $320,000 worth of plant and equipment. None of this equipment was ever delivered. Indeed it never passed the design stage. However, the press release did draw attention to Attunga's scheelite.

The Sydney Morning Herald quoted one director of Attunga, Louis Harmanis as saying: "We don't need a geologist to prove there's a million tons of ore under this hill — you can see it." When the results of DDH1 and 2 were released to the press in mid-January, they were given natural prominence. Endurance's capital moves of August, combined with the "new blood" in the boardroom and the current rumors that Endurance would be used as a vehicle for the backlisting of Attunga, had focussed attention on the stock. The Sydney Daily Mirror, whose finance pages were then frequently devoted to tipping "hot" stocks, was particularly enthusiastic.

One journalist given inside information was the finance editor of the Mirror, Cliff Riggs. On December 22, Riggs printed assay details which were not made public until three weeks later. On January 19, he wrote about rumors concerning Endurance (which were certainly rife at the time) and quoted "reliable sources". Asked by the inspectors what they were, he said: "A reliable source is someone you heard say it three times, in this sort of journalism".

The inspectors deplored Riggs' journalistic standards, but there are some points in his favor. A significant element of the public is interested in stock-market tips and rumors and in those days the Mirror was deliberately trying to cater for them by printing the current tips and rumors. One may deplore tips and rumors but they are very much a fact of life around the market. There is also an element of snobbery involved. One can gain prestige by passing along an occasional tip at a discreet downtown luncheon and half an hour later wax indignant about a tabloid evening newspaper trumpeting tips to its readers. In Riggs' tipping of Endurance, he tipped the stock on assays he believed were genuine. The only time he held shares in Endurance himself was in May 1969 when he did a quick trade of 700.

One point which was not evident to the press or the public, but which must have been known to the syndicate was the parlous financial state of Attunga Mining. At June 30, 1968, the company had been insolvent, with a deficiency of nearly $300,000 in net tangible assets. The company's biggest debts were $337,000 owing to shareholders, including $165,000 in unsecured loans. The inspectors were unable to verify these loans. By next balance date in 1969 Attunga had become a subsidiary of Endurance and was consequently subject to the rigors of public company audit. The auditors, Greenwood, Challoner & Co, heavily qualified the Attunga accounts. Greenwood, Challoner said they were unable to state whether the balance sheet and other accounts were properly drawn up in accordance with the Companies Act to give a true and fair view of Attunga's state of affairs. The auditors listed seven reasons for their opinion. The seventh read: "During the year ended June 30,1969, the company made various disbursements described as the repayment of the unsecured loans by shareholders as shown in the balance sheet on June 30,1968. The company has not been able to supply us with sufficient documentary evidence to enable us to satisfy ourselves that all or any of such disbursements were a valid application of the company's funds."

On the market, the shares of Endurance had hit $10 by the end of January 1969 and were the talk of the town. Endurance, which had been capitalised at a mere $100,000 at its low point eight months before, was capitalised at $15 million at its high point in early February. A substantial reason for this frantic activity was that on January 16, Endurance's directors had succeeded having the stock listed in Sydney. It had opened at $8.10 in an otherwise quiet market. Turnover of the stock was enormous. A total of more than 3.8 million shares was traded in January for a total value of more than $30 million. As the figure of 3.8 million shares was more than double Endurance's issued capital at the time, it is obvious that traders were turning over the stock very quickly and in large volumes. It is evident also that some of the syndicate must have been selling in this month or there would have been no supply of scrip to the market. Velocity is the best measurement of speculative interest in a stock and by any normal standards speculation in Endurance had reached fever pitch in January.

The market had been further spurred by the news that one of Australia's biggest and most respected mining companies, Peko-Wallsend Ltd, was considering a bid for Attunga. Peko, which had just gained control of the King Island scheelite deposit after a bitter fight with Consolidated Gold Fields of Australia, had considerable expertise in mine evaluation and development. On January 24 Peko decided not to go ahead with the offer, although it retained a strong interest in the prospect.

On February 3, 1969, the directors of Endurance met and agreed to issue 800,000 shares to acquire the whole of Attunga Mining's issued capital. The offer was 8000 Endurance shares for every share in Attunga. In what can only be described as a welter of paper, they also decided to make a one-for-one issue of bonus shares and following this bonus to make a further one-for-seven issue to Attunga's former shareholders at $1 per share, consisting of 25 cents as the fully paid value, plus 75 cents premium. The result of all this activity was that the former shareholders in Attunga now held some 32 percent of the stock in Endurance and Endurance owned the scheelite deposit. The Attunga mine had gained backdoor listing. If the original syndicate had held their shares and taken up their full entitlement in the earlier issue of contributing shares they would have now held about 26 percent of Endurance. However, several of the syndicate members had been selling parts of their holdings on the way up and their total holding would have been below 20 percent, possibly well below. The syndicate had formally disbanded on November 14, 1968. It was now every man for himself and they could buy or sell Endurance as they saw fit.

Simultaneously with the announcement of the takeover of

Attunga, the Endurance directors announced an ore-reserve figure for the deposit. The company's statement read: "The company's geologists have reported that indicated reserves of scheelite amount to approximately 500,000 tons of 1.4 percent WO3 and the board are hopeful that with further drilling reserves will be established well in excess of that figure."

This statement was wrong. Worse, it was allowed to stand uncorrected long after directors must have known it was wrong.

The importance of this cannot be over-emphasised, because exploration companies are valued almost entirely on their ore reserves or their prospects of finding ore. They generally operate at losses and have only some working capital and plant for assets, so that earnings, dividend yields, net tangible asset backing and other yardsticks used to value industrial shares are meaningless. A company with only $100,000 in assets can discover an orebody worth $10 million or more, which means its shares will be very sensitive to changes in the grade or tonnage of the ore.

The only possible basis for the reserve estimate of 500,000 tons was a weekly report by Hall Relph's site geologist, John Holland, in mid-January, which said in part: "... the footwall dips an equal amount giving true thickness of approximately 35 ft. Reserves of the order of half a million tons of 1.4 percent ore are indicated by this thickness and the results of DDH2 (value — $20 million). There are, however, some anomalies in the results from diamond drilling."

This report was made by Holland to his principals and the estimate was substantially qualified by other statements he made. His report was sent by his principal, Hall, to Endurance. When Paul Murray proposed to include the estimate in a public statement, Hall said he had said, "'Over my dead so-and-so body,' or something like that"[2] Nevertheless, Paul Murray included the ore reserve estimate in his February 3 statement to the Sydney Stock Exchange. The inspectors concluded that the statement was false because it omitted the qualifications and gave an untrue impression. Further, Murray had not even mentioned the qualifications to the board. The inspectors found that Maguire had been as responsible as Murray for the events that had occurred.

Endurance shares hit their peak of $10.60 shortly after the reserves and the Attunga takeover were announced. Then they began to fall sharply. A major reason for the drop would have been the considerable dilution of capital resulting from the proposed issues, but successive poor drill results also beared the market. Nevertheless, the buyers who were still in the market had no way of knowing from the company's confident statement of February 3 that the reserve estimate was not soundly. based. The calculations had taken in the results of DDH2 and the assays on this hole had been wildly high.

One problem was a lack of communication between the Endurance and Attunga boards. Paul Murray said afterward that he had been unable to get records or even a balance sheet from the company. One example of the lack of communication is that Hamilton had recommended to the Attunga board on February 3 that a check should be made of the DDH2 assays. The Endurance board knew nothing of this recommendation, according to Murray, and instituted their own checks in March, on the recommendation of geologist Mark Foy.

Meanwhile, a third diamond drill-hole had been completed at Attunga on January 22. This hole had intersected tungsten grades of only 0.09 and 0.1 percent. These cores were despatched for assay on January 24, 10 days before the announcement of the ore reserves. Even if the laboratory had not reported the assays back to Endurance by February 3, the field geologists would have had a fair idea of the results from on-site logging. One characteristics of tungsten is that it fluoresces under ultraviolet light. When drilling for tungsten it is standard practice for the field geologists to check the cores as they arrive from the drillers. Under an ultraviolet lamp the tungsten will fluoresce blue, while a yellowish glow indicates powellite. A high powellite reading is bad because it indicates a high molybdenum content in the ore.

Molybdenum is an impurity in tungsten and buyers will impose penalties in the form of lower prices on miners delivering tungsten containing more than 0.1 percent molybdenum (usually called 'molly'). Attunga's scheelite had a disturbingly high powellite reading, indicating molybdenum could have run as high as five percent in concentrate. At this level it might have been worthwhile extracting molybdenum also, if the orebody had been large enough to justify the extra cost of separation.

This on-site logging of cores is important because it provides a check on laboratory assays. When the assays are returned to the field geologist he can compare them with his own, necessarily rough, field readings and see whether they are "in the same ball park". It was because the DDH2 assays disagreed so violently with the on-site readings that the geologists pressed for check assays.

The results of DDH3 were published on February 10. The Melbourne Stock Exchange, puzzled by such a poor assay report only a week after the bullish reserves estimate, queried Endurance. Paul Murray, who a week earlier had said the board hoped further drilling would establish even higher reserves, replied that insufficient information was available to be able to comment on the significance of DDH3. The inspectors concluded that the results of this hole must have been known, from on-site logging, to Holland and almost certainly to Hall, Paul Murray and Maguire at the time the ore reserves were published on February 3. Actually DDH3 appears to have been drilled as the first of a number of holes to try to delineate the shape of the orebody, in which case the tungsten assays in the hole had only limited significance.

Normally the first diamond drill-hole sunk on any prospect is put down in the most promising spot to see if ore is there. If the results are ambiguous, another one or two holes may be drilled to find the ore. If the first two holes find good enough ore values over enough depth (on really hot prospects, one hole can be sufficient), the next holes sunk are to delineate the orebody, i.e. to work out its shape. These holes may produce low readings or may even strike barren gound. Without full data on the positioning of the drill holes and the angles at which the drills are sunk, the average investor has no way of evaluating the drilling program.

The market, confronted with the task of digesting the huge increase in Endurance's capital as well as the poor drill result, marked the shares down to $7.50. Endurance had begun its long downhill run.

It should be recorded that a geologist from one large, respected company came to a similar conclusion to Holland on ore reserves at this time. On January 22, Mr P. Forwood of North Broken Hill Ltd wrote a report to Endurance following a visit to the Attunga orebody. He estimated 650,000 tons of 1 percent grade, giving a total value of recoverable mineral of $14.6 million. He concluded: "Using this limited reserve figure, it is doubtful whether a big company such as North could proceed with mining even if it had discovered the deposit without any purchase premium". He had reservations about DDH2, saying: "I feel that hole No 2 was exceptional. I firmly believe that another hole yielding such widths of high-grade material is not likely to be achieved. For estimation of grade it would probably be best to take hole No 1 viz, 0.87 percent WO3". He added that Attunga contained enough reserves to be mined, "but whether these reserves are sufficient for a large operation, or for only a small operation, remains to be proved by drilling".

Summing up, Hall, Holland and Forwood all had misgivings about DDH2, but as long as either the syndicate members or the Attunga group were in control of Endurance nothing was done to qualify the bullish reserve statement of February 3, including the expressed hope that further drilling would increase reserves.

Now followed a curious period of drift. Maguire and Paul Murray were busy on other fronts. In April 1969 they floated Reef Oil NL, a spin-off of another of their oil exploration companies called Basin Oil NL. Negotiations were begun for a farm-in with Peko-Wallsend at Attunga. All that was announced to the exchange was the results of the drill reports as the geologists tried to delineate the orebody. The low values encountered resulted in Endurance shares drifting ominously, sinking to $2.30 by the end of May.

During this period, the vendors of the Attunga mine, now major shareholders in Endurance, were becoming restive. They now challenged the Murray-Maguire control. On June 11 Brudenell-Woods wrote to Maguire and Paul Murray demanding board representation and threatening a proxy fight. The response was immediate. Both Murrays, Maguire and O'Neill resigned in the next two days. Paul Murray and Maguire had just received, on June 10, the results of the check assay program. The checks, by three separate laboratories, proved conclusively that the grades for DDH2 had been grossly overstated.

A new Endurance board was formed including Edgar, Harmanis, Graham and Brudenell-Woods. They requested an immediate and full geological report from Hall and Relph. This report was delivered orally to the new board on June 25 by Mark Foy, a Hall Relph geologist. He told them that proved ore reserves amounted to only 220,000 tons averaging 0.35 percent tungstic oxide. He also told them of the high molybdenum content.

This was very bad news. It was also crucially important to the prospects of the company and its shareholders, because this was an estimate made on fuller and more accurate information than had been available for the previous statement of ore reserves. Attunga was the predominant asset of Endurance and the recalculation of reserves had slashed its value. But the previous ore reserve statement of 500,000 tons was allowed to stand and Foy's report was never released to the public by this board. The general public was allowed to go on thinking that Endurance had 500,000 tons of ore when the latest calculations showed the figure to be less than half that tonnage.

The next three months saw a wave of selling as the syndicate and the Attunga vendors, including new members of the board, unloaded substantial portions of their holdings. During this time clients and partners of Clinton & Co sold $2 million worth of stock on the market. Murray and Maguire sold $845,000 worth and Peter Murray, $726,000 worth. The inspectors estimated the combined profits of the insiders, whom they defined to include such people as Murray and Maguire's dentist and a man who bought a house from Brudenell-Woods.

Schedule G of the inspectors' report calculates the profits made by the most prominent participants, including securities on hand at cost, were: Paul Murray and Maguire and an associated company, $778,000; Hall and Relph, $483,000; Brudenell-Woods family and associated companies, $471,064; Peter Murray, $400,818; Clinton & Co, families and associated companies, $348,323; Baker, $247,206; Graham, family and associated company, $201,942; O'Neill, family and associated companies, $191,159; and Dr Furber and associated company, $184,434.

Under this heavy selling, it is not surprising that the price of Endurance should have slumped to 44 cents by the end of September. This slump occurred despite the news that Peko was now definitely looking at the prospect. On June 26 Peko agreed to pay for an exploration program and erection of plant in return for a half-interest in the deposit. As Peko was one of Australia's most respected mining companies and had proven expertise in scheelite mining, a joint venture with Peko ought to have enhanced the Endurance share price.

Another "outsider" who had become interested in Endurance was a tough entrepreneur named Walter Shapowloff. Shapowloff had a substantial holding in A.G. Healing Ltd, which, through Goodwins interest in Attunga, now had a large holding in Endurance. On August 26, Endurance announced that Brudenell-Woods and Harmanis had retired from the board "due to pressure of their personal business commitments" and Shapowloff had been appointed.

Two week later Shapowloff was in the Endurance office

going through the files to familiarise himself with the company when he stumbled upon the Foy report downgrading the reserves. Shapowloff was shocked and immediately released the report to the Sydney Stock Exchange. The market promptly slashed $1.2 million off the value of the company and the exchange called for an investigation. Brudenell-Woods explained that the report had not been released because the board considered it "would be equally as misleading as the previous report".

Meanwhile, Peko had been evaluating Attunga. In November 1969 it issued a detailed report. Peko's geologists said that on their calculations drilling of the first 23 holes suggested total mineralisation of only 13,600 tons with grades ranging from 2.15 to 3.02 percent tungstic oxide. On these figures the Foy report may have been described as misleading but it was not as misleading as the February 3 reserves statement. On Peko's estimates, molybdenum grades ranged from 0.18 to 0.87 percent. Attunga was, in summary, high-grade but too small to be viable. Peko also reported that no large (ie, greater than 500,000 tons) orebodies occurred in the Attunga area but they did not preclude the presence of small (around 30,000 tons) orebodies of high grade (perhaps five percent) tungstic oxide. These would be economic if found.

The inspectors conceded that there was some support for the view that, because of conflicting reports as to the reserves and grave errors in the assays, the directors could have been justified in their non-disclosure of the Foy report. However, the inspectors concluded that the directors had not acted in good faith because they had sold shares on the basis of the same report. The inspectors believed the report should have been published and condemned the directors' sales as reprehensible. They reckoned that the profit of nearly $5 million was probably the largest example of insider trading to be investigated in Australia.

Nor did Hall and Relph escape unscathed. "In our opinion their vast share trading activities in Endurance, a company to which they were consulting geologists, was improper," the inspectors said. "We examined both gentlemen closely about this matter and neither of them saw anything wrong with their conduct. They admitted to us that if they had used inside information to trade in the securities of Endurance this would have been improper. In our opinion they did the very thing which they themselves admitted would have been improper."

The report concluded: "In summary, the trading in

Endurance securities was not on a fair-and-honest market. It was a false market created and manipulated by persons intent on making personal gains at the expense of the uninformed public. The information on which the public relied was false in almost every form. One cannot but be concerned for the plight of the investing public, especially the small investor."

Maguire, in a letter to the Financial Review, dismissed the findings of the inspectors as "preposterous and utter nonsense". He said that at no time when he sold shares had he been in possession of information which was not public knowledge. "I did make a substantial profit selling my Endurance shares on a rising stock market during a mining boom, and for which I offer no apologies, but I emphatically deny that the means by which I achieved this profit were illegal or immoral," he said.

Brudenell-Woods also wrote a letter saying that profits on sales of Endurance shares by the original Attunga shareholders were "a normal business transaction, which any person under similar circumstances would have undertaken." He added that the figures quoted were far in excess of the true amounts received.

It must be pointed out that insider trading was not illegal at that time, unless fraud was involved. The Securities Industry Act, making insider trading illegal, was not enacted until another year had passed.

When the inspectors were appointed, they naturally focussed on the highly misleading assays from DDH2. They concluded that the drill cores had been salted. The following quotation is from paragraphs 219, 220 and 221 of their report.

"During August, Attunga was gaining excellent results from drilling which could have been and probably were anticipated some weeks before being officially reported owing to assessing by ultraviolet lamp. September saw the first liquidity problem with Attunga and the problem grew worse until the company was insolvent in October and November, 1968, with $54,157 outstanding in accounts payable and an overdraft of $3111. Loans made by Furber, Edgar, Harmanis and Brudenell-Woods totalling $164,832.98 had all been expended.

"However, September, October and November saw exceptional drilling results from DDH1 and 2 (which cannot be substantiated) obtained and with the option being granted to Endurance in December the liquidity problem of Attunga was relieved by the receipt from Endurance of $50,000 in cash as part consideration for the option. In December, DDH2 showed exceptional tungsten values on assay, which, as we have already stated, were subsequently checked by six laboratories showing errors of up to 4000 percent. The result was the rise in price of the shares in Endurance to over $10 in January. Attunga was able to sell off some of its shares in Endurance allotted as part-consideration for the granting of the option, restore its liquidity and retire $164,832 of directors' loans.

"Is it a coincidence that Attunga should find exceptional tungsten results (results that are seriously in error) just when it was in severe financial difficulties and certainly proceeding along the path to liquidation? Is it an accident that the assays are wrong? Is it a coincidence that Endurance immediately after being taken over by the syndicate expanded its capital step by step with the emergence of tungsten results from Attunga?

"We have already expressed our opinion that the samples sent for the assay from DDH2 or the assays were fraudulently tampered with by persons unknown."

DIGRESSION 1: The Art of Mine Salting

Mine salting is a long-established practice which has always carried a dash of glamor of the old Wild West. Salting reached its height of popularity in gold mining during the 19th century. A prospector would buy or otherwise obtain some gold dust, a shotgun and some cartridges. He would remove the buckshot from the cartridges and substitute the gold. When fired at a stope (face) underground the gold would splatter widely, embedding itself as fine particles in the rock. The result would look like auriferous ground and with luck the mine could then be sold at a high price. It is doubtful whether any experienced prospector or mining engineer was ever taken in by this device, but certainly a few mines were sold this way. Salting was mainly an American practice, but folklore alleges that it used to happen on the Victorian golfields too.

Considering the proliferation of sharp practice which occurred during the 1969-70 nickel boom, there were comparatively few rumors of salting. Perhaps this reflects not so much the morals of the promoters as their frequent lack of technical skill. Salting requires some expertise and usually the complicity or active assistance of whoever is working at the mine site. Salting also involves tampering with pieces of crushed rock amid the flies, heat and dust of the Australian outback and this was not a congenial pastime in an era when money could be made in perfect comfort, shuffling pieces of paper around O'Connell Street.

Gold mines have always been the easiest to salt. This is because gold is very valuable and widely available. The fact that gold is a high-priced metal means that it is economic to work very low-grade mines. An underground gold mine today would be economic at about half an ounce of gold to the ton of ore. An open-cut would be economic at two to three pennyweights to the ton (there are 20 pennyweights to the ounce). Anyone who could afford a few ounces of gold and scatter it judiciously among the ore going into the concentrator could boost the assays of the mine by 50 or 100 percent. A six-pennyweight mine is a far more attractive proposition than a three-pennyweight mine.

Obviously the people in the best position to salt a mine are the owners and the field workers. They have access to the site and if their holdings in the mine are substantial they have a strong motive.

When a prospect is in its exploration stages, the trick is to salt the drill cores. This can be done quite simply by adding whatever element is desired to the core samples being sent in for assay. Usually there is enough outcrop material around to suffice, but even if it is not, samples of metals such as copper, nickel or tin are easy to obtain. Not much need be added to boost an assay from, say, 0.5 to 2 percent. With gold, a little dust can be injected into the laboratory bags after they have been sealed.

The one proven case of mine salting in the nickel boom was that involving Leopold Minerals NL. On March 19, 1971, this small company announced that it had struck nickel averaging 5.33 percent over a 25 ft intersection at Nullagine in Western Australia. This was an excitingly high assay, reminiscent of the rich discovery values reported by Western Mining in 1966.

The shares took off. From Thursday's close of $1, they soared to $5.50 in Friday afternoon trading before closing on $4.90. When trading opened the following week, Leopold dominated the market, hitting a peak of $8.80 on March 22 in frantic trading. Turnover was enormous, with more than a quarter of the company's issued capital turning over in four days of trading. Then on March 23 the stock was suspended halfway through the morning session at the request of the chairman, Alan Treloar.

There were elements of humor, provided you weren't one of the bulls who had bought at more than a dollar. For one thing, the Nullagine prospect had been shown in the prospectus as a uranium prospect. Leopold was supposed to have been drilling for uranium when it struck nickel instead.

When Leopold took off, the shares had been listed in Perth and Sydney but not Melbourne. The Melbourne Stock Exchange requested their listing and took the unusual step of listing them on a Wednesday — a day earlier than usual. The result was that Leopold created a record which will probably stand for all time for the shortest listing ever on an Australian stock exchange. The stock was suspended just 95 minutes after being listed, which meant that Melbourne brokers had been allowed to participate in the Leopold boom for an hour and a half. The main loser was the company, which had paid $600 to list in Melbourne. This worked out at $6.30 a minute.

To professionals, there had seemed something wrong with the Leopold run from the start. The run occurred when the rest of the market was in the horrors (Minsec had crashed the previous month), which meant that Leopold took off very much on its own. The Perth Stock Exchange were sceptical from the start and asked Leopold how the nickel occurred. In Sydney, partners and clients of Constable and Co, the brokers who had floated Leopold, were very heavy sellers. David Constable was senior partner in the firm and his brother Robert was on the Leopold board.

Robert resigned on March 24 and issued two letters revealing doubts about the assays. Constable said that, despite sustained efforts and a visit to the field on March 23, he had been unable to see any part of the 25-ft strike core. He had been told that all the core had been pulverised except for two small pieces and those could not be located by the company. To anyone familiar with drilling procedure this sounded suspicious.

Leopold immediately became the subject of more official investigations than any other company in Australia. By the end of the week it was being probed simultaneously by the NSW Companies Office, the Perth CIB, the WA Government, the Senate Select Committee on Securities and Exchange and the Perth Stock Exchange. Brian Cutler, a director of Leopold who had made more than $300,000 trading the stock, was requested to appear before the Senate committee. He gave evidence on April 6 but during the dinner break the following day he caught a flight to Sydney. The Senate committee issued a summons to get him back and complete evidence.

On April 22, the 30-year-old Cutler was arrested on a charge of making a false statement with intent to deceive. Three weeks later, still unable to find the missing core, the company put down a deflection drill-hole under police supervision at Nullagine. The hole, providing core samples at the same depth as the original and only three feet away from it, showed there were no significant nickel values in the section where Leopold was supposed to have 5.33 percent.

Cutler pleaded not guilty, denying he had salted the drill-core. In a long unsworn statement from the dock in the WA Supreme Court, Cutler said he had lost an enormous amount whatever the outcome of the hearing. His exploration consultancy, Geodrillers Pty Ltd, had had to close, his staff had been scattered and his arrest had destroyed any hope he had of starting a consultancy again.

The jury found Cutler guilty and he was sentenced to 3½ years' jail. Exactly how the drill-core had been salted will never be known. Unofficially, the most popular theory was that some concentrate from Kambalda had been added to the core. Only a handful or so would have been needed to produce the readings.

Cutler's real crime was misjudgment. The assays were so high that they attracted suspicion. The timing was poor also, because by March 1971, when VAM and Minsec had just crashed, there was a scent of scandal permeating the whole securities industry and the call for action and reform was strong. Cutler was a natural scapegoat.

The only other time that salting was publicly alleged was in the Endurance report. The inspectors' finding was based largely on the fact that the timing of the high DDH2 assay had been extremely convenient for the vendors of Attunga. Given the enthusiastic share dealing of all parties involved and the excessively high assays this was a natural suspicion. But a closer examination of the facts suggests that the overstatement of DDH2 was due to laboratory error and the drill-core was not salted by anyone.

To explain this conclusion it is necessary to go a little further into the methodology of assaying and the characteristics of tungsten. Scheelite is very friable and the drill cores break and powder easily. It is very easy for them to become contaminated during on-site inspection and where economic ore grade can be below the one percent mark, quite small doses of contamination can boost the assay significantly. Even some of the oils used in drilling can produce fluorescent readings.

Drill cores are sent from the site as pulverised rock inside a bag. At the laboratory, a small sample is taken from each bag and is mixed with liquid to become a pulp. Some of this pulp is then tested for whatever metal is being sought. DDH2 was originally analysed by Geological and Mineralogical Laboratories Pty Ltd. They used a wet gravimetric analysis. In this form of analysis, the assayer begins with a sample of given weight, say 10 grams, and refines it chemically. First, sulphuric acid is applied to remove soluble sulphates. At this point the solution may still contain insoluble sulphates of the alkaline earths such as barium, strontium and calcium. These have to be cleaned off very carefully by leaching, a process that has to be repeated three or four times. If this is not done very carefully the heavy elements remaining in solution will distort the result of the final weighing and give an inaccurately high reading for tungsten.

After the check assays with three other laboratories had revealed much lower values from DDH2 (particularly in the lower-grade material), Foy returned to Geomin's laboratory. As was normal procedure, the unused parts of the pulps had been stored for future checking. Usually they are kept until the company orders them to be thrown out. If the drill-hole had been salted, these pulps would still have been salted. A number of them were checked, this time by colorometric analysis, and recorded much lower tungsten assays than before. They had not been salted. The most logical conclusion therefore is that there was an error in the original tests.

The risk of error in DDH2 was heightened by the fact that the ore was in a metamorphosed limestone, which meant that there would have been considerable calcium in the samples sent for assay. Unless the calcium was removed meticulously it would have increased the tungsten reading.

Geomin, which has earned a high reputation for its work, has always denied any error. It is fair to record that the results of the check assays also showed wide variations between the other three laboratories, although none of the readings were as high as Geomin's. Considering the volume of work done in laboratories on assays at this time and the speed at which results were demanded by companies, it is perhaps surprising that there were not more errors.

As this book went to press in the last half of 1978, a small share market boom was running in diamond miners. It should be pointed out that in some respects it would be easy to salt a diamond prospect.

Very briefly, diamond prospecting consists of searching for

kimberlite pipes, which are the host ground for diamonds, putting an auger drill-hole or two down in each pipe to test for "indicator minerals" and, if the indicators are there, bulk sampling of the surface gravel of the pipe.

Diamond mines can be economic at very low yields. The economics of each mine, and each country, will vary but in broad terms a mine would need to produce about 10 one-carat stones in every 100 tonnes of ground treated. Of these 10 stones, three would need to be of gem quality. A stone may be of one carat or more but not of gem quality because of shape, color, cracks or other flaws. As a carat is only 0.2 grams, or one-one hundred and fortieth of an ounce this means that a diamond mine can be viable on less than a gram of diamonds to every 100 tonnes of gravel mined. Where economic concentrations are as low as this, salting is easy and relatively inexpensive. Half a dozen half-carat to one-carat stones scattered on the ground to be sampled will vastly enhance the ground's potential as a diamond prospect.

Ideally, the diamonds should be characteristic of the locality. In northern NSW, for example, Bingara stones are known to be small in average size and very hard. However, where virgin or near-virgin territory is being explored little or nothing is known of local characteristics and salting would be easier.

In another respect, diamond salting would be most unlikely to fool an expert. Almost invariably, prospectors are put onto the scent of diamonds by what are called the indicator minerals, notably chrome diopside, magnesium ilmenite and pyrope garnet, which have chemistry peculiar to kimberlite pipes. Any prospector or company who announced a diamond discovery where none of these or other indicator minerals had previously been found would be viewed with heavy suspicion. This might not apply, however, to an alluvial prospect, where water has flushed the diamonds from their original home in the pipe.

The situation would become trickier if a pipe bearing some indicator minerals but no diamonds were salted. This is because, statistically, it takes a great deal of bulk sampling to prove the viability of a prospect. The rule of thumb is that a pipe cannot be said to be viable until 3000 carats have been recovered at a satisfactory rate of yield. This would entail the sampling of 300,000 tonnes of ground. The unreliability of an early high showing would therefore only slowly become apparent. Such salting would be difficult to prove, because it could happen in nature that a few diamonds could occur on the top of an otherwise barren pipe.

A final thought. Endurance and Leopold were by no means the only companies to report high assays or reserves which later had to be reduced dramatically. We shall see further examples in this book. In the hit-and-miss stages of early exploration some errors are inevitable. It takes a great deal of investigation and considerable technical knowledge to separate reasonable errors, compounded by optimism, from criminal attempt to defraud. However, after allowing for the necessarily imprecise nature of early exploration, some results looked a little peculiar. There were probably one or two other cases of salting during the boom which went unpunished because they were not investigated. There was one case where some shares ran strongly but the assays were not so inordinately high as to arouse an investigation.

Salting, it would seem, requires a fine hand.

---

[1] Where one or more inspectors are appointed to carry out a special investigation into a company under Part VI of the Companies Act (as Lockhart and Mealey were), they have the power and opportunity to probe the affairs of the company and all connected with it in considerable detail. They can demand all books and records relevant to their inquiry, and examine under oath anyone whom they suspect can aid their investigation. Such inspectors therefore have far greater access to facts that can ever normally be gained by the public. When tracing an intricate web of corporate manipulations, these inspectors are engaged in highly sophisticated detective work. Whether they reach the right conclusions from the facts they unearth depends, of course, on their own judgment.

[2] Inspectors' report, p. 48, para. 231.

[3] The main objection to first-come bids is that they can open and close very quickly, giving the mass of the shareholders no opportunity to participate. At present the law is being redrawn to try to force bidders to keep their offers open for a set period, perhaps 30 days.

CHAPTER FOUR

The Oil Floats

For Australia, 1969 was the last of the really good years. Our economic problems, as New Years Day dawned across the Pacific, were slight and arousing little passion. Kenneth Davidson, economic writer on the Australian newspaper, was at loss for any substantial criticism in his January 2 preview. Under the heading "Signs Reveal Sound Year for Business", he wrote: "In terms of demand, of production and of employment it looks like being a good year. The economy enters 1969 nicely balanced on the knife-edge of full employment without undue inflationary pressures".

The share market obviously agreed, because it had been bubbling right through the Christmas-New Year break. Almost the only other news was the play in the Third Test between Australia and the West Indies. The Financial Review headlined its market story "A Bullish Cocktail for 1969" and said: "The Sydney share market which reopens today should set the pattern for a continued upsurge across the boards with the emphasis — as experienced in the 1st days of 1968 — on the oil and mining sections".

The news was overwhelmingly favorable. One of Sydney's top broking firms, Ord, Minnett, T.J. Thompson and Partners, issued an optimistic circular predicting that the Sydney mining index would at least double by 1975. The firm said that in the six years to June 1968 the earnings of seven heavyweight mining stocks (BH South, Mount Isa, CRA, EZ, Mount Lyell, New Broken Hill and North Broken Hill) had grown at a compound rate of 21 percent a year while their share prices rose at a rate of 36 percent a year. "In view of the fact that company earnings can be expected to grow at a compound rate of at least 20 percent over the next seven years we consider the strong share-market rise in 1968 well justified," Ords said.

The managing director of International Nickel (Australia) Pty Ltd, Mr D.K. Couts, writing an article for the Financial Review, declared that Australia's worldwide image was changing. "Within the next decade, it will be known for its minerals more than its wool," he said.

Investors were thinking along the same lines, reflected by surging turnover on the Sydney Stock Exchange. On January 7, turnover reached 4,776,000 shares — only 100,000 short of the previous record volume set in September 1968. Good news continued to flow in from the field. On January 10, Amad NL announced that it had secured exploration rights to the whole of the Sudan and BHP announced a nickel find at Mount Monger in W.A. The Amad announcement underscored the fact that the Australian boom was going international. Even small companies were exploring overseas. The BHP find was bannerlined across page one by the Australian, which said: "Stock markets took it that BHP was close to entering the booming nickel industry". In the event, BHP never developed its find at Mount Monger (about 12 miles northeast of Kambalda) and Amad did little in the Sudan, but both announcements reflect the promise and excitement of the time.

The enthusiasm was not confined to Australia's shores. In London the financial press indulged in a welter of favorable comment on Australian shares over the holiday period, crowned by a lengthy and bullish article in the Investors' Chronicle, Britain's most influential investment magazine. The Chronicle, said 1969 could well bring a rise in prices for Australian mining stocks and was particularly bullish about Metals Exploration NL, which was then developing the giant Greenvale nickel mine in partnership with Freeport of the U.S.A. This was another case of misplaced enthusiasm. The soaring costs of Greenvale, combined with depressed world nickel prices in the 1970s, have made the project a financial disaster.

However, at the opening of 1969 no such clouds darkened the sky. Dr Lewis Weeks, discoverer of the Bass Strait oilfields, estimated that Australian oil reserves amounted to about 2500 million barrels of which some 93 percent was under the strait. On these figures the strait reserves were 50 percent higher than the reserves which had been announced by Esso-BHP only a few days before.

Another factor reinforcing optimism was the short lead-time between discovery and mine development in those days. One example was the railway to the Pilbara's Mount Newman,

Australia's largest iron ore mine. The final spike was driven by Western Australia's Premier, David Brand, on January 22, 1969. The 265-mile railway had been completed in the truly remarkable time of 18 months. Such speed in development encouraged a belief in quick rewards.

In these expansive times, investment analysts took a bullish view of price-earnings multiples. The p/e ratio is one of the standard measurements of a share's value. If a stock is earning 10 cents a share after tax and is selling on market at $1, it is on a multiple of 10 times earnings. This ratio is especially useful when comparing shares of companies in the same industry. Thus if the shares in three breweries are selling at 12 times earnings and a fourth is only on a multiple of eight, the fourth is the best value to buy unless there is some good reason for it being marked out of line with the others.

The great problem with p/e multiples is that they change with the fashions. In the good times of the early 1970s, the accepted p/e ratio for a good industrial stock might be 15 times. In the crunch of the recession four years later, this shrank below 10. By mid-1978 it would be around 12. As p/e multiples are largely an expression of confidence, they are likely to fall at the same time share prices are falling. This produces a doubly dramatic effect. If a company was earning 20 cents a share when the going multiple was 20 it would be selling at $4. If its earnings dropped to 10 cents at a time when multiples had shrunk to 12 it would be selling at only $1.20.

Take Mount Isa as an example. In January 1969, the Sydney brokers, Rudd, Bain, Macdonald & Co, adopted a range of p/e ratios from 25 times to 35 times in projecting future share prices for the company. Rudd, Bain said these ratios were 'conservative in relation to past experience". In January 1969, Mount Isa shares were selling at $7.50 on earnings of 11.2 cents per share, putting them on a multiple of 67 times earnings. So Rudd Bain were certainly being conservative. But by March 1978 earnings were 15.2 cents a share on expanded capital and Mount Isa (now MIM Holdings Ltd) was selling at $1.70, putting the stock on a p/e of just over 11 times.

One index of the times was the sudden upsurge of interest in the workings of the stock exchange. During January, the Sydney Stock Exchange had enrolled a record 352 students for its course in stock-exchange practices and procedures, an increase of 140 on the previous year. (The mortality rate among these students must have been severe; the course took four years and any student graduating in 1973 would have found himself in an industry racked by corporate crashes and stockbroker insolvencies.)

The good news continued to roll in. At the start of February Alcoa landed a $US250 million contract to supply alumina to the British Metal Corporation and Bougainville secured a 15-year contract to supply 950,000 tons of copper in concentrate to seven Japanese smelters. Contracts such as these emphasised the fact that the heavyweight miners and the small, speculative companies were living separate lives. The great mining discoveries and developments which had created the boom were made by the heavyweights, whose shares had now become expensive by Australian standards. If these companies announced higher profits, new discoveries or large developments, the announcements often had little impact on the share market. The heavyweights led the market into boom conditions, but the trading was increasingly in the smaller stocks.

The reason for this can be summed up in a single word, leverage. Anyone buying shares in Conzinc Riotinto of Australia Ltd at $20 would have to see them go to $40 before he could claim a 100 percent gain. A small miner which had few assets, however, could easily rise from 20 cents to 40 cents. Indeed, a small stock could rise from 20 cents to $2, in which case the gain was 900 percent. Also, CRA had total assets of $500 million, which meant that a discovery worth $20 or $30 million was of little significance. But to a tiny company such as Endurance, the news that Attunga could be worth $20 to $30 million sent the shares through the roof.

An amusing example of the relative immobility of the heavyweights occurred at the CRA annual meeting in March 1969. Late in the meeting, in reply to a question, the chairman, Sir Maurice Mawby, revealed that the group was in a consortium which had discovered two new nickel deposits in WA. CRA had 26-2/3 percent of the consortium, with Anaconda of the USA holding 60 percent and New Broken Hill holding 13-1/3 percent. The partners had found one orebody of 1 million tons of nickel averaging 3.5 percent at Redross and another of the same size averaging 1.2 percent at Widgiemooltha. As soon as Sir Maurice had announced this two share traders at the meeting bolted from their seats and raced for the lifts. Although the meeting was on the top floor of the 36-storey CRA building at 95 Collins Street, the high-speed lift service would have delivered the pair downstairs to phone their brokers before Sir Maurice had answered the next question. For all this fast footwork they may still have lost money, because CRA moved down 40 cents to $20.80 that day.

Speculative stocks were by no means the sole province of shady, disreputable traders. Companies which would normally be regarded as staid pillars of the establishment were participating. In March, Boral Ltd announced its participation in the float of Narla Minerals NL. Boral took four million of the initial issued capital of 12 million 50c shares paid to 10 cents each and the rest were underwritten by Ian Potter & Co. Narla had been formed to jointly explore several diverse mineral claims with Newmont Pty Ltd, a subsidiary of Newmont of the USA. The most noteworthy feature of the prospectus was that it conformed to US practice by declaring boldly on the cover "This issue is speculative".

To companies the size of Boral there was a natural incentive to get a stake in mineral exploration. The outlay was tolerably small and the rewards could be great. The Australian Gas Light Co was another large company to join the ranks of the explorers. With an eye on the possibility of discovering its own reserves of natural gas, AGL participated in the float of New South Wales Oil & Gas NL. This float was launched by the Planet Oil group. NSWOG issued 10 million 50 cent shares at par with one transferable option attaching to each share. AGL, which was already a large shareholder in Planet, took 500,000 shares together with the options. NSWOG was to acquire all the Planet oil and natural gas interests in NSW plus 4000 square miles of adjoining territory in Queensland.

In the hunt for big equity partners the most outstanding coup was by Dr Kon Panikol, a former chemist who had become one of Australia's busiest mining entrepreneurs. Panikol floated Genoa Oil NL to the public in November 1968. Shortly afterward one of the top advisers to Crown Agents of the UK met Panikol during a visit to Australia and said that Crown Agents would have liked a stake in oil exploration. He couldn't have mentioned it to a better man. Within three months Panikol had accommodated Crown Agents by forming Hartog Oil NL as a spin-off of Genoa. Crown Agents invested $1.1 million in the Hartog float and the terms of the agreement provided that their ultimate investment could rise as high as $5.5 million. In practice, the Crown Agents financed Hartog's exploration in two promising areas of Alaska. Compared to other investments by Crown Agents in Australia, Hartog did not do too badly. The amount Crown Agents dropped in Hartog would barely have paid the interest on their property disasters here, and the potential rewards in oil exploration were greater.

March also saw some old gold miners back in prominence. On March 13 shares in A1 Consolidated Gold NL, an old company which had become a vehicle for boom prospects, suddenly jumped from $2.40 to $6.20. The rise was on a rumor that the $3 shares would be split into 20 cent units. The company denied the rumor in the afternoon but the shares only fell to $5.80. In theory, of course, a split adds nothing to the value of the stock, but speculators were banking on the theory that there would be opportunities for high-leverage gains in a lower-priced stock.

The run on shares in Cocks Eldorado Dredging NL had even less logic behind it. This was another old-established company which had fallen upon hard times. The shares ran from $1.20 to $1.75 in a few days as three broking firms vied for stock. It was noted by one bemused commentator that all Cocks owned was a disused dredge, a few investments and its listing. The buying had been by Sydney entrepreneur Boris Ganke, who turned up a few days later at Cocks Eldorado annual meeting with one-third of the issued capital. The meeting, held in a small Melbourne office, was a strained affair, even though it was suggested that "someone should tell a joke to relieve the tension". Chairman C. Wallace Smith, himself a prominent Melbourne broker, said the company was negotiating for areas in WA.

The highlight of the market each week was Thursday morning. This was when new floats were listed for the first time, providing they had been approved by the listing committee of the home exchange. Speculators and shareholders would wait eagerly in the public gallery to see how the new companies made their debut. Usually they would enter the market at handsome premiums, giving big killings to the stags.

A stag in market parlance is someone who takes shares in a float and sells them soon after listing. During the boom this was a painless way of making money, and there was intense competition to be on a broker's "float list". Other things being equal, clients with the largest and most active trading accounts received preference. Sometimes brokers would reserve large quantities of scrip for their own families, employees and friends. On other occasions large volumes would be reserved for the promoters, becoming in effect a hidden vendor consideration. However, the underwriting broker had to place some stock with the genuine public to achieve a spread of shares sufficient to meet exchange listing requirements.

A lucky client of an underwriting broker might be offered 500 shares at 50 cents each in a float. The shares would cost $250 to the client, but they might come onto the market a few weeks later at 75 cents. The client could stag the shares immediately and take a profit of $125 less tax. A big trader would not normally confine his account to one broker but would have accounts with several, thus multiplying his chance of getting into floats. Speculators who bought the shares upon listing also had a chance of making money, because frequently a stock would languish in the market for a month or so while the stags sold and then might go for a run.

As floats were such a dominant feature of the mining boom, it might be best to describe the mechanics of floating publicly. All companies when first formed are private companies, although they may change their status very quickly into public companies. To form a private company it is necessary to choose a name, which is approved and registered by the Corporate Affairs Commission of the home state, a memorandum of association setting out the general aims of the company and articles of association setting out the way in which the company shall be run (voting power of shares, election of directors etc.). Upon lodgment of the necessary forms and declarations with the CAC the company is legally formed.

Normally it is necessary to have at least two shareholders upon foundation. If they have one $1 share each the private company thus formed has an issued capital of $2. Its authorised capital, which is the amount it may issue without a further resolution by shareholders, may be several million dollars or even hundreds of million. To become a public company it will need a greater number of shareholders. Let us say a company was being formed to invest in mining stocks and was seeking to begin with $1 million issued capital. The promoters would seek the money by issuing a prospectus to the public, inviting applications for shares. A prospectus is a legal document which is required by the Companies Act to give subscribers essential information about the company, its assets, its directors, the capital structure and other specified facts. The aim is to give enough information to enable a subscriber to make a reasonable decision on whether he should invest. The making of a false statement in a prospectus is properly regarded as one of the most serious offences which can be committed under the Companies Act, and the penalties include jail. It was for a prospectus offence that Stanley Korman was sent to Pentridge, although to be fair to Korman he was convicted on a prospectus upon which no subscriber lost money.

In the boom, when the public was avid for mining floats, brokers used sometimes to reserve large proportions of an issue for themselves so that they could stag it at big profits after listing. One effect of these large reservations for vendors and others was to make the scrip more tightly held. The float of Australian Consolidated Minerals NL in October 1969 was a prime example of this.

ACM was formed with 17 million 20 cent shares, which was quite a large issue by boom standards, and the issue was underwritten by two Perth broking firms, D.J. Carmichael & Co and Saw, Cambridge & Brannelly. Of the 17 million shares, 6.8 million were reserved for prior subscription by shareholders of Metals Exploration NL, Western Queen (1936) NL, Mountain View Gold NL and West Coast Holdings Ltd, four companies which had agreed to sell ACM various minerals claims and leases for the float. The remaining 10.2 million were "offered for public subscription".

The underwriting brokers took 4.2 million shares for distribution, of which 550,000 was retained by the brokers, their families and associates. Another 3.2 million shares were taken by Patrick & Co in Sydney. Six share trading companies associated with that firm took two million of that allocation. The remaining 2.7 million was allocated to directors and promoters of ACM.

March and April of 1969 saw a spate of oil company floats, some of them huge. March 6 saw the listing of Vamgas NL, an offshoot of the diversified mining group VAM Ltd. Vamgas' 50 cent shares opened at 80 cents and closed the day on 84 cents. Vamgas had a 10 percent interest in the Gidgealpa-Moomba gasfield which would eventually supply natural gas to Adelaide and Sydney. In the case of this float the faith of investors was not unjustified, because Vamgas survives to this day as a partner in an operating gasfield. Some speculators went further on March 6 and also bought shares in VAM, which rose 42 cents on the day to close at $6.22 with a market capitalisation of $25 million. VAM was an ambitious group put together by Frank Close, a mining engineer. What the speculators did not know was that the downside in these shares was $6.22. Eighteen months later VAM was in receivership and their shares were worthless.

Nickel-boom prospectuses sometimes left room for doubt about their legality. The few that were ever subjected to an official inquiry of any sort revealed material facts which had not been disclosed to subscribers, and doubtless there were more that went undetected. The most notable feature of boom prospectuses was the high value placed upon oil or mineral prospects which were unpromising. But it is not an offence to buy something at a high price, even something valueless, and as long as all known details are set out in the prospectus the public cannot say they were unfairly induced to buy. But in 1969-70, when the market was boiling, the trick was not to be able to analyse a prospectus but to get hold of one, because subscribers to almost every float had the opportunity to make money, sometimes multiplying their original subscription.

In the case of the investment company quoted above the prospectus would be "empty"; that is, the company would have no assets to begin with. Most mining and oil companies were formed with either possession of some mineral claims or an intention to acquire them. A promoter might have a block of claims he valued at $500,000. Typically, the company would float with the stated intention of buying the claims from him. He would be paid either in cash or shares. The shares, taken at par value, would be called vendor shares and were usually restricted from trading for a set period.

There were also a great number of spin-offs, sometimes called hive-offs, where an existing public company floated another. An example of both vendor considerations and the spin-off can be seen in the float of Southern Cross Exploration NL. This company was floated in August 1969 with the aim of an issued capital of 15 million shares. Seven million 50 cent shares paid to 10 cents each were taken up by directors and their nominees. Half a million fully paid 50 cent shares were issued to vendors. These shares were the consideration for the purchase of a private company called Flinders Mining and Exploration Co NL, which had a copper prospect in South Australia, a base metal prospect in Queensland and nickel claims at Scotia and Nepean. (It was almost obligatory during the boom for new companies to have a nickel claim somewhere in their prospectus.) Of the remaining 7.5 million shares, one million were reserved for subscription by shareholders of the Brisbane-based investment company Alexanders Ltd. The other 6.5 million were offered to the public. All these 7.5 million were 10 cent contributing shares. The company would therefore have an issued capital of $1.7 million, of which $250,000 were vendor shares and the rest issued for cash. As the directors included several from Alexanders, this was an Alexanders spin-off.

Formation expenses could be heavy. One of the largest was the commission paid to the underwriter, who was usually a broker. An underwriter signs a contract promising to raise the amount required by the prospectus, say $2 million. For his risk and services he receives a commission. This was normally around the one percent mark, but could vary. In the case of Southern Cross Exploration quoted above, the big Sydney firm of Ralph W. King and Yuill took a commission of three-quarters of one percent on 7.5 million shares being issued to the public, ie $5600. If King and Yuill were only able to place five million of the shares they had underwritten there would have been a shortfall of 2.5 million shares which the brokers would have had to take up themselves for a cost of $250,000.

However, there was no shortfall on the Southern Cross issue, nor on any mining float at the height of the boom. But there were some big shortfalls in ones that came in late, from 1971 onward.

On March 27 Hartog listed and rewarded stags with an instant 300 percent gain. Perhaps buoyed by the news that Crown Agents had invested in the company, the 10 cent shares sold at 40 cents in Melbourne before closing at 36 cents.

The market was kept on the boil by the constant news of yet more floats in the pipeline. On March 28, for instance, the Australian reported three upcoming floats. Endeavour Oil had registered a prospectus at the Victorian Companies Office and directors had announced a public float would be made within a few weeks. In Sydney, a company called Combined Resources Ltd was planning to float as a mining investor. It was overshadowed by news of the impending float of Harbourside Oil NL. Harbourside would have 19,990,000 50 cent shares issued at 10 cents each with a free option attached to each share. A spokesman for the underwriting brokers, Johnson and Wadell, said applications already exceeded the number of shares available. John Bovill was to be chairman of the five-man board, whose directors included Alexander Barton. A press release said Barton was responsible for the formation of Harbourside and also for attracting overseas capital to the venture.

Harbourside was large, but there were larger to come. The number of oil companies formed in the boom was far smaller than the number of mining companies but the average oil company had much larger issued capital. In terms of numbers of shares and options issued, these companies were among the largest ever floated in Australia.

Although there had been plenty of losses by traders of oil stocks, their faith in the companies had not been entirely misplaced as the leading explorers of the Rough Range boom — Ampol Ex, Woodside and Santos — had all gone on to find oil or gas fields. This explains much of the enthusiasm for oil floats in 1968-9. Oil was still being discovered in Bass Strait and it seemed possible that Australia might be as rich in oil as it was in minerals. The gloom of the 1970s, with its monotonous dry holes and poor tests, was still distant.

This optimistic mood supported a phenomenal chain of oil floats in 1968 and 1969. In the 12 months to the end of April 1969, oil-exploration floats raised more than $33 million. All this would not be spent on exploration, however, because formation expenses had already snipped more than $2 million from the total. In this period 15 major new public companies were launched with a staggering total of 350 million issued shares and options. The most profitable business at the time must have been printing scrip certificates. Neil Spiers on the Sydney Morning Herald tried to calculate one day the acreage of scrip created by the new companies and was disappointed to find that all the pieces of paper laid side by side would only cover a few square miles. He had expected it to cover New South Wales.

Formation expenses tended to be high. Basin Oil raised $1 million cash, out of which underwriting costs took $125,000 and a further $18,000 went in preliminary expenses. Thus $143,000, or one dollar in every seven, went before the company spent a cent on exploration. Genoa raised $3 million of which $80,000 went to a company called DTR Services Pty Ltd. DTR was owned by two directors, Dr Panikol and Gabriel Lorentz, and by Pat Burke, who joined the board shortly afterward. The $80,000 paid for legal, accounting and other services, including professional consulting fees, plane fares and accommodation. In addition $200,000 went to cover other preliminary expenses, including Ord's underwriting fee.

On top of these considerations, Genoa directors and their nominees were reserved 7.8 million of the 15 million shares in the float.

The floats were also obtrusive enough to begin diverting the attention of Federal Cabinet from such weighty matters as the

SEATO Treaty and the future of Malaysia. In the case of some of the oil floats it was an open question whether the prime motivating factor was the discovery of hydrocarbons or the avoidance of tax. Under the tax laws as they then stood, subscriptions to calls on oil company shares and money spent exercising options were tax deductible. The initial and praiseworthy objective of this concession was to stimulate the formation of companies that would search for oil in a continent that badly needed it. By April Fool's Day 1969, however, the authorities were becoming increasingly concerned by the fact that the concession was stimulating the formation of oil companies without any corresponding increase in the discovery of oil.

The apparently insatiable public appetite for oil exploration floats was largely explained by their appeal as a tax shelter. The shares were normally issued in a form to make them tax deductible. Thus, instead of offering shares fully paid to 50 cents on subscription, they would be offered as 50 cent shares paid to only 10 cents. The further 40 cents to be called would be tax deductible. Sometimes the shares were issued at one cent on application plus a nine cent call payable immediately, which meant that 90 percent of the money subscribed could be claimed as a tax deduction. Share options were also deductible when exercised. This led to a rash of combined share and option issues, as companies offered shares with options attached. Sometimes the options were free but sometimes they cost one cent each. Subscribers did not even care if the shares listed at a discount on par, because losses on the shares were also tax deductible. Thus on the same share in the same issue a subscriber could claim a deduction for a call paid and, after selling, a deduction for any loss made. This double deductibility began to come under heavy fire, particularly from the Treasury.

The financial editor of the Sydney Morning Herald, Tom Fitzgerald, calculated that if all the known option holders in Australia exercised their options on the same day, the Federal Treasurer would wake up next morning to learn that his Budget was down by $65 million in revenue. At that time, eight companies accounted to 108 million of the options.

Until April 1969, the prevailing convention had been to limit option entitlements to a ceiling of one per share. This barrier was broken by Ian Sykes[1] founder of the cut-price petrol chain XL Petroleum and a notorious convention-breaker, in the float of Petrocarb Exploration NL. Petrocarb issued a prospectus for five million $1 par shares with one-and-a-half options attached to each share. The other novel twist to this float was that it had no underwriter. The prospectus was available from the office of XL Petroleum or from XL's petrol stations in Melbourne, Albury and Geelong. It conjured up the spectacle of a motorist pulling up in the driveway and saying: "I'll have five gallons of super, a pint of 30 grade and 2000 Petrocarb."

The peak of the oil floats was in April 1969 when two of the biggest floats Australia had seen in terms of share scrip hit the market simultaneously. This was on April 17, when Reef Oil NL and Harbourside Oil NL listed on the same day in Sydney. Reef Oil was a Murray and Maguire float, later to become a minor partner in the Cooper Basin. Harbourside was from the Alexander Barton stable. The two companies between them had 60 million shares and 40 million options.

A wave of paper hit the trading floor as turnover on the Sydney Stock Exchange reached 8.3 million units. The great bulk of trading was in the new stocks. Nearly 1.6 million Reef shares were traded plus 1.4 million options. Harbourside turnover was nearly 1.9 million shares and 1.4 million options. This meant that 6.4 million units were traded in these two stocks alone in one day. Volume was so huge that some of the larger houses put extra operators on the floor. A few had one specialist each in Harbourside and Reef. Considering the volume of stock the prices were good. Reef opened at 13 cents and fluctuated between 12 cents and 15 cents before closing at 15. The options closed at 11 cents. A stag would have shown a 160 percent profit on the 10 cent shares and one cent options. Harbourside closed at 10 cents, while its options closed at seven cents. As these shares were issued on the same terms as Reef's, a stag would have shown 60 percent profit.

A year before there had been 45 oil companies listed in Sydney and 48 in Melbourne. Now there were 64 in each city. Meanwhile, turnover figures in the two capitals had changed dramatically. A year earlier trading in oil stocks had been roughly even, with 2.4 million shares turning over in Sydney in the third week of April 1968 alongside 2.1 million in Melbourne. By April 1969 Melbourne turnover had increased slightly but Sydney's had rocketed. In the week Reef and Harbourside listed Sydney's oil turnover, including options, totalled 15.8 million units. Oil and mining floats had made Sydney the dominant exchange.

Another half a dozen oil floats were still in the pipeline,

including the biggest of them all. On April 29, Harbourside registered Bounty Oil Ltd as a spin-off. It had an authorised capital of 200 million 10 cent shares. Alexander Barton was on the board.

---

[1] No relation to author.

CHAPTER FIVE

Indigestion and Interest Rates

The oil floats were giving the market a bout of indigestion. Apart from the oil explorers and a steady flow of small mining floats, there had been large rights issues by some of the biggest companies in Australia. The public paid $44.5 million in application money during March for share issues by BHP and the Bank of New South Wales. In April, CGFA and British Tobacco were raising another $30 million. A further 14 companies were making smaller issues, bringing the total raisings to $50 million that month. The flood of new money in the equities market was being dried up by masses of paper.

This checked the market, although nothing had happened to daunt confidence. Patrick & Co, who by then had become the largest broking firm in Australia and specialised in mining stocks, published a comprehensive review entitled "Australia — 1975". The firm concluded that the future appeared full of promise, particularly for the corporate sector.

The bullishness of the small promoters was reflected in the real estate market. As the newly-floated companies set up their headquarters, they opted for impressive offices. One of the favorite addresses was the Australia Square building in the heart of downtown Sydney and handy to the Stock Exchange in O'Connell Street.

This was high living in both senses of the word, because Australia Square was then Australia's tallest building and, in the days before the property boom had bequeathed us an office glut, was charging top rents of $7 to $8 per square foot. The Genoa-Hartog group were on the 43rd floor in combined offices. Don Hoult, writing for the Sydney Morning Herald, estimated they were paying about $18,000 rent a year. Alexander Barton had a luxurious suite of offices on the same floor. The only people higher in the building were Lend Lease Corporation, who had erected it, and the revolving Summit

Restaurant. Barton and his son Thomas were regular diners at the Summit.

Other tenants included Canadian Superior Oil, Kennecott Exploration and Union Oil. In contrast with these American giants on the tenant list was Dominion Mining NL with an issued capital of $1 million. Allstate Exploration, which had not even listed at that stage, was also planning offices in Australia Square. The broking fraternity was represented by Ord Minnett, which spread over three floors. After the boom collapsed, Ords prudently retreated to the Royal Insurance building alongside the Harbour Bridge approaches, nearly half a mile from the exchange.

Another industry thriving in boom-time conditions was the charter aircraft business. The most organised operation was run by Business Jets Pty Ltd from Mascot. For $1975, Bizjets would take you on a five-day tour of selected mine sites in a Falcon fan-jet which seated 10 (the seating was adjustable for smaller parties). The plane zoomed its passengers around the Outback in air-conditioned comfort at 540 miles per hour while a hostess supplied food and drinks. It was by far the most comfortable way of travelling to Australia's remote mines and for overseas visitors who wanted to see as much as possible in a short time, the fare was bargain price.

Typically, a flight would begin in Melbourne pick up passengers in Sydney then visit Weipa, Mt Isa, Juno (the Peko-Wallsend gold mine at Tennant Creek), Port Hedland, Dampier, Mount Tom Price, Mount Newman and Kambalda. In addition, tourists would fly over the Moura open-cut coal mine in Queensland, BHP's Groote Eylandt manganese mine, Mary Kathleen, Barrow Island and Mount Goldsworthy. The passengers were usually geologists, mining analysts, stockbrokers and the occasional journalist or professional investor.

At the other end of the charter aircraft industry was the espionage business. As small exploration companies began working on prospects in remote areas, they found themselves plagued by spies and trespassers. Leopold Minerals' experience was typical.

Even before announcing its infamous "find", Leopold had been under constant surveillance by spies in light planes and helicopters. A senior geologist with Geodrillers Pty Ltd, Mr. D. Harrison, told the Rae Committee[1] that people from other mining companies flew over the site regularly. Their planes had been recognised. Sometimes the spies moved in closer. On one occasion a stockbroker's agent had been found hiding in the undergrowth near the drill site.

Companies tried to keep trespassers off their ground by imposing security. This security was often rough and ready, as at Trough Wells where International Mining Corporation used barbed wire and shotguns. Aerial spying was faster and more comfortable than crawling about in the bush. It could also be quite informative. Outcrop mineralisation and gossans can be seen clearly. The amount of work done and the delineation of the possible orebody can be estimated from the number and position of trenches and drill-holes. The West Australian inland air is dry and clear in the early morning before heat haze develops and a skilled geologist with a good camera can glean a lot of information. If the racks of drill-core trays are not covered, he can estimate the footage drilled and might even be able to detect how much of it was in ore. There is no record of any outraged exploration boss ever taking a pot-shot at a low-flying aircraft, although some may have tried.

One example of this aerial snooping[2] is given in the Rae Report which described, with an air of faint incredulity, the methods used by Burrill and Associates. Geoffrey Burrill and Roy Jones were partners and directors of Burrill and Associates, one of the top geological consulting firms of the boom. The firm produced a monthly bulletin called "Mineral Exploration in Western Australia" for which subscribers paid $1000 a year. It had about 30 subscribers in Australia and overseas.

One means by which Burrill and Jones gathered information was by flying over exploration sites. An essential part of the operation was to look for black sludge coming from drill-holes, as this could indicate mineral sulphides.

Burrill's newsletter had a deservedly high reputation, combining technical expertise with brevity and giving its subscribers an excellent precis of what was happening on exploration sites all over WA. It was probably the most reliable information produced by any boom commentator. Here is a short extract from their January 1970 bulletin:

"Zone A — Map 8: Area 1 — Mt Keith.

"Metals Ex and Freeport are continuing their program at Mt Keith. They have two diamond drills near the north end of the property. Recent results are believed to be very erratic and they are having difficulty correlating their drill results with the geophysical anomalies.

"Area 2 — Windarra.

"(a) Poseidon NL now have 3 diamond drills and 2 percussion drills on their property. All these rigs are concentrating at the north end of the property.

"(b) Conwest have now gridded their property and carried out a magnetometer survey. It is thought probable that they will be drilling within the next 2 months.

"(c) North Deborah have commenced a program of gridding and a magnetometer survey. It is not considered that this ground has any merit."

Sometimes the bulletin was caustic, such as the comment about the Kambalda area in February 1970: "Kambalda Mines have reported that their rig is now at a depth of 1,400 ft and aimed at a target area of around 2000-3000 ft. It is considered that the possibility of this hole encountering nickel sulphide mineralisation is roughly one in a billion. It would be most surprising if they ever hit serpentines let alone nickel sulphides."

Back on the market, Mineral Securities Australia Ltd had announced on May 6 1969 it had gained control of two prominent beach-sand miners, Cudgen RZ Ltd and Consolidated Rutile Ltd. Mineral Securities, usually called Minsec, was one of the glamor companies of the boom. It had emerged as the biggest share trader Australia had ever seen. Minsec had made huge, fast profits trading in mining shares and was now building up a diversified group of mining companies. In a quick coup it had now gained control of 42 percent of Cons Rutile and 63 percent of Cudgen. Minsec had previously held significant minority stakes in both companies and had now upped its holding in a single deal.

Both the deal and its announcement were typical of the big, dramatic style of Minsec. From its incorporation with an issued capital of just under a quarter of a million dollars in 1965-6, Minsec had mushroomed. By bold and brilliant share trading it had amassed huge profits, which had been reinvested to acquire controlling interests in a string of medium-sized mining companies.

The market marvelled at both the size and skill of the Minsec trading. Brokers vied for their business, which had been a large factor in boosting turnover on the Sydney Stock Exchange way above that of its traditional rival Melbourne. Minsec had sprung from a geological and mining consultancy named Kenneth McMahon and Partners. Its founder and chairman was Kenneth McMahon, a likable and highly competent mining consultant who had the vision of forming the first

Australian-owned major mining house since the rise of the Collins House group in the 1900s.

McMahon's vision was shared by many. With his technical expertise, his financially patriotic statements and his hail-fellow-well-met demeanor, he was widely liked and admired. The managing director and co-founder of Minsec was Tom Nestel. Dark, reserved and a Presbyterian lay preacher, he was a complete contrast to McMahon. He was widely respected, however, for the way he had financed Minsec's rise to stardom. It was Nestel who ran the share trading which provided Minsec's financial sinews.

In addition to Cudgen and Cons Rutile, the Minsec stable included Petroleum Securities Ltd and Oil Investments Ltd, which were largely share traders, Amad NL, Pexa Oil NL, and the Aberfoyle group. The Aberfoyle companies — Greenbushes Tin NL, Paringa Mining and Exploration Co Ltd, Ardlethan Tin NL, Cleveland Tin NL, North Australian Uranium Corporation and Golden Plateau NL — were mostly tin miners but had interests in copper, wolfram, gold and other minerals as well.

The share market was looking patchy. Stags in Petrocarb and Flinders Petroleum NL enjoyed good listings on May 6. Flinders 5 cent paid shares (it had a complicated capital, but this class of share was the best guide) ranged between 13 cents and 15 cents. Petrocarb's 10 cent shares closed on 15 cents with the one cent options at 11 cents.

At the other end of the mining spectrum Mount Isa announced it would double production by bringing its northern leases, later called the Hilton mine, on stream. This would make Mount Isa the biggest silver-lead-zinc producer in the world. Australia was already second only to Russia as a lead producer and was the fourth largest zinc producer.

For years it had been a standing joke around the market to say that some company had "found another Mount Isa". Now Mount Isa had done exactly that. The market mulled over the news for the weekend and following Monday then put the shares up by 44 cents on Tuesday. Other heavyweights showed only small gains and then fell back to their original levels. Interest was flagging, particularly in London. Only a discovery by one of the small miners could, it seemed, provide the necessary shot in the arm.

If the small miners were not discovering anything, it was not through lack of pegging effort. The Western Australian Minister for Mines, Arthur Griffith, told the annual meeting of the WA Chamber of Mines on May 28 that his department was having trouble meeting the demand for prospecting land in WA. Almost 9000 claims had been lodged in 1968 compared with 3000 in 1967.

In June the market went lower still, hitting its lowest point so far for 1969 on June 10 when the Sydney non-ferrous metals index plunged to 3656.55 points.

This was a reaction to a sudden rise in US interest rates. In the first week of June the US prime rate — the minimum bank lending rate to the biggest and most credit-worthy customers — jumped a full percentage point to 8½ percent. In Europe, Eurodollar rates moved up sharply in sympathy. With three-month Euro rates at 11½ percent, London money was diverted away from Australian mining stocks into fixed interest securities on the Continent.

In theory, rising interest rates have a dampening effect on the share market. This is because fixed-interest returns are seen as competing with share yields as an investment attraction. If share yields are low and interest rates are rising an investor is presumed to switch money into fixed interest to avail himself of the higher returns. It also used to be the theory that share yields should always be higher than fixed-interest rates, because of the greater amount of risk involved in holding equities. This has not been truly applicable in Australia for the past 10 years, as share yields have generally been lower than returns available on debentures. There is a continuing argument among market analysts and economists on the relative merits of equities and fixed interest securities, and the point at which investors should switch from one to another.

In June 1969, the diversion of London funds was blamed for a setback in the share market for industrial and mining heavyweights. The general tone of the market dragged speculatives down too. The impact of the interest rate rise on speculatives was secondary, because a speculative miner normally does not pay any dividends and therefore has no yield.

The Australian authorities had also been edging up official interest rates. In the three weeks to June 10, the Reserve Bank had been more active in open-market bond-selling than at any time in memory, driving the rate on long-term non-rebatables up from 5.44 to 5.8 percent. In the bond market, this is a large move. In a society where low interest rates had been stable for more than a decade, it was a substantial shift. The Reserve Bank normally retains a Sphinx-like silence when asked about its market operations. Sometimes it will issue a statement of Delphic ambiguity. In June 1969 not only was there a policy of inscrutability, but almost nobody was scrutinising because the attention of the financial sector was monopolised by the share market. News of the long-bond rate-rise gradually filtered through from the bond dealers, which meant the market took weeks to feel the impact.

News from the field was still good. WAPET struck a promising oil flow at Dongara, raising false hopes for what turned out to be a gasfield. The end of June was marked by the opening of the Mount Newman iron-ore mine, a ceremony which was one of the most impressive testimonials ever paid to Australia's mineral prospects. The opening was officially performed by the Governor-General, Sir Paul Hasluck, in a ceremony transmitted by satellite television to London, Tokyo, New York and Australian capital cities. It was a fittingly great opening for a great mine.

On the economic front, the Commercial Bank of Australia's monthly economic review opined that any tendency toward boom conditions in the Australian economy may have already been checked. The lack-lustre market was given a shot in the arm by the listing of Carr Boyd Minerals Ltd on June 5. The 20 cent shares ranged from $1.80 to $2.50 before closing the first day at $2.35. Although the company had raised only $1.2 million in issued capital, it had been capitalised by the market at $14.6 million at the peak of trading that day. Carr Boyd had a group of mineral claims at Carr Boyd Rocks, near Scotia. Their claims surrounded a very promising prospect being explored by Great Boulder and North Kalgurli.

Not many other floats had the same glamor and new listings began to come under pressure. Normally when a company was listed the underwriting broker would try to set the market on the opening day. A broker who had underwritten an issue of 10 cent shares by a newly-floated company might decide to "set" the market by buying and selling around the 25 cent mark for the first week or two. The broker saw this as a duty to both the company and to his own clients who had taken the shares, but the practice is sometimes indistinguishable from market rigging. A lot depended on the strength of the broking firm. Patricks might spend $50,000 net in the first week of a stock's listing to set its price where they thought it should be. In companies that had no earnings and where net tangible asset backing was only a fraction of their share price, this could be a highly subjective exercise. In the final analysis the broking firm was subject to market forces, because if it misjudged what the market would bear and over-pitched the price it could face heavy losses. Even Patricks could not ignore these market forces. Some underwriting brokers did not try to establish a post-listing market and some were too small to do so effectively.

Proponents of a free market object to this "setting", but in practice it would seem preferеble to what can happen when the market is left to work untrammelled. On June 12 the market was called upon to absorb three new mining and oil companies — Platina Developments NL, Ashburton Oil NL and Murumba Oil NL — with a total issued capital of 86.7 million shares. They came on in a doggy market and operators watching the abnormally heavy trading in Murumba sensed that something strange was happening. It quickly became apparent that the stock was being taken over on-market.

The arithmetic of the raid was simple. Murumba Oil had spent little on exploration. The 10 cent par value shares were backed by nearly that amount in cash. If their market value fell much below par they would be vulnerable to a raid. Anyone paying six or seven cents for a Murumba Oil share was buying a piece of scrip backed by nearly 10 cents cash. As the Murumba Oil shares were widely spread and there were no significant vendor considerations, there was no one with a large enough stake in the company to rally to the defence.

The raider was Mining Finance Corporation Ltd, the recently-formed mining investment arm of Slater Walker Securities (Australia) Ltd. In just one week, Mining Finance acquired a 45 percent stake in Murumba Oil, which amounted to control. Turnover in the stock had been huge. In the first two days trading a total of 10.3 million shares and 3.2 million options had changed hands, generating magnificent brokerage. The stags, happy with their tax deductions, had been prepared to sell below par because they could still show a profit. Double deductibility had become a two-edged sword for company promoters. Mining Finance had not been floated publicly at this point, although its prospectus was released soon afterward.

Two week later another cockleshell launched into the whirlpool as Weatherly Oil NL made its debut. Weatherly had 35 million 10 cent shares and no exploration areas or farm-ins, which meant that nearly all the $3.5 million was still in kitty. In the first days trading 14 million shares changed hands, or 40 percent of issued capital. Of these 7.5 million shares were bought in Melbourne at six and seven cents.

Weatherly, which had been floated by a Texan oilman of that name, had been raided by Genoa and Hartog. The two Panikol companies had bought 13.5 million shares in Weatherly within a week of its public listing, giving them effective control. After a short court fight, the Panikol group established their control in mid-July, and immediately made their presence felt with two announcements. The first was that Weatherly might be given the right to buy an interest in Genoa leases at Moonie and offshore of Western Australia. The second was that contracts had been terminated with three former Weatherly directors, J.E. Weatherly, D.W. Wales and Peter Murray. The contracts, ranging from four to seven years, had been for $20,000 to $36,000 a year.

Overall the market was still stagnant. The Sydney non-ferrous metals index[1] had opened the month at 3986.46, slid to 3656.55 at June 10 and was still around the same level a fortnight later. Then the index jumped 250 points on June 25 to breach the 3900 level and rose again on following days.

This performance should be treated with some cynicism. June 30 is balance date for many investment companies and institutions in Australia, and some of them are not above pushing stock prices up at that time to improve the quoted market value of their portfolio as it will appear in the next balance sheet. On one memorable occasion a large Melbourne broking firm pushed up the whole preference share market on June 30 by entering a series of higher bids for stocks in this generally neglected area of the market. The large number of mining spin-offs during the boom meant that the parent companies had a strong interest in seeing the shares of their listed subsidiaries and investments at a healthy price at balance date.

As the financial year drew to a close the comparatively depressed state of the share market was blamed on London and short selling. As both these influences have a mysterious and malign image in market mythology, this might be an opportune time to look at them.

DIGRESSION II: London and Bear Country

As 1968-9 closed, the financial editor of the Sydney Morning Herald was reflecting on the way in which London was influencing key Australian mining stocks. He referred to them as "the Earl's Court set of Australian stocks ... that colony of shares which have become part of the every-day London investment scene". After looking at recent gyrations in this market, he concluded philosophically: "Our share market has become a cork on the ocean waves which Britannia no longer rules but rides. We will just have to get our sea-legs and put up with it."

The size and nature of the London influence on Australian share markets has been the subject of continuous argument throughout the years. London is not, of course, monolithic in its attitude. There are buyers and sellers in London just as there are in O'Connell Street. Also, much American, European and other foriegn buying is channelled through London, obscuring the picture still further.

London's attitude toward investment in Australian equities can alter for reasons beyond Australia's influence. As we have just seen London buying dried up in early June 1969 because of a rise in New York bank rates. In the mid-1970s London investment in Australia stabilised because of a reluctance to commit funds through the dollar premium pool.

During the boom, London appears to have been a heavy net buyer of Australian stocks, particularly mining heavyweights. Unfortunately, there is no way this can be quantified with any accuracy, because Australian stock exchanges do not declare the amount of trading on behalf of foreign principals, and it would be almost impossible to define what constituted a foreign order anyway. There is even less information available in England because the London Stock Exchange does not disclose turnover volume. All it reveals, quaintly, is the number of "bargains marked", i.e. the number of transactions. This is not very helpful because if there were seven transactions in a stock during one day the actual number of shares going through could be anywhere from seven shares to the whole of the company's issued capital.

London trading has always been strongest in the largest Australian companies. The first prerequisite of what might be called the normal London institutional market is that the institutions can get stock in large volumes. This is why London has historically concentrated on companies such as BHP, CSR and Woodside, where there is a large issued capital and also a speculative interest. London Institutions also have a natural inclination toward companies with strong British links such as CRA, Bougainville, Comalco and the ANZ Bank. Among the mining heavyweights, Western Mining, Mt Isa and

Peko have been strongly traded in London. In 1969-70, Minsec and Metals Exploration could have been added to this list. In stocks such as these the turnover in London has often been greater than in Australia. There is enough scrip on hand in these stocks in London for the British market to have a life of its own, independently of Australia.

A London broker or his client, wishing to buy shares in Australian companies, had the choice of buying on one of the Australian exchanges or buying in London, where a number of Australian stocks were listed. The London board did not list them all, but generally any speculative issue which attracted sufficient interest was soon quoted there.

On Australian exchanges a broker buys or sells shares for his client by dealing directly with another broker. In London he has to go through a jobber to get them. A jobber specialises in dealing in a particular group or groups of stocks. He might be a jobber in building companies, brewers, retailers, etc. During the boom there were up to half a dozen jobbers specialising in Australian stocks.

The disadvantage of the jobbing system is that it adds an extra layer of cost to the business of trading shares. In addition to paying the broker's commission, the client pays for the jobber's "turn". The advantage of the jobbing system is that there is always a market.

A client may want, say, 5000 shares in a small, obscure listed company. His broker will approach the jobber and ask for a quote on the company for 5000 shares (he does not have to reveal whether he is buying or selling). The jobber might quote 90 cents and a dollar meaning he will sell the stock to you at a dollar or buy it at 90 cents)in practice the quotes would be in sterling). If the price is right, the broker can complete the deal by buying 5000 at $1. The jobber may not have the stock on hand, but he is bound to deliver at the price he has quoted. If the client had to buy on an open market, as in Australia, he might not have been able to pick up the shares in one parcel, or may have forced up the price. The jobbers protect themselves by widening their quotation on larger parcels. If our hypothetical buyer had wanted 50,000 shares, the quote would have probably been 85-$1.05. The jobbers are particularly wary on thinly quoted stocks and are rarely caught offering bargains.

Under the London Stock Exchange system, every second Tuesday following the end of the stock exchange "account" is settlement day for jobbers. All shares bought and sold during the account period of two weeks are delivered on settlement day, and cash is paid for purchases.

Settlement of Australian transactions soon began raising problems in London. The vast increase in the number of companies and the volume of scrip on issue, together with the huge increase in trading volume, put huge pressure on the companies' share registries. In a market where the same piece of paper could change hands several times in one day and where share registrars were forced to hire inexperienced staff to cope with the flood of work, errors and delays were inevitable. London brokers were soon complaining that they could not get scrip in Australian companies. Delays of months were common and there were several delays reported of more than a year. In the case of some speculative companies which later collapsed, the share registry was never straightened out and never will be.

If the London Stock Exchange caught up with the 20th century and disclosed turnover figures we would know the size of trading in the London market, but it would not necessarily follow that this was the same as the amount of foreign buying and selling in Australian shares. Much London and other foreign trading is done through Australian brokers on Australian markets. Conversely, an Australian broker or institution or even a well-connected individual can telephone a London broker and buy stocks on the London LSE. During 1969-70, overnight traffic was very heavy in both directions. Even small Australian brokers would keep one partner or senior operator back late at night to check the opening London market and place orders if necessary.

There was also a great deal of arbitraging. An arbitrageur is one who makes his profit when the price of a stock gets out of line in different markets. If a stock is on offer for $2 in Melbourne and is simultaneously being sought at $2.10 in Sydney, an operator on the Melbourne floor can phone his contact on the Sydney floor and buy and sell simultaneously until the price differential disappears. Arbitraging is normally only possible for brokers because the key requirements are direct access to the trading floor and freedom from broking commissions.[4] In 1969-70 when prices were moving hectically, arbitraging could be highly profitable when wide margins opened between different markets.

Brokers would also take stock firm as the Australian markets closed and offer it in London that night. This was done whenever a line of stock appeared late in a company which had some appeal to London. This was not arbitraging in the strict sense of the word. The broker was holding the stock as principal and if London opened lower he would suffer a loss (or he could hold the stock and cross his fingers).

From talking to brokers who were active in London trading during 1969-70, I have come to the following imprecise conclusions:

1. The Australian market generally took the lead from the London market.

2. The jobbers were the most powerful force in the London market.

3. The bulk of the trading done in London was probably on behalf of Australian interests.

The first observation is interesting from a psychological point of view. The impact of London on the Australian market was out of all proportion to the volume of stock traded there. If the London market closed higher, the Australian market would open next morning at least at the London level. This could be on quite small turnover at times. It could occur that 50 Australian stocks might be quoted in London and that on a given night there might be transactions in only twenty of them. Yet when the market closed the whole 50 would be marked up. Next morning Australia would open at the same level.

In stocks such as Woodside there was enough scrip in London for that city to have a genuinely independent market. But in the speculative stocks with tiny issued capitals, London was making a market on scrip that did not exist there and frequently in companies which it knew little or nothing about. Why Australia should follow the London lead in such cases is a phenomenon more in the psychological sphere than the financial. Perhaps we can write it down as one more manifestation of our cultural cringe.

The second observation raises the position of the jobbers. Visiting London brokers to Australia have for years gently pooh-poohed the notion that the jobbers are a powerful influence in the market. A myth, they say. The poor jobber is just making his turn and has acquired totally overblown status in Australian demonology, they say. I have been told this so often that at one stage I believed it. Since researching this book I have reverted to my original opinion. Australian brokers who had to deal nightly through the jobbers consider them to have been extremely clever and the most powerful single influence on the market.

For one thing, the jobbers can effectively override the system by failing to settle each fortnight. Their best excuse for failing to deliver Australian scrip was the all-too frequently valid argument that there were share-registry delays. Where a jobber was short in a stock, the position might not appear as a short in his books, but as a sale for a client. Where a jobber was long he could roll the position over through a buy-back agreement with a bank. Certainly jobbers appeared able to hold positions — particularly short positions — for much longer than 14 days.

In a market with very few jobbers, they could easily combine to their mutual advantage. There is no way this could ever be proved, but if two jobbers were specialising in Western Mining and they had lunch together, they could quickly work out what stock was in the market and what orders were on hand from Australia. They could then move the market wherever they wanted to in the afternoon to suit their books. The same could apply to any other stock.

Midway through the boom, one Australian broker managed to see the book of Brothers, a broking firm which was also a jobber in Australian stocks. The broker was amazed to see the jobber's book totalled $20 million. By comparison, Australia's biggest share trader, Minsec, had a trading portfolio totalling $10.9 million at June 30, 1970. That one jobber had twice the trading clout of Minsec.

The third point, on Australian trading through London, is impossible to be specific about. The partner who handled London trading for one very large Australian firm through the boom years says that in more than half the trades the name on the other side of the transaction was an Australian broker or nominee company. The Australian-based nominee company could have represented either Australian or foreign interests — no-one will ever know — but in the bulk of transactions the scrip never left Australian shores.

Not only are there clever jobbers; there are a few clever Australians too, and some of them could use the London market to set the scene for the Australian market. Presuming an Australian broker, promoter or individual had established himself as a client of a London broker, he could ring the broker at midnight and ask for a quote on 50,000 shares in a speculative company. The reply might be $1.30-$1.50. The Australian could then buy 50,000 at $1.50, effectively taking out the market. Later that night, just before closing, the Australian could ring London again and check what the jobbers were quoting. The company would now be a little higher, say $1.50-$1.65. The Australian would then mop up whatever the jobbers had at $1.65, paying a little more for the few thousand still left in the market.

That morning in Australia, institutions and brokers would scan the Reuter overnight report and see the company had moved up 15 cents in London to $1.65 on reportedly heavy trading. The first hour of a bull market generally being the strongest each day, the stock would probably open around $1.80. Our Australian would promptly unload. This was a particularly easy ploy where the Australian concerned held a large existing position in the stock, such as a vendor interest. The profits were worth the sleepless nights.

This presence of Australians in the London market, combined with the ability of the jobbers to make a market without scrip, would explain a phenomenon which has to date passed without comment. Although London was often reported to be buying Australian stocks heavily, when the boom subsided there did not appear to be any significant increase in UK ownership which had come about through purchases on market. This is especially true of the heavyweight mining stocks, which were traded strongly in London throughout. London institutional buying is certainly very heavy at times, but the evidence points to the conclusion that much of the trading was simply that — trading. If this is true, the London market could have gone on merrily if the companies and the scrip never existed at all.

THE BEARS.

In popular imagery, the bear is another sinister figure in the stock market. For the benefit of those still uninitiated to the market menagerie: A bull is someone who buys a stock believing it will go up; having bought the stock, he is long in it and hopes to sell it for a profit later. A bear believes stock will go down, so he sells it and becomes short in the stock; he covers his short position by buying back in, he hopes at a lower price. There are other denizens in the market menagerie including dogs (chronically hopeless stocks), nits and geese (foolishly optimistic investors) and pigs (speculators who hang on too long instead of taking a profit or cutting a loss), but for the moment we are concerned only with bears.

The problem with short selling is that the bear's total risk is unquantifiable. A bull who buys a Poseidon share at $1 knows that the most he can lose on the share is $1.[5] A bear who shorts

Poseidon at $1 does not know his total risk, but it would have turned out to be $279 per share.

Because of the size of the risk, the bears are usually people with plenty of capital. In the Wall Street crash of 1929-30 the market was beared heavily by big operators such as Mike Meehan and Joseph Kennedy (father of John F.). This led to the image of short sellers as big, wealthy villains who formed a pool to belt the market down and wipe out the values of stocks in which small shareholders had invested their savings.

A layman is often unable to see why someone should be able to sell something he doesn't own. The usual answer is that this happens in all commodity markets and shares are just another commodity. Defenders of short selling say it is a balancing device which helps to thicken and stabilise the market. A short sale is always expected to be covered by a purchase within a limited time (it used to be a fortnight) but there is nothing to stop a bear from shorting again as soon as he has covered his original sale.

In the boom, many small traders who enjoyed early success and developed a little market sophistication began attempting short sales. They quickly discovered that like anything else in the market it depends on timing. Indeed timing is crucial to a bear because he has to cover within a short time so that he can deliver scrip to the original buyer.

There are legends about people who shorted Poseidon when it rose from $1 to $5 and had to cover frantically as it soared upward. I heard of one trader who was short at $5 and lost so much money that his broker had to make him a sub-underwriter in the Laverton Nickel float to get his account square.

When caught in this kind of bear trap, the classic advice is to buy double and hope it keeps rising. This bold course of action calls for strong nerve because when it fails the results are horrible. The Melbourne broking firm of John T. Martin went broke applying this strategy. The firm had a client who had sold Leopold Minerals contributing shares short between 48 cents and $1.06. The last short sale was completed just before Leopold announced its 5.33 percent nickel. The contribs soared to $7.80 within a few days. The firm's clients were short 48,000 shares and under stock exchange rules the firm would be required within three days to meet the responsibility for supplying any Leopold scrip which their clients would not or could not deliver.

Martin and his partners decided to cover and then — after a

nervous lunch — to buy. As covering the short cost $200,000 which the firm did not have, it would have probably gone out of business anyhow and so was justified in trying to recoup its loss. It was a choice between certain death and possible life. The firm went into insolvency with a deficiency of $1.4 million. Its overdraft had reached $670,000 before the bank refused further accommodation.

One group in the boom who were widely suspected of shorting the market were the London jobbers. Some of them may have stayed short for months in particular stocks while pleading scrip delays.

In Australia, the father and mother of all bear traps was in Antimony Nickel NL. This company listed in February 1971 with an issued capital of seven million shares split into 10 cent fully paids and one cent contributing. In the depressed market after the Minsec collapse of February 1971, the Antimony Nickel shares came on only a little above par. A total of 15 brokers began short selling the stock, which was around 20 cents at the time. It was an insane gambit, even without the benefit of hindsight, because the maximum profit would only have been 19 cents a share and the risk was unlimited. What the hears did not know was that one of Australia's most brilliant entrepreneurs, Gordon Barton,[6] was simultaneously going long. Barton's group had been allotted 1.8 million shares in the float, taken up by the Australian Mutual Growth Fund, which he controlled. After listing, Barton began buying with the aim of taking over the company.

Early in March he realised that he could not gain control because the directors and vendors had at least 51 percent already. But by that time Barton also realised that some of his own buying had been from short sellers. He was in a position to squeeze them, which was also his only way of getting out of the stock at a profit.[7]

Barton bought more stock and the price began rising dramatically as scrip dried up. On March 12, Antimony Nickel reached 40 cents and a fortnight later they were $2.30. At one point they hit $3 and Barton was talking of selling back to the bears at $6 or $10. By March 19 the directors had 51 percent of Antimony Nickel, Barton had 44 percent and small shareholders who had not yet sold had about 10 percent. The bears' short sales had resulted in the stock being sold 105 percent.

There was a minor epidemic of these oversold stocks in the market at the time. Northland Minerals Ltd had two groups each claiming to own 51 percent and Metana Minerals NL was reportedly oversold 104 percent, but Antimony Nickel presented the most intractable problem.

The Sydney Stock Exchange suspended the stock on the grounds that there was not an orderly market. The fact that several Sydney brokers were short in the stock, including seven of the 10 firms represented on the exchange committee, was considered by cynics to be the overwhelming reason for the suspension. The exchange chairman, Jim Cooper, whose own firm was not short, stoutly denied any such base motives. Cooper repeatedly asserted that the exchange was trying to establish an orderly market in the stock, but was reluctant to define what constituted an orderly market.

This led to a Kafkaesque confrontation with Barton after the stock had been suspended. Barton's version of the meeting, as told to the Rae Committee, was as follows:

"I asked Mr Cooper why he had delisted the stock and he said that it was done in the interests of preserving an orderly market in the shares of Antimony Nickel. I asked him what, in his view, constituted an orderly market and he said, 'Only experience could tell you that.', and said that if I wanted to know what an orderly market was I would have to refer to my brokers and he took it that they would be able to advise me on this matter. I turned to Mr Clinton, and to Mr Wolfe from A.B.S. White and Co, who had also come to this interview, and asked them what, in their view, constituted an orderly market and they said they were prepared to be guided by Mr Cooper. Mr Cooper then said that in his view any broker of experience would know what was the right thing to do in a set of circumstances and he did not propose to say any more than that."

At this point the directors of the company, who wanted to retain control, had agreed with Barton that they would not sell without first informing him. If the shorters had to cover therefore, they would have had to buy either from Barton or from the small pool of scrip still left in the hands of the original shareholders. The price would be high in either case and could have been ruinous to one Sydney firm.

In the event, the suspension of the stock meant that they never had to cover. To cut a long story short, Barton cancelled the contracts on the eve of a block of scrip becoming available to ease the position. After a court case, Mr Justice Street upheld Barton's right to cancel. The brokers were off the hook but the episode had left a nasty taste in the public mouth and the NSW

Government outlawed short selling. As it was already banned in Melbourne, this was the death knell for shorting in Australia, although it lingered on in Brisbane for some time. Antimony Nickel was never relisted.

Shrewd operators can still go short in Australia by buying put options. A put option, which can be bought through a broker, is the right to sell, say, 1000 Woodside at $2 at any time within the next three months. The buyer of the option might pay a premium of 25 cents. Then if Woodside drops below $1.75 at any time during the term of the put option contract, the holder of the option can exercise his option by selling to the party on the other side of the contract (called a writer) and cover by buying Woodside on market. If Woodside shares rise, or do not fall far enough to cover the premium, the holder of the option simply lets his contract lapse at the expiry of the period, losing the premium. The opposite of a put option is a call option, which is an option to buy a stock at a set price within a set period. Another form of option is the double, under which the buyer can exercise if the stock goes either up or down. This is a particularly risky form of option for the writer, particularly in volatile mining stocks, and it is difficult to find anyone prepared to write doubles. The biggest doubles writer in Australia in the early 1970's was the Melbourne broking firm of W.J. North & Co, which went broke writing them.

Bears can also use a technique called a sale against the box. A trader may have 1000 shares in a stock which he believes is over-priced. He sells and the price goes down. He can either deliver the scrip or, if he believes the price may bounce back, cover the original sale by buying back. This is a sale against the box. The Rae Committee discussed the possibility of outlawing sales against the box, but it is difficult to see how this could be done or how the market would benefit from such a ban.

---

[1] The first chairman of the Senate Select Committee on Securities and Exchange was Sir Magnus Cormack, but for convenience I have referred to it as the Rae Committee throughout.

[2] This aerial spy work would appear to be within the law. There has been an interesting judgment by Mr Justice Griffiths in a UK case, Berstein of Leigh (Baron) v. Skyview & General Ltd. Lord Bernstein had sued for trespass a company which had taken aerial photos of his estate. Mr Justice Griffiths said in his judgment: "The balance

(between the rights of the landowner and the general public) is in my judgment best struck in our present society by restricting the rights of an owner in the airspace above his land to such height as is necessary for the ordinary use and enjoyment of his land and the structures upon it, and declaring that above that height he has no greater rights in the airspace than any other member of the public".

[3] Throughout this book I have referred to the Sydney non-ferrous metals index as a gauge of the market in mining shares. However, this index only reflected the movements in a handful of heavyweights. At June 1969 the stocks in this index were BH South, CRA, EZ, Mount Isa, Mount Lyell, New BH and North BH. A better barometer of the speculative market was the Patricks Speculative Index, which showed the market capitalisation of 30 selected mining stocks, none of which paid dividends. Poseidon and Tasminex were excluded from the Patricks index to avoid undue distortion. A comparison of the two indices is shown on page 146 at the end of Chapter 10.

[4] A broker does not charge himself commission.

[5] Presuming he buys a fully-paid share or one in a no-liability company. If he buys a part-paid share in a limited liability company he is also liable for the amount of unpaid capital on each share if the company makes a call. And if he buys a share in an unlimited liability company his risk is unlimited, as unit holders in Dr Michael Garretty's Dividend Fund discovered to their grief.

[6] No relation to Alexander Barton, whose name was Sandor Buchalter before he Anglicised it.

[7] Barton even constructed a mathematical model to determine how far the price would rise if he kept buying.

CHAPTER SIX

The Lull

The 1969-70 financial year opened bearishly on the stock exchanges. The indices slid backward as the month progressed. At the start of July this appeared to be due to heavy London selling, with large turnover in mining stocks. For four consecutive days trading on the Sydney Stock Exchange was running at about 2.5 million mining shares per day. This was nowhere near as high as the turnovers achieved in the scrip-rich oil stocks, but was the highest turnover yet seen in mining shares. As July progressed, the market continued to slide amid pre-Budget jitters. London was still reported to be selling heavily.

An additional factor depressing the market was the rise in domestic interest rates. At the start of the month the Commonwealth Government announced the terms for its July cash-and-conversion loan, which included a hike in the long-bond rate from 5.4 to 6 percent.

As ever, the speculatives were the least vulnerable to monetary pressure. In the week after the higher rate was announced, the most notable performance on the market was that of Scamander Mining NL, which jumped 16 cents in one day to a high of 88 cents on rumors that it had tendered successfully for nickel claims in the Carr Boyd Rocks area.

Carr Boyd Rocks had become the latest fashionable "address' for mining companies. When the boom became more frantic, the search for addresses, in the case of some companies, had a higher priority than the search for minerals. These were the companies where the concern for the share price exceeded the concern for exploration. There was no real 'rush' to Carr Boyd because the most promising area had been pegged by Great Boulder-North Kalgurli and by Carr Boyd Minerals. Nevertheless it was the first of a string of fashionable addresses which later included Windarra, Mount Venn, Murphy Hills, Trough

Wells and a host of lesser areas. For some reason, addresses at successive new "hot" areas excited more speculative fever than those near proven mine sites such as Kambalda, Scotia and Nepean.

The 16 cent gain on Scamander to 88 cents represented a rise of more than 20 percent. On the same day, Carr Boyd Minerals moved only from $2.38 to $2.45. Patently, it was far easier to make a 20 percent gain in a lower priced stock. This was the magic of leverage.

Another point worth mentioning about Carr Boyd is that there was a pegging dispute between Carr Boyd Minerals and Great Boulder-North Kalgurli over one area of interest where the claims adjoined. The dispute was settled out of court, in Carr Boyd's favor, but not before Great Boulder-North Kalgurli had sunk a few drill-holes in the contested area. In mid-July the partners announced they would open a mine in their area at Carr Boyd. They simultaneously announced — unsportingly — that the drill-holes sunk in the contested borderline area had produced negative results. Shares in Carr Boyd Minerals dropped 18 cents to $2 on the news.

This was one of the first serious over-pegging disputes of the boom and the precursor of many more. Over-pegging reached plague proportions in Western Australia, particularly in areas where interesting new discoveries were being made. Some over-pegging was accidental and some caused by incompetence, but some was deliberate. At best, a prospector might overpeg a neighbor's claim as a precaution in case the neighbor's claim was faulty in some respect. At worst, over-pegging was a form of extortion in which the over-pegger could either drive out the original claimant (where a big company deliberately over-pegged a small prospector) or demand cash in return for waiving his right to fight a lengthy and expensive court case.

Some of the disputes were authentic Wild West material. Leopold had one intruder who drove up at night in a four-wheel vehicle with the lights out and began moving claim pegs. They caught him red-handed and sent him packing.

The situation was not made easier by Western Australian mining law, under which a claim could be pegged for any minerals not being sought there already. This meant that the same piece of ground could be pegged more than once. A first prospector might peg it for gold, a second for iron and a third for nickel. This provided a loophole for unscrupulous prospectors and companies. A company may have already pegged a promising area for nickel, copper, gold and cobalt.

They could be over-pegged, legally, by someone else who said he was looking for tungsten or some other mineral. The over-pegger would then proceed to look for nickel. No one ever found viable nickel by this method, but it is fascinating to speculate on what might have happened if anyone did. What did happen was considerable friction between prospectors and rival companies. In the long run the only people who made any money out of these stratagems must have been lawyers.

Professional claim peggers made fortunes during the boom. One such specialist was 31-year-old Campbell Cooper, who eventually floated his own company, Western Compass Minerals NL. Cam used to operate in a converted five-ton military weapons-carrier. This tank-like vehicle carried a diamond drill, percussion drill, a motorcycle and a gyrocompass. It had a range of 1500 miles on a full tank and carried 60 gallons of water. On arriving at a site, Cooper would work out where he wanted to peg, take a bearing by the compass, calculate how many chains he had to go in each direction, then drive off in a straight line. "Unless trees are six inches or more in diameter, the vehicle charges straight through," he said.[1]

Some peggers did not bother whether the ground was prospective for minerals. If a company contracted them to peg claims, they would peg claims — quite legally — in whatever area was desired. The peggers were well paid for their work and the company could boast of claims at a fashionable address.

Another ploy used to create interest in a stock was to report preliminary exploration results such as surface sampling and costeaning (trenching) in terms of parts per million. Thus a company might report that it had recorded 80 parts per million nickel and 75 parts per million copper on one of its prospects. There is no evidence that any such announcement ever induced worthwhile speculative buying in a stock, because any investor could work out for himself that 80 parts per million was only 0.008 percent. As mentioned in chapter two, nickel comprises 0.009 percent of the earth's crust, so on an average an investor could have found these sort of values in his own backyard. The measurement of surface samples in parts per million is a standard geological practice, but like many other practices it is capable of misuse.

As July progressed the market drifted downward. There was no lack of good news of the solidly reassuring type. BHP announced a modest lift in profit from $47.2 million to $50 million in 1968-69, and there was the usual debate among analysts about what the steel giant's real profit was after adjusting for its unique method of charging fixed-asset utilisation. More to the point was Hamersley's profit jump to $11.8 million in the June half, indicating that the iron miner would top $22 million in calender 1969 — a year ahead of its prospectus forecast. Western Mining reported new nickel finds in the St Ives area of the Kambalda field.

Geoffrey Blainey's colorful history of Australian mining, *The Rush That Never Ended,* was being published in its second edition and was winning a lot of space in the book review columns. As well as telling the exciting and sadly-neglected history of the industry, the book carried the message that mineral discoveries in Australia had always been due to a combination of incentives and intelligent searching. The message is still true.

Blainey's optimism was outstripped by that of G. Paul Phillips, executive director of the Australian Mining Industry Council. Speaking in Melbourne, he predicted that the coming decade could witness the discovery of 12 new nickel mines in Australia. He added that a calculation made by one of his colleagues indicated that iron-ore reserves in Australia could total as much as 100 million million tons — "enough to supply the world for the next 140,000 years".

Even further along the optimism scale was Frank Close, founder and chairman of VAM Ltd, who on July 11 held one of the era's classic press conferences. VAM was a company which had mushroomed with the 1968 boom and now had a stake in a number of mineral ventures, the main one being the Gunpowder copper mine in western Queensland. The press conference was to discuss the development of a gold prospect on the Burdekin River. The newsmen present dutifully noted that VAM and its partner, Amad NL, might make a working profit of more than $2 million a year out of the prospect and that the area was now being explored intensively. Good assays had recently been reported from the field, where exploration had begun six months ago, and it had even been said that there was gold in the trees.

The newsmen paused. Gold in the trees?

Yes, the trees, said Close.

Close said his attention had been drawn to the Burdekin four years ago by a prospector named 'Red' Kennedy who had been addicted equally to pig-shooting and fighting. Kennedy had told him that there was gold in the trees along the Burdekin and that he had washed grains of gold out of sand caught in the paperbark trees to prove it. VAM shares rose 10 cents that day to $8.50, close to their top of $9 for the year to date. Their bottom had been $2.70.

The gold-in-the-trees story raised a laugh at the time, but the market did not know that VAM began making an undisclosed share issue just over a fortnight later. The company raised $676,600 by "placing" 100,000 shares on market through the Sydney broking firm of Ralph W. King and Yuill. The market and the investing public were not informed of what had happened until after the shares had been sold.

Now the price of a share — in theory — is the worth of the company divided by the number of shares it has on issue. If a company is believed to be worth $3 million and it has three million shares on issue, they should be worth $1 each. If it has six million shares, they will be worth only 50 cents. The value an investor places on a company can be calculated in many ways and always has some subjective element in the calculation, but the number of shares on issue should be a matter of record. In the case of VAM, the company's share capital was being increased without the investing public being informed until afterward.

The Rae Committee inquired into the placements and found that King and Yuill had begun placing the line of 100,000 on July 29. These shares were sold on market — many of them in London — to buyers who had no way of knowing that they were buying previously unissued stock.

This placement accounted for 22 percent of VAM's turnover in the seven weeks that King and Yuill were pumping the shares out into the market. The so-called placement is one reason why the shares dropped from $7 to $5 in that time. It was not until September 19 that VAM announced it had made a placement through King and Yuill.

Normally, a placement is an issue of shares made privately to one or more institutions or private investors. It is a fast way of raising capital and can be used as a defensive manoeuvre when a takeover is threatened. However, it also dilutes the equity of the existing shareholders. However, one normal characteristic of a placement is that the recipients know it is a placement. In VAM's case the shares were sold on market.

New shares issued by a company in a private placement are not usually allowed to be sold on a stock exchange until the exchange approves their quotation. The Rae Committee found that "in the case of this private issue of VAM shares, the company, with the acquiescence of the broker, deliberately withheld information about the placement in the belief that this would enable the new shares to be sold to the misinformed public at a higher price than would have been possible had there been immediate and proper disclosure".

While the speculatives were showing flashes of life, the market in general was continuing down. Pre-Budget jitters had pushed the non-ferrous metals index to 3766.09 by the end of July. The market stayed around these levels until the Federal Budget was brought down on the night of August 12. Despite the fact that the Federal Treasurer, William McMahon had been arguing for restraint, the Budget was expansionary. This was interpreted as being the result of pressure from Prime Minister John Gorton, who was facing an election at the end of the year. The only factor in the Budget directly affecting the market was the removal of double deductibility on oil-and-mining shares. These shares actually rose slightly next morning, while the rest of the market showed little reaction. The non-ferrous metals index was drifting, and shed 90 points by the end of August.

The stags were having varied luck. They made gains of up to 375 percent when Westmoreland Minerals Ltd (another Alexander Barton float) listed on August 7, but later in the month North Flinders Mines NL came on at par. Offshore Oil NL, entered the lists on August 21, was the first oil stock to come on since the Budget had removed double deductibility. It had also removed a lot of charm from oil stocks, as Offshore showed when its four cent paid shares sold for 1½ cents on the first day.

More significantly, the free market price of nickel was soaring overseas. On July 11, the miners of International Nickel had walked out at the giant Sudbury mine in Ontario. The strike, over a new three-year contract for the miners, showed no sign of ending. At that time Inco produced some 50 percent of the non-Communist world's nickel. Consumers, who had undergone cutbacks in 1966 due to a similar strike, now faced another scarcity of nickel.

At the beginning of 1969 the free market price of nickel had been about £1500 a ton, but by mid-March it was £2000. When the Inco strike began it had risen to £2730. It rose to £3600 in the next few weeks and then miners from Falconbridge, the other giant Canadian producer, joined the strike. The free market price soared to £4250 a ton.

American and European consumers were more concerned by

the scarcity of the metal than the price. The producers' price was only £986 a ton, but supplies were rationed. The soaring nickel price naturally enhanced the earnings and prospects of existing Australian mines, although only part of their output was sold on the free market. It also provided a strong incentive for the burgeoning number of Australian explorers.

In Australia, the non-ferrous index firmed steadily in September, reflecting renewed confidence in the heavyweights as Mount Isa announced a record profit and BHP agreed to sell 150 million tons of ore to Robe River. Among the floats, one of the better performers was Gippsland Minerals NL, the 25 cent paid shares of which opened in the market at 50 cents. The enthusiasm for Gippsland was particularly strong in the firm of W.J. North & Co. Having underwritten the company at 25 cents a share, the company sent out a circular before Gippsland listed recommending that clients buy at 38 cents as soon as it hit the boards. There was "an air of quiet confidence surrounding this exciting newcomer," North said.

Not everyone was confident. The chairman of Duovest Australia Ltd, leading Melbourne stockbroker Arthur Goode, told his company's annual meeting that no immediate short-term lift could be anticipated in market prices. He was wrong, but he could hardly be blamed for failing to foresee what was about to happen.

On September 25 an obscure Adelaide-based exploration company suddenly jumped 33 cents to close at $1.48 in brisk trading. The big roller-coaster ride had begun.

Poseidon was moving.

---

[1] Interview with John Mitchell, the Age, November 1969.

CHAPTER SEVEN

The Poseidon Adventure

On Friday September 26 Poseidon moved forward another 42 cents to $1.90. After a weekend of feverish rumors, it hit $5.60 by the close of trading on Monday, then moved up to $7 on Tuesday. It was headed for the wide blue yonder and was destined to take the whole mining market with it. In its lifespan of four months, the Poseidon boom was the wildest phenomenon ever recorded on Australian stock exchanges. During that time, the public knew very little of what a handful of insiders had been doing in the stock. The facts came to light years later.

A major reason for Poseidon's spectacular rise was that it was a classic tightly capitalised stock. When its price took off in the last week of September, Poseidon had only two million shares on issue, of which 1.1 million were fully paid to 20 cents and the remainder paid only to 12 cents. At market levels of $1 for the fully paids, the company was capitalised at only $2 million. When a company with such a small number of issued shares and such a low market capitalisation strikes a major orebody, the shares can rise meteorically.

The gambling spree was apposite in one respect: Poseidon owed its name to a racehorse, which used to run in all purple. Poseidon is the name of the greatest three-year-old ever to race in Australia. He won the AJC and Victoria Derbys and St Legers, and the Caulfield and Melbourne Cups all in one year — 1906. Among the punters who backed him in the Melbourne Cup were a group of Italian miners. Later, working a claim 100 miles south-east of Tennant Creek, they named it the Poseidon mine in fond memory of their big killing.

The Poseidon mine was a wolfram deposit in the Hatches Creek area. After many vicissitudes, the mine passed into the possession of a syndicate headed by E. Broughton Jensen in 1952. Jensen deserves a special niche in Australian corporate history. He was one of the founders not only of Poseidon NL, but also of Woodside (Lakes Entrance) Petroleum NL — two of the greatest mining exploration companies Australia has ever seen. Jensen formed a company to mine the wolfram deposit and named the company Poseidon after the mine. The other members of his syndicate were Colin McArthur, an Adelaide stockbroker, and Gordon Campbell, a solicitor.

Poseidon No Liability started badly and continued the same way. The price of wolfram fell and for the next decade the share price was never more than sixpence.

"We didn't have very much cash to start with," Jensen said.[1] "Wolfram could not be sold at that time and we were just trying to keep the mine going to pay the workmen. I worked the winch without salary for months, although I was a director of the company."

By 1966, Poseidon was a destitute company. Its shares were quoted at two or three cents. In July of that year a Sydney entrepreneur Boris Ganke wrote to the largest shareholder, Pioneer Mines NL, offering to buy 70,000 shares at 2½ to 3 cents each. The offer was refused. Ganke was attracted at the time to a number of small mining companies because he felt they could be rejuvenated. Since the discovery of nickel at Kambalda six months before, mining appeared a promising area of investment.

Ganke pursued his object, travelling to Adelaide in February 1967 for an auction of forfeited shares in Poseidon.[2] He bought 50,000 for two cents each, out of the total of 237,000 shares which had been forfeited by shareholders for non-payment of calls. Ganke was the biggest single buyer at the auction. Later Ganke acquired another 50,000 shares, bringing his total holding in Poseidon to 100,000. This represented roughly a quarter of the company's issued capital at the time and entitled Ganke's group to a seat on the board. A.C. Scott was nominated as Ganke's representative.

At the Poseidon annual meeting in December 1969, many of the Adelaide broking establishment who supported Shierlaw and the board appeared to be under the impression that Ganke was an interloper who was trying to muscle in on Poseidon. In fact, Ganke had been in the company well before Shierlaw.

The company was still tiny when Ganke became its largest shareholder in 1967. Paid capital totalled only $82,491 at balance date that year. But after 15 years of hand-to-mouth existence, Poseidon's fortunes changed early in 1968 when it bought an option over some nickel claims at Bindi Bindi, about 120 miles north of Perth.

The claims had been bought by a syndicate headed by Shierlaw. A condition of sale with the original vendors had been that the syndicate would put the claims into a listed company. Poseidon, small, broke and Adelaide-based, was a logical choice. The syndicate took up 400,000 shares in Poseidon, some paid for in cash and others as vendor shares in consideration for the Bindi Bindi claims. The deal could be described as a partial reverse takeover, because at the finish the syndicate had 22 percent of the issued capital and the vendors had 15 percent. Shierlaw became the largest shareholder, with a total of 150,000 shares in various family holdings. The deal was consummated before Ganke, resident in Sydney and involved in a network of other transactions, knew it was happening.

Shierlaw, a rough-hewn individual with a shock of black hair and a down-to-earth manner, had graduated as a mining engineer from Adelaide University in a course interrupted by Army service in World War II. After completing the course he went to work for North Broken Hill at the Barrier, where he earned a reputation as a good mining engineer and also did an accountancy course.

He later set up as sole proprietor of the Adelaide broking firm of N.C. Shierlaw and Associates, but continued to take an active personal interest in mining and prospecting. He spent a lot of time in the bush inspecting mines and claims, and had sunk many a cold beer while chatting with prospectors. In the urbanised profession of stockbroking he had a refreshing smack of the great open spaces, always looking out of place in an office. He was a throwback to the earlier type of miner-cum-broker who had traded on the Bendigo and Kalgoorlie exchanges in their heydays.

The rough exterior covered a shrewd brain. One adversary said later, without rancor: "Under that rough and ready, good-old-Australian front, you've got to realise he's a cunning bastard."

In early 1968 the Bindi Bindi claims looked promising. The Shierlaw-Poseidon deal was mutually beneficial. Poseidon badly needed an injection of life in the form of a promising new mining prospect and Shierlaw was in search of a public company. Such marriages are occurring constantly between syndicates and small companies, in both mining and industrial stocks.

Poseidon's share price jumped from 10 cents in December

1967 to 63 cents in mid-January 1968 on the strength of the deal. Eventually it ran to $1.35 by the end of January and the Perth Sunday Times tipped it as the stock of the year. They had the wrong year, but Poseidon put on a surprising performance in the 1968 stock market nevertheless, hitting $3.45 at its peak. This share rise made it possible to inject fresh funds into the company. The Adelaide broking firm of F.W. Porter managed one placement of 105,000 shares at a price of $1.30 per share in June 1968 — a price which would have been thought unattainable only a year before.

Poseidon took more than 700 surface samples from Bindi Bindi then percussion drilled the claims in May. They discovered four mineralised zones, dipping vertically, with average values of 0.6 percent nickel and top values of 0.9 percent. The combined strike length of the zones was 1200 ft. The grades were too low to be minable, but they were not negligible. They had warranted investigation.

Meanwhile Shierlaw had hired for Poseidon a full-time prospector named Ken Shirley. A Tennant Creek man and an old friend of Shierlaw, Shirley had a caravan and a wife who was prepared to tolerate the hardships of the Outback. Shirley's nomadic prospecting took him to Mount Windarra, 15 miles north-west of Laverton. There he noted a sinuous belt of banded iron formation, part of which was exposed and the rest hidden under sand. Amid the loose rubble of this formation, Shirley discovered inconspicuous gossans, the ironstone caps of a sulphide lode. He pegged 41 claims along the belt, which stretched seven miles, and covered all of it except a short section already pegged by Selection Trust of the UK. Shirley deserves great credit for his prospecting skill because Windarra still ranks as one of the best nickel deposits found in Australia.

The claims covered 10,800 acres, roughly in the shape of a battered exhaust muffler, with the pipe at the top. The richest area, named the Shirley zone, was toward the neck of the muffler. Poseidon, which had just under $200,000 cash at June 30, 1968, originally sought to farm out an interest in the area rather than carry the full exploration costs. Shierlaw approached Consolidated Gold Fields of Australia, who turned down Windarra saying it was no good. Other majors were equally apathetic. So Windarra was inspected by Poseidon's own consulting geologists, Burrill and Associates Pty Ltd. Burrill had been hired early in 1969 because Poseidon had been dissatisfied with the geologist who had worked on Bindi Bindi.

At Windarra, Burrill found surface sampling of the gossans

registered values of 0.5 percent copper and 0.7 percent nickel, together with associated platinum. Burrill and Associates described this to Poseidon as "very encouraging" and "intensely interesting". Their work also indicated anomalous areas which warranted drilling.

They were impressed enough by the claims to ask Shierlaw if they could buy shares in Poseidon. The Poseidon directors agreed and the geologists bought 10,000 Poseidon in April 1969 through Shierlaw's firm. The shares rose from 60 cents to $1.20 while Burrill and Associates were buying. The geologists' buying appears to have caused the rise.

On April 29 Poseidon directors released the Burrill report on the Windarra surface samples. In reply to a stock exchange query about the rise in the price of the shares, Poseidon directors said they knew of no reason. They did not mention the share buying by Burrill and Associates. Much later, in evidence to the Rae Committee, Burrill conceded that the rise in share price had probably been caused by his firm's buying.

As the shares gradually sagged back below the dollar level, work continued at Windarra. After surface sampling, the prospect was costeaned (trenched). In the costeans they had values up to about 0.8 percent nickel. In association with the visible copper oxide on the surface, this was the type of mineralisation the geologists had been seeking. By the end of August when costeaning finished Burrill considered it "an absolute top prospect".

Poseidon then began percussion drilling about half a mile south of Mount Windarra. Percussion drills are very similar to the pneumatic drills that can be seen at roadworks. A percussion drill is cheap to operate but penetrates only to shallow depth. It brings up the rock in pulverised form, making exact measurement of ore zones impossible. Diamond drills are the opposite, being expensive but capable of penetrating to greater depth and producing smooth core samples which allow geological formations to be measured with precision. It is normal in exploration to use percussion drills first to establish whether the prospect warrants diamond drilling.

By September 1969 very little was known outside South Australia about Poseidon, and not much more was known inside the State. In the moving market of 1968, the April flurry in the stock had been forgotten. It had been dormant in the market since then. Little had been reported about Windarra, which was on an unknown area far from such fashionable addresses as Kambalda and Carr Boyd. For the first half of September the shares stayed around 80 cents. Then they began moving in rhythm with the exploration program, although Poseidon made no announcement to the stock exchanges except for one on September 12 saying percussion drilling was beginning. Poseidon firmed to the dollar mark on September 13, reaching $1.07. The shares retreated to 95 cents, then rose to $1.15 in Sydney at the close of trading on September 24.

What was happening in this and the following week has since been the subject of an exhaustive Senate inquiry, but some of the facts are still cloudy. On the field, the first hole (PH1) caved in at a depth of 45 ft. It was restarted as PH1A, drilled through oxide zone material and was lost at a depth of 145 ft. On September 23 a batch of samples from the drill cuttings of these holes was flown to Kalgoorlie, about 200 miles south. Burrill sent a letter with the samples asking the laboratory, Geochemical and Mineralogical Laboratories (W.A.) Pty Ltd — better known as Geomin — to carry out an analysis for nickel and copper by atomic absorption spectrophotometry (AAS). Burrill marked some samples as specials and asked for them to be analysed immediately.

Geomin phoned Mrs Burrill in Perth at 11.30 am on September 24 with the assays from the special samples. The nickel content of these samples was about one percent. The nickel content of one sample was greater but unspecified. Unfortunately, these assays, while encouraging, were relatively meaningless. The presence of nickel values in an oxide zone does not necessarily mean there will be worthwhile nickel in the underlying sulphides. The formal Geomin report did not arrive until October 3. That said the average values in PH1 had been 1 percent nickel and the average grade in PH1A had been 1¼ percent.

One person who had been taking keen interest in the activity at Windarra was Keith Biggs. A shrewd entrepreneur, although he had limited education, Biggs saw an opportunity to make money on Poseidon. He was a prospector, investor and businessman, owning the butcher shop at Leonora. He was a smallish share trader and saw little downside risk in buying into a tightly capitalised company which was about to drill but whose shares had not yet moved.

Biggs owned the pastoral property on which Windarra was found, and had given some rock samples of his own to Burrill for analysis. Perhaps because he was so close to Poseidon, he decided to take a bigger punt than usual. He bought a "shelf" company called Granby Pty Ltd as a family investment company, apparently on the same day that he decided to buy Poseidon shares. He began picking up a line of 10,000 Poseidon shares on August 29.

When he heard the first rumors of sulphides at Windarra on September 24, he ordered another 15,000 shares through the Perth broking firm of Hartley Poynton & Co, who bought them on September 25. This second parcel of 15,000 shares may well, have been the trigger that detonated the whole Poseidon boom. Norm Shierlaw, who had been keeping a nervous eye on strong Perth buying of Poseidon, plunged into the market himself on the 25th, and the boom started running.

Poseidon's historic hole, PH2, was started by the contractor, Selective Drillers Pty Ltd, at 10 minutes past noon on September 23. It reached 140 ft by 6 pm, when drilling stopped. The drillers did not know it, but they were only five feet away from what was to become one of the world's most famous orebodies.

Soon after drilling began at nine next morning, September 24, Burrill saw a change in the color of the sludge coming out of the hole. An expert geologist who had worked for years with Falconbridge, Burrill could see nickel and copper sulphides in the dark sludge. He was very excited, because if nickel is prominent enough to be seen with the naked eye, the concentration is at least three-quarters of one percent, which means it is approaching mine grade.

As Burrill told the Rae Committee: "If you can see nickel sulphides you can be pretty certain it is at least three-quarters of one percent, but I would not have an idea whether it was one, two, three, four or five percent". He could also see the chalcopyrite (copper sulphide), another exciting aspect because all the big deposits discovered in the world in sulphide orebodies to that date had contained copper with the nickel.

These massive sulphides were sent to Geomin on September 25 with instructions to carry out an urgent analysis for copper and nickel by AAS. This was a fast test, but inaccurate for readings of more than one percent. On Friday September 26 Burrill learned that 30 of the 36 samples he had sent in from PH2, including all those covering the massive sulphide zone, had assayed greater than one percent nickel. The copper readings were also highly encouraging. Geomin had to complete further tests in its Perth laboratories before accurate assays could be produced. Either over the weekend or on the following Monday, Burrill rang Geomin and asked for the assay readings of PH2. The results were excellent, with the nickel content ranging between two and five percent, accompanied by favorable copper values. As anything over two percent was mine grade, Poseidon was beginning to look exciting.

The excitement had already reached the stock exchanges, where there had been heavy buying of Poseidon shares. The stock had closed at $1.15 in Sydney on September 24, the day of the discovery. The next day marked the real start of Poseidon's trip to the stratosphere. As trading was heavy in a stock the scrip of which was scarce in some exchanges, the price varied substantially between the main exchanges. The shares rose from $1.15 to $1.48 in Sydney, from $1.12 to $1.53 in Adelaide and from $1.20 to $1.55 in Perth. The latter prices are those at the close of trading. During trading the stock went as high as $1.65. The Melbourne exchange was closed for Show Day, and the channelling of its business into the Sydney and Adelaide exchanges may have helped push the market up generally.

The Adelaide Stock Exchange asked directors of Poseidon why the shares had risen. The reply, released to the exchange at 1 pm, was brief and uninformative. Later it was also alleged to be untruthful. Directors said: "The drilling commenced on September 15, 1969, but due to technical difficulties only a limited footage has been completed. No assay results are available. Therefore the board is unable to explain the sharp increase in the price of the shares."

A few assay results had been released to Mrs Burrill on September 24, as mentioned earlier. But by themselves they were relatively meaningless. Roy Jones, a partner in Burrill and Associates, had flown to Adelaide that night and discussed the "whole Poseidon situation" with Shierlaw.[3] However, Shierlaw has always said that he knew nothing of the strike until the following Saturday when Burrill phoned him from the field.

Nevertheless, Shierlaw and Associates began buying Poseidon shares heavily on September 25 — the day after the strike and Jones's visit to Adelaide. The firm's purchases of 32,000 Poseidon represented more than half the Adelaide turnover of 48,000 that day and about one quarter of the total Australian turnover of 115,800 (a total in which there would be some double counting). The buying of Shierlaw's own firm would have been one reason for the price rise.

Shierlaw and Associates bought 11,600 shares for NCS Securities Pty Ltd, a share trading company in which the Shierlaw family held a controlling interest; 2700 shares for B.R.Lewis Pty Ltd, the family company of Basil Lewis, who would later become chairman of Samin Ltd and a director of Poseidon; 4000 for Burrill and Associates; 3600 for Jones; and 10,100 for the Shierlaw and Associates house account. The house account shares were held for resale in London and 5000 of them were resold that night. Shierlaw's firm and associates had been remarkably active in a stock there was no reason to buy.

Another buyer on the 25th was Geoffrey Hynam, chairman of the Perth Stock Exchange, who bought 2100 shares. His wife, who held a nine percent stake in Burrill Investments Pty Ltd, had bought 2000 Poseidon in late July. Hynam was a close friend of Jones and acted as his broker.

On Friday, September 26, the market ignored the directors' official statement and kept buying. The shares ranged between $1.40 and $1.94. There was also strong trading in the stock in London that night. One person who heeded the directors and sold this day was Dr Max Kimberley Anderson of Perth. If Poseidon's directors could have foreseen the trouble this sale would eventually cause, they might have considered a placement to pacify Dr Anderson, whose enmity was to have far-reaching effects.

On Monday, September 29, Poseidon directors reported that the second percussion drill hole at Windarra had encountered nickel and copper sulphides. "The assays of the samples will be published as soon as possible," they said.

This was enough to more than treble the price of the shares to $6.70, although in later trading they shed a dollar to close at $5.70. Poseidon's shine was beginning to be reflected by other stocks, particularly Metals Exploration NL, which jumped $1.10 to $6.50 for no apparent reason apart from the fact that it was an existing nickel producer (at Nepean) and had a number of other mineral claims in WA.

On Tuesday, September 30, the speculators were busy again, pushing Poseidon up another $1.30 to $7 at close. Poseidon was increasingly dragging other stocks with it, including heavyweights. CRA, Hamersley and Western Mining all moved up and the mighty BHP rose 50 cents this day to $13.95. There is perhaps nothing more remarkable in all this saga than the fact that a share run in a small nickel explorer, which had so far announced very little, should have added $65 million to the capitalisation of the largest company in the land.

At the other end of the scale, Mallina Mining, which had listed on 19 days earlier, jumped 85 cents to $1.60.

Two factors contributed to Poseidon's strength this day. One was strong London buying. The other was shorters covering their positions. A number of traders who had sold the stock short during its rise of the previous week, reckoning it a two-day wonder, now decided to cover by buying.

On October 1, Poseidon directors issued a historic report to the Adelaide Stock Exchange before the start of trading. The statement began: "Further to the report of the recovery of nickel and copper sulphides on September 29, the directors of Poseidon NL announce that the assays received to date of the first completed drill hole PH2 at Windarra, WA, are as follows:-

From-to in feet	Length in feet	Ni%	Cu%	Type of Ore
0-25	25	0.40	0.10	Leached Oxide
25-115	90	1.53	0.25	Oxide
115-145	30	1.60	0.40	Disseminated Sulphides
145-185	40	3.56	0.55	Massive Sulphides

The consulting geologists, Burrill and Associates Pty Ltd, quote that the mineralised zone has an indicated length of 1000 ft and minimum width of 65 ft".

After giving some other exploration details, the board also announced a placement of 500,000 shares to unnamed mining interests for a total of $2.8 million.

Both the assays and the placement were to come under fire. In the assays announced, the key figure was the 3.56 percent value given for nickel in the 40 ft of massive sulphides. The way the table was produced in the announcement, with the metal content given to two places of decimals, had the appearance of precise, careful calculation to the last 0.01 percent. This air of precision was quite misleading. No justification was ever produced for the figure of 3.56 percent and Poseidon had only limited assay information available at the time.

When the October 1 statement was released, the only specific knowledge available was the Geomin report to Burrill on September 26 stating that 30 of the 36 samples from PH2 had assayed greater than one percent nickel. At some time over that weekend he had phoned Geomin and had been given assay readings of between two and five percent. The laboratory claimed in evidence to the Rae Committee that as readings taken by AAS above one percent were inaccurate the figures had been rounded down to the nearest whole percentage point. It was on the basis of these figures and no others that Poseidon's directors and geologists announced an average of 3.56 percent for the massive sulphides.

Considering that it was such an important assay, signifying that Poseidon had struck a major mine and gone from rags to riches, and considering that directors and geologists bought so many shares on the strength of this drill-hole their memories afterward of how the figure was calculated were surprisingly poor.

In evidence, Burrill said: "Maybe it was sloppy but I think it was the right thing to do. For the first hole I do not consider it the least bit important providing it is in the right order and the whole thing is not a fake."[4] Shierlaw said: "It was four years ago and as far as I am aware it was done in the correct geological manner. Just do not ask me the details now. I could not tell you."[5]

Geomin's detailed assays did not reach Poseidon until October 3. These showed an average assay for the sulphide intersection of 3.17 percent. The directors did not correct their October 1 statement and the public never knew of the discrepancy until the Rae Report appeared in 1974. The size of the error need not be labored. A mine is not determined on one hole and 3.56 overstated the actual grade by only 11 percent so the error in reporting was not great. More important was the fact that this assay gave a misleading impression of Poseidon's richness. The eventual recoverable mine grade was to turn out to be nearly half the level reported and this had dire consequences for the company. This was the key initial statement by Poseidon, and apart from some disappointing drill-holes reported in mid-October, was all the market had to work on until November. By that time the shares were $40.

The second area of contention in the October 1 statement was the placement. On all the available evidence, the major, if not the only purpose of the placement was defensive. Shierlaw's family holdings amounted to about 150,000 shares at this point, representing less than 10 percent of issued capital. Turnover in the stock had been very heavy, and if any substantial percentage of recent buying had been by a raider, he could have lost control. The accompanying table, taken from the Rae Report, shows sales of Poseidon fully and partly paid shares in the 10 trading days from September 15 to 26.

POSEIDON TURNOVER

1969 Sept	Sydney	Melb-ourne	Adel-aide	Perth	Total	Price of f.p. shares Low/High cents
15 Mon	26,700	3600	24,900	7600	62,800	90-105
16 Tues	5900	2000	13,600	11,170	32,570	94-105
17 Wed	6600	4600	6000	3500	20,600	94- 96
18 Thur	29,700	6400	6900	1000	44,000	98-105
19 Fri	42,900	19,300	23,250	9600	97,050	105-115
22 Mon	53,800	18,100	25,800	2900	100,600	109-128
23 Tues	24,500	3000	7800	6040	41,340	115-120
24 Wed	12,300	7700	4300	closed	24,300	110-112
25 Thur	53,500	closed	48,800	13,500	115,800	115-165
26 Fri	44,200	14,100	40,800	8000	107,100	**104-194**
Total	300,000	78,800	204,150	63,310	646,260	

A total of 646,260 shares had turned over in two weeks, representing nearly one-third of the issued capital of two million shares. Shierlaw defended his buying on September 25 on the grounds that he feared a raid was developing on the company. The size of the turnover did give grounds for such fear. On Thursday morning, September 25, Shierlaw rang his Perth agent and asked who was buying. He was told: "The butcher, the baker, the candlestick maker". Keith Biggs was the Leonora butcher. Shierlaw now believes others close to the site were also buying.

The total turnover shown in the table would be inflated by double counting of transactions where a purchase or sale had been made in one exchange on behalf of a broker in another city. Even allowing a notional 20 percent for this, the total would still have been half a million shares or one-quarter of issued capital. Actually, no raid was occurring. The only likely raider was Ganke and he had sold half his holding — 50,000 shares — the week before the run started. This still amuses Shierlaw.

More important was the short position. Bears in Sydney had shorted Poseidon and doubtless accounted for a portion of the 300,000 turnover in Sydney during those two weeks as they covered again. These purchases cancelled earlier sales. Some covering undoubtedly occurred on the 25th and 26th in Sydney. In a bull market, it is dangerous to stay short over the weekend in a rising stock.

Nevertheless, the level of turnover was dangerous and there was a chance Shierlaw might lose his mine before he found it. By placing half a million shares in friendly hands he made a raider's task much more difficult. The recipients were not identified for some weeks but eventually proved to be North Flinders Mines NL (100,000 shares), Nobelex NL (50,000), Australian Development NL (50,000), Conwest Exploration Co Ltd (50,000), clients of London broker James Capel & Co (50,000) and Samin Ltd (200,000). Samin received the shares at $5 each and the others at $6.

North Flinders, Nobelex and Australian Development were all linked in various ways with Poseidon and its directors. Samin was the most controversial placement, because the company had not even been incorporated. Samin had no cash or assets, although it was in the process of formation and intended to reopen the old Burra copper mine. It was in no position to pay $1 million for a share placement. Eventually Samin paid for the shares by raising the $1 million from the E.S.&A Bank. When it floated soon afterward, it used most of the $1.65 million raised in public subscription to repay the loan. By that time Poseidon shares had risen substantially in value which meant that Samin came on the market at a huge premium. Samin shares, issued fully paid at par of 50 cents, sold for $20 on the opening day.

Shierlaw and Associates received $72,488 for underwriting the Samin float. The lucky recipients of Samin shares included Shierlaw's family companies (100,000 shares); Shierlaw and Associates (46,000); the wife of T.A.Hutton, chairman of Poseidon, and Hutton's private investment company (1800 shares); B.R.Lewis, his family and family companies (151,000); Burrill Investments (3000); the family of C.F.Wegener, another Poseidon director (2000). Poseidon itself received 200,000 Samin shares as part of the placement deal.

Nor were North Flinders, Nobelex and Australian Development rich companies. The placements represented large investments for them too. Shierlaw and the other Poseidon directors who had links with these companies were obviously convinced by the morning of October 1 that the shares were worth the $5 and $6 values put on them in the placements. It is worth noting that the shares had never been $6 in the marketplace before the previous day.

Burrill helped with the placements. He had some impressive overseas contacts, including Conwest, which took 50,000 shares purely on his recommendation. Burrill was on the investment management committee of the First Investors Mining Petroleum Fund, whose board included the chairman of Falconbridge, and Union Corporation, and a representative of Union Miniere. This was a body with powerful and international links. Its board and management committee included representatives from Canada, Belgium, South Africa, France and Australia. It was managed by Capel in London, but was based in Switzerland for tax reasons. Its meetings were held in Toronto and Geneva.

Burrill swore in evidence that he did not disclose any drill details to Capel when they took the placement. He said Capel would take whatever he recommended. An outsider may find this incredible, but mining investment can work in this loose way. If an adviser has a good track record — and Burrill's was excellent — a broker will buy blind on his recommendation. To probe for details takes time, and an opportunity may be lost in a moving market. To seek details may also compromise the adviser if he is in an insider position such as Burrill's. Once an adviser such as Burrill has built a position of trust, very large deals can be done on an unsupported recommendation. The legal fine print is signed later. This looseness and informality can lead to an occasional disaster, but its overwhelming virtue is the efficiency and speed it provides. Without such faith and informality, for instance, the whole short-term money market would be unworkable.

There was a conflict between Burrill and Shierlaw over the placements. Burrill believed that Poseidon needed a strong partner and suggested a stronger link with Conwest. Conwest's chairman, Martin Connell, met Shierlaw and discussed this but Shierlaw and the rest of the Poseidon directors wanted to go it alone. Conwest still took the placement.

History was to prove that Burrill was right and Shierlaw was wrong. As the months passed, the question of financing Windarra's development was to grow from a small cloud on a bright horizon into a big black problem. The company's failure to clinch a favorable deal with a big brother would ultimately ruin it. The October, 1969, placements squandered valuable equity which bigger companies would later have bid high for. The Samin placement would become a source of weakness, and expensive to remove. This is not judgment in hindsight. The faults of the placement were argued vehemently by Ganke at the time. Burrill saw the need for a big brother also. We will return to this problem later.

The question which generated the most heat about Poseidon

was whether the directors and geologists were insider traders in the stock. This was investigated at length, to the point of tedium, by the Rae committee. It boils down to the question of what the directors and geologists knew when they were buying. Any investigation of this involves a careful reconstruction of the sequence of events, an exercise reminiscent of Hercule Poirot putting together a timetable of suspects' movements on the night of the murder at Greystanes Manor.

The timetable of events in Poseidon certainly arouses suspicions.

September 24. — Windarra drill strikes massive sulphides between 9 and 10 am in PH2. Burrill, on site, recognises copper sulphide and is 'almost certain' he can see nickel sulphides. At 11.30 am Geomin in Kalgoorlie phones Mrs Burrill in Perth to tell her the average grade of oxide material in PH1A was 1¼ percent nickel. In the afternoon, Burrill's partner, Roy Jones, flies to Adelaide where he meets Shierlaw that night.

September 25.— Great activity in Poseidon shares. A total of 115,800 traded (the total includes some double counting) and the price rises from $1.15 to $1.65. Shierlaw and Associates prominent in market buying a total of 32,000 shares. In response to an Adelaide Stock Exchange query, Poseidon releases a statement at 1 am saying directors are unable to explain the sharp rise in the shares.

September 26.— Burrill receives AAS assays from PH2 and is told 30 samples had nickel values greater than 1 percent. Burrill phones Shierlaw Friday night or early Saturday morning to tell him.

September 29.— Poseidon reports that the second percussion drill-hole at Windarra has encountered nickel and copper sulphides.

September 30.— Burrill flies to Adelaide, meets Poseidon directors and phones James Capel to offer Poseidon share placement.

October 1.— Poseidon issues report before trading saying the company has struck 40 ft of 3.56 percent nickel. Directors also report placements of 500,000 shares. Poseidon rises to $12 on market.

Burrill said that he had been 18 miles from a telephone at Windarra and had been unable to contact Shierlaw until early on Saturday morning, the 27th. Jones said he did not know PH2 had struck nickel sulphides when he flew to see Shierlaw on the 24th. Shierlaw said he had gone into the market and bought heavily on the 25th, because he had seen heavy buying in Perth earlier in the week and "if it was good enough for them to buy, we should buy ourselves". Having bought the shares, he then booked them to various friends and associates who would support the stock.

This is a feasible chain of events. Shierlaw had a big personal stake in Poseidon and as a stockbroker would have been monitoring its movements closely as drilling started. A big buying surge for the stock in Sydney and Perth would have been a good enough signal for him to move in and buy also, whether as a trader or defensively against a possible takeover.

However, the credibility of the defence of Shierlaw and the geologists was undermined by the fact that when allegations of insider trading first arose, they misstated the date of the discovery at Windarra. When Senator Lionel Murphy raised the accusations of insider trading in the Senate, Burrill and Shierlaw issued indignant denials. Burrill said nickel had been discovered on September 26 and Shierlaw said it had been on September 29. As the date of discovery was crucial to the allegation of insider trading, it is hard to understand how they could have both got it wrong. In his reply to Murphy, Shierlaw had referred to "the discovery of nickel and copper sulphides at Windarra on September 29." Confronted by the Senate committee with evidence that the discovery was on the 24th, Shierlaw defended his original statement by saying, "the English is wrong".

Jones had also originally stated that the discovery date was September 26 and in a statutory declaration had misinformed the committee of the day he had travelled to Adelaide. Telling untruths to a Senator would be close to the ultimate political crime in Australia and whether their untruths were intentional or not, Burrill, Jones and Shierlaw all stood convicted. The Senate committee was particularly nettled by the fact that its members did not discover they had been misled until at quite a late stage of their inquiry. It was not until the end of 1972 that the committee uncovered evidence which cast serious doubt on the reliability of earlier testimony. They reopened the inquiry to examine Burrill, Shierlaw and Jones.

The Rae Committee concluded: "During the period leading up to the announcement of the discovery of nickel and copper sulphides the consulting geologists and one of the directors misused their official and fiduciary positions to buy Poseidon shares for private profit. The geologists also sold many of the shares at a time[6] when they had a greater knowledge of the prospects than the general public whom they had informed about the Windarra field. Another conclusion of wide significance is that the Windarra announcements, which gave nickel prospecting a new glamor and gave rise to a great wave of mineral company flotations and placement issues, can now be seen to have been misleading in significant respects and made on careless, to say the least, reporting of assays."

My own view, on the evidence available, is that the geologists and Shierlaw were gambling, albeit with the odds in their favor. They knew they were on to something that looked good, but they could not quantify it on the assays available. Shierlaw says he had always thought the discovery was on the 29th until he was confronted with the evidence at the Senate inquiry.

On the assays available to October 1, the directors and geologists knew little more, in specific terms, than the general public. Nevertheless, they were mining experts and knew that Windarra was increasingly looking like a good mine. Stock exchange reporting requirements are necessarily laid down in terms of specifics. A company making a bullish statement in general terms about its prospects when there are no specific facts to back it can get into trouble not only with the stock exchange but with the Corporate Affairs Commission.

In hindsight, Poseidon's board could have described the initial drilling and assays as "encouraging" or "promising". They could have advised shareholders not to sell pending an announcement, which would undoubtedly have put the share price up. They could have disclosed their own buying, which would have had a similar effect.

Throughout this book it should be remembered that in 1969-70 no Securities Industry Act existed to prohibit insider trading. Everyone in the securities industry today is acutely conscious of ethics, because ethics have been forcibly dragged to their attention. But in 1969 the new morality had yet to arrive.

A point in Shierlaw's favor is that the only shares he ever sold in Poseidon were $200,000 worth to make a donation to the Australian Mineral Foundation. His only tangible reward from Poseidon was a plaque on the wall of the Foundation building in the Adelaide suburb of Glenside.

Of course, the geologists and directors were obviously not the only ones buying Poseidon in the run-up to October 1. A glance at the share trading table shows that other people, who presumably knew even less about the assays and drilling progress, were buying too.

Readers will remember that it was on September 24 that Poseidon struck nickel and on September 25 that the shares first took off in Adelaide. As far as the Rae Committee could ascertain, at that stage Burrill had seen the nickel-copper sulphides but had not contacted Adelaide, the samples had been sent to Geomin but not assayed, yet despite the lack of information and communication the share traders in Adelaide and Sydney were buying heavily into the stock. This happened frequently during the boom to the extent that it is surprising that no apostle of extra-sensory perception has ever written a thesis arguing that share traders are abnormally endowed with telepathic faculties.

The truth is more mundane. On any site where geological work was progressing there were various employees of the company and the contractors. Many of these knew enough about mining to identify minerals and make a rough estimate of assays.

Some minerals, such as gold, copper, silver, iron and tin, can be easily identified on sight in certain ores. The richer the ore, the easier the identification. Burrill could identify sulphides in drill sludge and could even identify them from a low-flying plane. Tungsten concentrations, as we saw with Endurance, can be identified in the field under ultraviolet light. Radioactivity in rocks can be measured with a geiger counter or scintillometer, although it won't tell uranium from thorium or potassium, unless properly calibrated.

An experienced employee or trespasser with a little equipment could make a reasonably accurate guess of what a company had discovered. After work, the drillers might drive 20 or even 50 miles to the nearest pub for a beer and swap information with their mates. Barmen and publicans became great traffickers in information.

Even if the employees kept their mouths shut and simply telephoned or telegraphed orders to their brokers, PMG employees had access to the information. Burrill said about Poseidon: 'We were very, let us say, security conscious over the fact that sometimes when you were phoning you would hear 'click, click, click', and very often I used to go to Leonora to make my phone calls if I thought they were such that someone might be listening".

At one stage during the boom it was rumored that telegrams to eastern companies and brokers from WA were being delayed in transit while telegraphists got set on the market. Boards of companies often complained privately that they were the last people to know what was happening in the field.[7]

A whole irregular communications network was set up linking the nickel fields with the stock market, but it was almost invisible to the various inquiries held into share irregularities. These inquiries tended to concentrate on directors and brokers, putting what might be called the underground network into the 'too-hard' basket.

In the case of Poseidon, rumors had been circulating for some time. I was first tipped the stock, several weeks before the strike, on the floor of the Melbourne Stock Exchange, where Poseidon was thinly traded. Newspaper reports at the end of September, when Poseidon finally took off, referred to long simmering speculation in the stock. A few keen-eyed professionals had detected the optimism of the mid-year statements, and the fact that Burrill was personally supervising exploration work indicated that Windarra was of more than normal interest. The stock was due to fly as soon as the underground network scented that it had the real thing.

Possibly the trigger was Keith Biggs, who bought 9100 Poseidon between August 29 and September 16, then ordered another 15,000 on September 24. When his broker suggested this was a big punt, Biggs replied: "A punt's a punt, and if it does not come off I go broke."

Asked why he bought the shares, Biggs told the Rae Committee he had followed the exploration work closely and that it was common knowledge in the district that Poseidon thought highly of the area. He bought the first 9100 shares on this basis. The big order for 15,000 had been placed after he heard rumors that Poseidon had struck nickel. The news must have reached him very quickly because he placed his order on the same day — apparently during trading hours — that Poseidon struck the sulphides in PH2. He said: "One strong rumor — you may well laugh at me — was that the drillers had been seen in Laverton with sulphides on their boots and things." He also said the drillers had been seen ringing up.

The story has a lot of credibility to anyone who has ever visited Laverton. Before the boom it was a one-horse town in the middle of nowhere. As the boom got into stride it was still a one-horse town in the middle of nowhere except that more transients were passing through. The town basically consisted of one street with a few ramshackle dwellings. The only human life in sight might be a knot of black children staring at a motorist as he put water in the radiator and asked for directions into the low, parched scrub of the surrounding terrain. There were few local inhabitants, and in country where the towns are a hundred miles apart, everyone knew what everyone else was doing.

Communication with the outside world was limited. When Poseidon's directors wanted to contact Burrill at the drill-site they used to ring the postmaster at Laverton, who would then drive 18 miles to the geologist's caravan. Obviously this could only be done when the need for communication was urgent.

The brokers and investors in the eastern states did not know whether Poseidon had struck sulphides or received assays. But the same brokers and traders knew how the underground network operated and were constantly alert for any buying signals emanating from Western Australia. Monitoring them became a full-time occupation. Finally, there was the trading floor itself. Any alert operator who saw Shierlaw's firm buying Poseidon in quantity would doubtless have thought it a good idea to pick up a few shares himself.

The smarties had done well to get set by the end of September. On October 1, after Poseidon's inaccurate announcement of 3.56 percent nickel, the shares soared to $12 in hectic trading. On October the fully-paid shares hit a high of $17.70. The boom was off and running.

---

[1] Interview with author, October 1969.

[2] A holder of shares which are partly paid in a no-liability company has the option, if there is a call on the shares, of paying the call or forfeiting his shares at no cost. All forfeited shares must by law be subsequently offered for auction.

[3] Rae Report, ch. 2, 11.

[4] Evidence to Rae committee, p. 3011.

[5] Evidence to Rae committee, p. 3051.

[6] Between October 21 and 30, Burrill Investments Pty Ltd sold 3000 Poseidon shares for $103,922. Between December 31 and March 23, they sold further shares at prices ranging from $196 to $235. They ultimately netted nearly $1.4 million for 9500 shares which had cost them less than $10,000.

[7] BHP at one time tried to circumvent the 'underground network' by having its men in the field fly back to Melbourne and report orally. The market still had the news before BHP.

CHAPTER EIGHT

OCTOBER: The Fever Starts

October, 1969, was an election month, with polling day on October 25 to decide whether John Gorton or Gough Whitlam would be Australia's Prime Minister. If either candidate could have captured public imagination to the same extent as the share market did in the final month of the campaign, he would have won in a walkover.

Poseidon's October 1 announcement of 3.56 percent nickel had sparked a share boom even greater than that of the golden 1890s. Poseidon shares hit $13.30 on October 1 before closing at $12 for a $5 gain on the day.[1] On October 2 they climbed a further $6.40 to $18.40, closing at their peak. They moved back to $16 on the Friday as weekend profit-takers moved in, but recovered to $20.10 on Monday, October 6.

The market had been dominated by professionals at first, but small investors quickly caught the excitement and plunged into the market, clamoring for Poseidon and other stocks. So many speculators flooded the Sydney Stock Exchange building on October 2 that many could not reach the third level of the building where the public gallery was situated. Their problems were exacerbated by the narrowness of the Sydney public gallery, which covered only part of the trading floor, and the fact that pillars intervened between the gallery and the boards.

Overcrowding was also a problem on the trading floor, where operators had massed around the mining boards. One of the boards collapsed, but prices continued to be written on it while floor staff refastened the ropes.

Activity centred on Poseidon, but other companies' shares began to make sudden rises. The first to run were those rumored to be pegged alongside Poseidon at Windarra. The first to jump was Omega Oil NL, whose shares had been languishing at par value of 10 cents until September 30, when they rose three cents. On October 1 they rose another six cents, which meant they had almost doubled in price in two trading days. They reached 27 cents before directors released a bemused statement on October 3, listing Omega's prospecting areas and showing that none of them were at Windarra. The rumors — showing typical adaptability — then said Omega was about to negotiate a farm-in with someone else bordering Windarra. The shares receded slowly.

Mallina Mining, which had vaguely reported striking sulphide mineralisation while drilling in the Pilbara, rose 80 cents to $2.40 on October 2. Carr Boyd, gyrating wildly on the same day, rose from $2.28 to $3 before settling at $2.45 for a net gain of 17 cents.

One reason for the various runs was the "nickel shield" theory. This held that the Archean rocks running some 400 miles north-south from Wiluna to Norseman in Western Australia were rich in nickel. The theory proved to have considerable validity, as all Western Australia's operating nickel mines and several marginal and sub-economic deposits have been found in this belt. It was a theory much favored by speculators and company promoters because it provided a basis for arguing that any claim in the area was a potential nickel mine.

The speculative fever spread to other minerals as well. Al Consolidated, an old gold miner which had recently been taken over by interests associated with Boris Ganke, found itself caught in the action. The shares leapt $2.50 to $4 on October 2, apparently because the company had reported three weeks earlier that it had discovered "radioactive material" in its tin lode at Glen Esk. When the company reported on October 2 that assays were due from Glen Esk that night or the following day, the shares dropped to $3.10. The drop was symptomatic of a market which preferred rumor to fact.

On the following day, Friday, October 3, trading on the Sydney floor was frantic. The total turnover was 23.5 million shares and options, beating the previous record for a single day's trading by nearly five million. A better indicator of the scale of trading was the transaction count. There were 10,128 transactions recorded during the day compared to the previous high of 9600.

After an opening flurry the market was dominated by profit-takers. This also was typical of boom trading. Speculators who were showing a gain on their stocks would cash in on Friday rather than hold over the weekend, particularly if it was a long weekend in Sydney such as this one.

Trading was given a further shot in the arm by Queensland Mines Ltd, which ran from $10.80 to $15 on a statement announcing rich uranium values at Westmoreland in the northwest of Queensland. Westmoreland Minerals, a recent float from the Alex Barton stable which had claims next door, went from 33 cents to 90 cents.

Spectators again jammed into the Sydney public gallery. Two regulars came craftily prepared for the fray, carrying two fruit cases. They stood on these to get a clear view of the boards, and, whenever one left to phone his broker, the other guarded his case. On the trading floor sweating, dishevelled brokers and operators shouted themselves hoarse in a milling mob. They were haggard but cheerful at the end of the day. Only two weeks before many of the operators had been facing retrenchment because of the slack market.

That night in London, Poseidon shot past the £10 mark for the first time. Any British speculator who had bought on the previous Friday would have made 900 percent profit selling now.

While their London counterparts were happily engaged, members of the Melbourne Stock Exchange were attending their annual dinner where the guest speaker was John Gorton. Judging by their comments afterward, many must have wished they had stayed back in their offices to clear contract notes. Gorton was only three weeks from polling day, but knew he already had the votes of his audience. His speech was brief and superficial, shunning any suggestion of abolition of capital-gains tax, the only point which interested the brokers.

Monday October 6 was a public holiday in Sydney, which meant that much of its normal business was channelled through the Melbourne exchange. More than 19 million shares changed hands in easily Melbourne's busiest trading day on record. One Sydney broker flew his staff to Melbourne to help cope with the increased business from Sydney. At Patrick & Co in Sydney, the staff worked through the Monday, disregarding the public holiday as they strove to catch up with the backlog of paperwork. Patricks was now the largest firm in Australia, specialising in mining stocks. It employed its own resident geologist, Lionel Milligan.

Melbourne's day of monopoly of the share market was enlivened by the saga of the Harrietville Hot Rock. This episode centred around the personality of Jim Say, the chairman of the recently floated Gippsland Minerals NL. Say, a jovial, knockabout fellow with a colorful turn of phrase, was a scrap merchant in the Melbourne suburb of Malvern and was also a part-time prospector. When Gippsland floated, he and his wife had received 400,000 vendor shares, fully paid to 25 cents each.

On October 6 Gippsland shares suddenly jumped by $1.20 to $2.70 when the company announced that rocks found on its leases at Harrietville in the Victorian Alps had "disclosed the presence of uranium". Thoughtfully, the company invited journalists to come and see the discovery that day, with the result that the Melbourne Herald carried a photograph right across page one of the mountain road where the 'hot rock' had been found.

The discovery, it seemed, was one of those happy accidents that lend the mining industry its special fascination. Say had been driving along the Harrietville-Omeo road with fellow-director Bruce Westwood, a clothing manufacturer from Adelaide, and Westwood had been pointing their scintillometer out of the window. Suddenly the scintillometer, a device for measuring radiation, became active. Say's recollection of the episode went like this:

"Bruce asked 'Is this thing singing?'

"I told him 'It isn't Al Jolson'.

"I thought the machine had gone on the bugle, that there was something wrong with it. Then we went back and played it on the rocks and the needle nearly jumped off the scale."[2]

Dr D.E. Thomas, a former director of geological survey for the Victorian Department of Mines, pronounced the Harrietville hot rock to be the most promising uranium find ever made in Victoria. Dr Thomas, a geologist since 1932, was now a director of Gippsland. Nine days later the assays showed that the scintillometer reading had been mainly due to the presence of thorium, a mineral of little commercial value although there has been some discussion of the possibility of building thorium reactors as an alternative to uranium. The hot rock assayed 1.34 lb of thorium to the ton and 0.16 lb of uranium, well below commercial grade. Directors said they would begin diamond drilling the prospect immediately. Commercial uranium was never found, but then it had never been found anywhere else in Victoria either, so Dr Thomas has never been proved wrong.

The next day saw the heaviest share trading recorded to that date in Australia. A total of 35.8 million shares and options turned over on the Sydney Stock Exchange, eclipsing the previous record set only four days earlier. The total included 27.7 million units in oil stocks, of which 17.7 million were shares and options in Bounty Oil.

The enthusiasm for Bounty appears to have been based on some peculiarly roundabout reasoning. As mentioned before, the glamor of Queensland Mines had rubbed of on its neighbor Westmoreland. Speculators now reckoned that the cheapest way into Westmoreland was by buying Bounty shares and options, because these securities carried an entitlement of one Westmoreland share for every 20. As Bounty shares and options each went to five cents this day, it meant the buyers were paying the equivalent of $1 each for Westmoreland shares. Harbourside shares, which had the same entitlement to Westmoreland, went to 5½ cents on the same rationale. As Westmoreland shares were hovering around the $1 mark at the time, buyers regarded the Bounty or Harbourside shares they bought as a bonus.

Elsewhere in the market, the small Bendigo-based miner North Deborah Mining Co NL rose from 90 cents to $1.60, following an announcement from directors advising shareholders not to sell their stock. This kind of announcement is known as a 'don't sell'. Three days later North Deborah announced that it had acquired mineral tenements adjoining Poseidon and that it had raised $600,000 in a share placement through Ralph W. King and Yuill.

This placement, which was never investigated, was of 400,000 shares at $1.50 each. As the 25 cent paid shares had never reached $1.50 until a day or two earlier, the speed of the placement was remarkable. At the end of September the same shares had been only 30 cents in the market.

It will be remembered that King and Yuill had placed shares on market for VAM between July and September. The October placement for North Deborah may have been made the same way, because the Rae Committee discovered that King and Yuill made two subsequent placements on market for North Deborah in January 1970. Once again previously unissued stock which had not been granted quotation was being sold on the market. From the three placements North Deborah raised a total of $2 million, which may have consoled the company for the fact that the Rae Committee later described the January placements as 'illicit'. The shares reached $7 in a market burning with enthusiasm for North Deborah's Windarra claims. Considering the claims had cost only $1779 to peg, they must rank as one of the great bargains of the boom.

Being a neighbor of Poseidon was glamorous, but the glamor could be extended even further. Ten days after North Deborah's announcement of claims alongside Poseidon, the consortium of North West Mining NL, Haoma Gold Mines NL and Hill 50

Consolidated NL announced that they had "pegged claims which adjoin claims of North Deborah which were previously reported by that company as immediately adjoining the claims pegged by Poseidon". And presumably if anyone pegged alongside the Haoma-Hill 50-North West consortium, they in turn would be pegged next to the group next to the company next to Poseidon. Poseidon's neighboring companies among listed companies were gradually sorted out as Newmetal, North Deborah, Kia Ora, with Allstate and Target close by. Their shares ran, but none of them had nickel. Shirley had pegged well.

On the exchange floor, the drama of each day's trading was heightened by the opening ceremony. Before open-post trading began each day the list was called over and opening bids set in key stocks. In Melbourne, the operators would form a ring on the floor listening to the selected companies being called by the floor manager.

October 9 was a typical day. The call started with Barrier, at that time having a run of its own, and worked through the alphabet. Tension mounted on the floor and in the packed gallery as the call reached the Ps. Poseidon had closed at $23 the day before and many operators had orders to sell at $25 and $26.

As Poseidon was called, one operator boomed: "Buy Poseidon 30!" Rival operators with selling orders scattered to their phones to check whether they should sell or hold. In the confusion, the first trade was actually done at $28 but a few moments later the price had gone to $31 — $8 above its previous close. This bold bid by a single firm triggered rises across the board and another frantic day's trading had begun. In the gallery, the atmosphere was electric. It was as though everyone was watching their football team win a big game.

On the floor the operators would pack three and four deep shouting orders at each other and at the chalkies scribbling prices on the big green boards. Being in the midst of this crush was like being in a rugby scrum but not as comfortable. Mini-skirted girls acted as couriers, rushing messages between the operators and the booths at the back of the floor. One broker reported he'd had a client ring up and say: "Buy me anything."

That week saw three new listings which could not have been better timed. Mining Finance Corporation Ltd, the offshoot of Slater Walker Australia, made a solid debut with its dollar shares selling at $2.30. Base Minerals, from the Murray and Maguire stable, gave stags 100 percent profits as the 25 cent shares closed at 58 cents. Minefields Exploration, a float by

Melbourne entrepreneur Rene Lipton, also saw stags double their money as the 10 cent shares sold at 21 cents. The Minefields float gave old exchange hands a warm feeling of nostalgia. One of the vendors was John Woolcott Forbes, the veteran mining promoter.

The volume of trading was putting the exchange computers under strain. Sydney Stock Exchange's computer broke down for the first time in more than a year. Both Sydney and Melbourne were in the teething stages of their computers at the time. Computer breakdowns were frequent during the boom months, adding to the chaos caused by unprecedented volume and wildly fluctuating prices.

The next area to take the strain would be the share registries, who usually begin handling transactions about two weeks after they occur on the market. Like the brokers, the share registries had been retrenching office staff in previous months and were now uneasily aware of an impending paper avalanche. As one broker said: "A little old man and a frail girl will suddenly be hit by the most enormous flood of paper they've ever seen".

The reasons for concern can be seen by a comparison with New York. At that time Wall Street was handling a volume of about 10 million units a day. Sydney was now running at three times that level. Admittedly the average share price in Sydney was only $1 against $50 in New York, but the paperwork still had to be processed and Sydney's may have been larger than New York's at this time.

Concern was particularly acute at Patricks, which had done about 60 percent of the volume on the day 35.8 million shares had been traded. In value Patricks had only about 40 to 45 percent of the turnover because it traded in many low-priced issues. The firm's growth had been formidable but so had its back-office problems. Brokers were prospering, but they were also working under dreadful conditions. Switchboards were constantly jammed and lunch hours had ceased to exist. Many offices were working from 9 am to 11 pm without any meal breaks.

There was chaos in the field too. The rush for leases near Windarra had led to disputes between prospectors in the area. More than 250 claims had been lodged during the week with the Leonora Mining Registrar, S.J. Winchcomb. Prospectors were queuing in his office to lodge their claims and scores of applications were arriving in every mail. There had been 73 applications in the morning mail for Friday, October 10, alone. Some claim jumping and over-pegging was occurring, mostly through errors. According to some reports, over-zealous prospectors around Windarra were pulling out their rivals' claim pegs and throwing them away. There had been an explosion of pegging at Windarra. When Poseidon had announced its find there had been only three other mining groups in the district. Now there were 20.

The big news on that Friday was from Poseidon, which announced two more drill holes had been completed at intervals of 200 ft along the strike. The samples had been submitted for assay and another percussion drill had arrived at the site. Poseidon closed the day at $27.50.

By this time nothing was more remarkable than the volume of turnover in the stock, which had an issued capital of just over two million shares. A total of 1.7 million had been traded in the two weeks ended October 10 in Sydney, Melbourne and Adelaide. Not counting the 500,000 shares in the placement, this meant that turnover in 10 trading days had been the equivalent of 85 percent of issued capital. Obviously, there was very heavy short-term trading in Poseidon with professionals moving in and out of the stock in the same day. There was very probably some short selling in Sydney as well, which would have swollen the apparent volume of shares trading. If these bears did not cover quickly, they would have faced horrendous losses.

Poseidon's best friends were in London. The London market was drifting downward at the time and Australian mining shares had been the only bright spot on the boards. The London traders would have been watching the free market price of nickel, which had moved up to £5500 during the week. Falconbridge had imposed stricter rationing on customers and a rise in stainless steel prices was being predicted. Another hefty boost for Australia was provided by The Economist in an article which said the Australian mining share boom had now entered a long-term phase. "For the longer-term buyer the list of some of the items of mineral news announced last week should be sufficient to show that Australian minerals cannot be ignored by anyone who wants to buy a bit of the future," the magazine said. Two days later the free nickel price moved up yet again to the £6000 mark.

Sellers hit the Australian share markets in the third week of October, driving many speculatives down. This should not have surprised experienced market hands. Considering how far and fast the market had risen, some profit-taking was due. In this case the generally weaker market was a signal of disappointing results from Poseidon's third and fourth drill hole. The third hole struck 1.51 per cent nickel over 150 ft and the fourth struck 80 ft of "very weak sulphides" averaging 0.38 per cent nickel.

Unfortunately for the London market, Reuters Economic Service was having an off day. The first news reached London in a Reuter flash at 12.27 pm, saying that 1.51 percent nickel had been struck in Poseidon's third hole. Poseidon slumped from 280/- to 200/- on the news. It was not until 1.15 pm that the rest of the message arrived, and then it said that the fourth hole had assayed 9.38 percent. Traders surged madly toward the Poseidon jobbers and the price roared to 250/- as operators tried to buy. The losers were the lucky ones, because then the message came through correcting the assay to 0.38 per cent. The stock slumped to 175/- and Reuters promised an investigation into the mistake.[3]

Back in Australia, the Financial Review rang some stockbrokers for their opinions on the election. One broker summed up the mood by saying: "I've enough on my mind at the moment without worrying about a change in government — God forbid". Another, asked about the impact of a Labor Government on the market, said: "Well look, Whitlam can't stop minerals being discovered, can he?"

On October 16, when Poseidon had closed at $27.50, Tom Nestel of Minsec had appeared in a television interview on the ABC's This Day Tonight program. As managing director of Australia's greatest and most successful share trader, Nestel's views on the market were regarded with awe. Nestel said he thought that Poseidon's recovery in trading had been a most impressive performance, because it could have dropped sharply if the shares had as usual followed the London trend and newspaper comment. Nestel calculated on rough preliminary figures that Poseidon could be valued at $80 per share, which meant "they are definitely not overpriced at $30", he said. That morning Minsec had bought 5000 shares at around $20 in the early market. When they reached $80 a few weeks later, Nestel's reputation was even further enhanced.

In the next few days the market returned to form. Scamander, which had mineral leases in Tasmania, jumped 23 cents to $1.01. Hastings, which earlier in the month had reported "strong mineralisation" on its prospects in northern NSW, ran from 25 cents to 70 cents.

Meanwhile, the Melbourne traders were having frenzies of their own in Pursuit Oil NL, which was drilling a well named Hindhaugh No 1 near the small coastal resort of Torquay. On October 16 the telephone lines were running hot between

Melbourne brokers and Torquay as Pursuit was rumored to have struck oil. On the trading floor the cry "Buy 10,000 Pursuit options" was heard all day. The options rose from 2½ cents to 8 on a turnover of nearly 750,000, while the shares rose four cents to 12 on 290,000 turnover.

A fisherman who had been out at sea in a lobster boat rang a broker and said he had heard voices from what he swore was the Pursuit rig saying: "Get into Pursuit — you'll be right, mate". The broker pointed out that it was a good story but the well was being drilled onshore. When the excitement had died it was discovered that Hindhaugh No 1 was being drilled alongside a creek and some of the lubrication oil from the sump had seeped into the water. Local boatmen, seeing the slick on the water, had jumped to the conclusion that Pursuit had struck oil and rushed off to phone their brokers.

At this stage of the boom, strange small companies suddenly swept into prominence. There was a flurry of panic on October 21 when a little-known stock called Sturts Meadows Prospecting Syndicate NL took off from $1.50 to $6.50. The stock was listed only in Adelaide which gave it the glamor of association with Poseidon while leaving traders in the eastern states short of information.

The rumor was that Sturts had struck 3 percent nickel at Kanowna, near Kalgoorlie. Eastern brokers went into a flat spin. Not only had they missed a 500 percent move in the stock, but most had never heard of it before. They hurriedly searched claim maps. This only proved what some brokers had suspected uneasily for a time — that the claim maps were out of date. Claims were being pegged and over-pegged so rapidly in the boom that there was no way the maps could ever get up to date. Spies were a better source of information.

Eastern brokers began phoning Adelaide with orders for Sturts shares, fearful that they may be missing another Poseidon. Then directors of Sturts (which had been suspended in July, 1968, for lack of precise information about the shares) said they knew of no reason for the run. It appeared to be "only a belated recognition of the real prospects of the company", they said. But speculators had become cynical since Poseidon's "no reason" announcement and the statement by Sturts' directors sent the shares higher.

Adelaide buying then switched to companies bordering Sturts. The eastern brokers, who hadn't known who Sturts were in the first place, were really floundering by now. In Sydney an obscure stock named Tennant Creek Metals rose 26 cents to 75 cents when someone discovered it was prospecting with Sturts at Feysville and reckoned it might have a piece of Sturts' Kanowna action as well. One problem was that not many brokers knew where Feysville was either. Days later it was discovered that Noranda, the operator at the Kanowna prospect, had begun drilling without bothering to tell Sturts' directors.

Another to run that day was Minefields Exploration, which gained 15 cents to 55 cents on rumors that it had struck nickel at Soansville — a slightly more familiar address. The story spread around the Melbourne floor about the operator who had sold his mother out of Sturts the day before at $1.30 and was afraid to go home that night.

Thursday saw the stags return in force as Newmetal Mines 10 cent shares listed at 70 cents. As the options came on at 50 cents, stags were showing an 1100 percent profit — one of the best in the boom to date. The market had been heartened by the Western Mining annual meeting, where the chairman, Sir Lindesay Clark, told shareholders WMC's nickel reserves at Kambalda now stood at 15.6 million tons of ore containing 575,000 tons of nickel. On current free-market prices, this put a total value of $1250 million on the mine.

That Saturday John Gorton was returned as Prime Minister with a smaller majority. The market rejoiced at a Liberal-Country Party victory, pushing Poseidon up to $34 the following Monday. The market's attitude was not entirely political. There was a rumor about that the latest drill holes would show better results. Poseidon showed continued strength over the next few days, finishing the month at $37. The word was that London was buying again.

Amid all this preoccupation with money-grubbing, the poet Judith Wright wrote a poem to The Australian entitled "The True Religion". It was a protest against the mining industry and its tenor may be gauged from the following stanza:

"Your Reverence for nature is obviously fake
It never made a dollar while look at what we make
The Good Works to our credit stand as angels bright
And show that Heaven chose us to prove the miners right."

The final line was: "The buck's the true religion".

[1] Unless otherwise stated, I have used Sydney prices for share movements. The boom was so frantic that prices were often significantly out of line at the close of different exchanges. Any differentials which persisted next morning would be wiped out by arbitraging.

[2] Interview with Philip Evans, Melbourne Herald, October 6, 1969.

[3] As one who worked two years for Reuter in Fleet Street, I can testify that the old newsagency prides itself on accuracy and gets very pained about mistakes. Investigations are sombre affairs, sometimes ending in dismissal.

CHAPTER NINE

NOVEMBER: Arithmetic Month

While it lasted, the boom was an exhausting, but marvellously exciting, phenomenon to those involved. One woke every morning wondering what new sensation the day would bring; and sensations were occurring nearly every day. We were all players in a great game, picking stocks, seeking rumors, watching the boards. The atmosphere on the seething exchange floors — sometimes lit with arc-lights for the benefit of television cameramen — was frenzied and at times electric. Players were getting rich and players were going bankrupt and if you were a player either thing could happen to you before the day was out. And unlike other games this one ran for five days a week, not counting the verbal speculation and telephone calls to London that occupied the nights and weekends. Later the boom would become a shameful thing and a cause for recrimination and moralising. At the time it was intensely exhilarating. Those who participated can be censured, but they can also be envied.

A share boom, according to an old stock exchange saying, is a lighted match being passed from hand to hand. No one ever put it better than John Maynard Keynes, who said: "It is, so to speak, a game of Snap, of Old Maid, of Musical Chairs — a pastime in which he is the victor who says 'snap' neither too soon nor too late, who passes the Old Maid to his neighbor before the game is over, who secures a chair for himself when the music stops. These games can be played with zest and enjoyment, though all the players know that it is the Old Maid which is circulating, or that when the music stops some of the players will find themselves unseated."[1]

When Poseidon was running wild in 1969, it seemed that anybody could be rich. A cadet covering the stock exchange for a Sydney newspaper made enough punting Offshore Oil options to buy himself a Jaguar, with a roof-rack for his surfboard. Later, of course, it was sold to pay debts to his brokers, but he had few regrets. "Fancy being paid to go in there every day," he said.

The mood was typical. It was very much a young man's boom. A Sydney Morning Herald survey concluded that the typical share trader was a man in his twenties who was now reinvesting his paper profits. This trader would have started with about $1000 which he could afford to lose. He would have bought one or two of the lower-priced mineral stocks in quantities, prepared to lose his stake if the price dropped and he had to sell. In other words, he was trading on margin. If he sold at a profit he had a credit with his broker and he used that as further margin, multiplying the scale of his trading. To some extent this was a by-product of the antiquated registry system of the time, because he would have had to wait a considerable time before receiving a cash settlement for the sale. The bigger buying orders coming through in this way were strengthening the whole market.

In the previous chapter, it was noted that on several days the market was knocked back by profit-takers. But what did they do with the profits so taken? With the market still looking good and Poseidon rising again after each of its setbacks, the logical thing to do was to reinvest in the market. The market was gaining strength from its own rises, and the margin system.

The impact of this enlarged trading was seen particularly in the oil stocks, where low prices were combined with virtually limitless supplies of scrip. Consider the volume of trading in Sydney in the week ended October 31 in the following stocks:

Ashburton Oil	2.2 m. shares	3.7 m. options
Harbourside Oil	0.9 m. shares	1.4 m. options
Murumba Oil	0.9 m. shares	2.9 m. options
Offshore Oil	7.0 m. shares	7.5 m. options
Omega Oil	0.8 m. shares	1.5 m. options
Reef Oil	0.9 m. shares	1.4 m. options
Target Petroleum	4.4 m. shares	6.0 m. options

The trading in mining shares was smaller but more dramatic, because the tighter capitalisation of the mining stocks created bigger jumps in price. Three months earlier, the daily average turnover in mining stocks had been a little over a million shares. By October daily turnover had reached four to five million. On some days it was six million.

Value of turnover had risen even more spectacularly. The combined value of mining and oil shares traded in Sydney had been only $28 million in the month of August. In October it soared to $168 million and in November and December was just below $200 million.

One miner which had been singled out for attention was Newmetal Mines Ltd, which had pegged areas next to Poseidon's eastern flank at Windarra. By the end of October it was turning over more than a million shares and options per day, an amazing volume considering that it had only six million shares and six million options listed. One block of 200,000 Newmetal was sold in London on November 3. In Australia the shares jumped 45 cents to $1.55. The company's shares were only paid to 10 cents and had listed a mere seven days before. A subscriber who had put $100 into the shares in the float would now have had a holding worth $2930, including the attaching options.

Monday, November 3, saw Sydney turnover reach yet another record with 37.9 million units being traded. The transaction count reached 14,000. The action was again heaviest in the oils.

The Melbourne Stock Exchange shut again on Tuesday, November 4, but this time for what was generally accepted as a sound reason — the Melbourne Cup. Newspapers reminded their readers that the great Poseidon had won the race as a three-year-old 63 years before. The cup was run in cold murky rain, and was appropriately won by the topweight Rain Lover. The great race seemed to have become infected by the drama that was sweeping the stock exchanges. The favorite, Big Philou, was sensationally withdrawn only 39 minutes before the start. He had been scouring badly in his stalls and was found to have been nobbled with a purgative called Danthron. The race was won in gallant fashion by Rain Lover, lumping 9.7, which became only the second horse in history to win two successive cups.

With Victorian punters away at the track, the market was having a rough day. Volume was again huge — this time 37.3 million units, but for the first time big turnover was occurring on a falling market. Many speculators saw their gains of the past few days wiped out. Ironically, one of the few stocks to perform well was Endurance, which posted a 17 cent gain to $1.10 on rumors that Peko was going to appraise Attunga. It was Endurance's last kick. From here on the stock was headed for oblivion.

Falls in the market continued during the rest of the week. The one bright spot was the listing of Westralian Nickel NL. The 20 cent paid shares traded between $1.50 and $1.35 on the first day.

The grief of speculators was not assuaged by an acidulous editorial in the Financial Review. The paper said: "The increasingly prevalent notion that the stock market is little more than an extension of poker machines, lottery tickets and horse races is hardly calculated to encourage a relationship of mutual respect between company directors on one hand and the shareholders who own the company on the other. In a situation where everyone is out for what he can get the first casualty may be the unlamented bellhop, whose ambition is exceeded only by his ignorance."[2]

By this time the brokers were becoming restive about the volume of trading and the associated paperwork. In Melbourne Ian Potter & Co imposed a $500 minimum limit on orders. A number of brokers had set similar limits, but Potter's move was seen as significant. Not only was it one of Melbourne's big three firms (with J.B.Were & Son and A.C.Goode & Co), but its senior partner was Sir Cecil Looker. Sir Cecil was chairman of the Melbourne Stock Exchange and of the Australian Associated Stock Exchanges (AASE). Potter's move gave a semi-official imprimatur to what had already been happening.

The move resuscitated an argument which has never been satisfactorily resolved. The brokers said their offices were being clogged with small orders. If a client traded the minimum parcel of 1000 in an oil stock selling at three cents, the total value of the transaction would only be $30. Brokerage would be a mere two percent, making it uneconomic to handle the paperwork. Brokers pointed out that the time and trouble over paperwork was much the same whether an order was for $100 or $100,000, which meant the larger clients were subsidising the small ones' costs.

This is true enough, but the logical extension of the argument would be to charge everyone the full cost of the paperwork, plus a margin for profit, on every trade. Such a flat charge would be a fleabite to institutions but would freeze small investors out of the market. One of the stock exchange's ostensible purposes is to encourage people's capitalism, under which Mrs Bloggs of Moonee Ponds can invest her $100 as easily as the AMP Society can invest $1,000,000. If Mrs Bloggs is being shut out of the market in favor of more lucrative orders, the image of people's capitalism takes a long step backward. Before the boom and after, when times were slack, brokers would fill small orders and the exchanges would solicit small investors. On this occasion the brokers who closed off access to very small investors were offset to some extent by other brokers who took orders from the "discards", hoping to lay the foundation for a larger long-term clientele.

In practice the exchanges are always nervous about imposing such undesirable things as order limits, higher brokerage charges or flat order-fees. High trading costs drive away investors — except in strong bull markets — with the result that the brokers gain little or nothing.

Another danger is the formation of a second market. If brokers' charges and conditions become too onerous, someone may set up a lower-cost market which bypasses them. Since the Securities Industry Acts of the early 1970s prescribed the various stock exchanges this possibility has faded. But in 1969 there was a live and fairly healthy Second Market, being run under the auspices of Dr Michael Duhan Garretty.

It was a thin market with few clients on either side. There might be two buyers in a stock and no sellers or vice versa, but the costs were low. There was no trading floor. You rang the Second Market and were given quotations over the telephone. In 1969 this market was enjoying a small boom of its own because it had built up a fair trade in pre-listed stock. Stags who had received shares in a float now did not have to wait until the stock was listed. They could sell it straight away on the Second Market. The promoters, in at least one case, used the Second Market to frame a market for their stocks. The financial press would check the Second Market just before the stocks listed and quote the prices as a guide to stags.

The Second Market no longer exists, but there is still an occasional danger of an alternative market. The biggest permanent threat is from the institutions and bank nominee companies, which hold large share portfolios. With modern communications, it would not take much organisation for them to form an electronic market between each other, trading shares by computer and telephone. Any such arrangement would be a deadly threat to the exchanges.

While the brokers were trying to cut down their paperwork in November 1969, the share registries were also having problems. The flood of paper threatened in October had now arrived. Brain Noble & Campbell, who handled the Offshore register, had been deluged with 10,000 transfers to mark at the end of October and had managed to process 4700 of them before the end of the month. Hungerford Spooner and Kirkhope, who handled the Bounty, Harbourside and Westmoreland registers, had an even more pressing problem because Harbourside and Bounty were being heavily traded and simultaneously having entitlement issues of Westmoreland scrip. Hungerfords were working nights and weekends to keep the registers up to date.

The brokers were trying to cope with a sharp reversal in manpower demand. Just before the boom they had been retrenching employees. In August two large Sydney firms had made substantial retrenchments, with Ords laying off 25 workers. In Melbourne between 400 and 500 people were retrenched between May and August, including part-time workers.

Now brokers faced a sudden demand for staff again. Some tried to rehire their old staff, but most took the view that hard times might come again so they soldiered on with a reduced workforce. Patrick & Co employed a large number of part-time moonlighters, who worked from 5 pm until at times midnight. Many firms followed this example. Such night work was lucrative but had unsuspected hazards. In Melbourne one operator who had been working back late knocked off and found his car had been stolen.

The boom began to recover its feet on November 10. Minefields, still drilling at Soansville, gained 20 cents to close at 78 cents, after reaching 90 cents in trading. Hawkestone was running on rumors that its joint venture with Longreach Metals had struck 1.5 per cent nickel at Weebo, north of Leonora. This is a good example of a patently untrue rumor. In the previous month Hawkestone had said percussion drilling at Weebo had been disappointing and they did not intend to diamond drill. Yet the rumor monger now had them making a nickel strike, complete to one place of decimals. Any market that could run on this type of rumor was still very much alive.

Poseidon came back into the ring with an announcement on November 12 that a further drilling report would be available soon. The shares leapt to $51 — their highest level yet — before closing at $50. The next day the market was given a further shot in the arm by the most spectacular listing of the boom — Kia Ora Gold Corporation NL.

Kia Ora had two million 25-cent fully-paid shares and three million contributing shares, paid to 10 cents each. The company had been formed by Brian Hill, the 32-year-old executive editor of the Port Lincoln Times.

Kia Ora had 22 claims pegged on the boundaries of Poseidon at Windarra. It had pegged on both sides of the northern tip and on the south-eastern corner. In addition, Hill's Welcome Stranger syndicate had pegged a large portion on the northwestern border of Poseidon. Hill's syndicates had been on the spot early at Windarra but time would prove they had no nickel. Shirley really had pegged well.

On the Thursday the shares listed, Kia Ora directors issued a circular revealing that the company now held 285 claims — three times the total shown in the prospectus issued a few weeks earlier. Many of their Windarra areas had been over-pegged but the company was confident it had legal title to all the claims. Directors also told shareholders in the circular that more than $1 million would be spent on Kia Ora's claims, under farm-out agreements and at no cost to the company.

It was the day the Port Lincoln Times went to press, but it is doubtful whether the executive editor, who had 300,000 shares in Kia Ora, was concentrating fully on the job. Kia Ora had been selling at $3.50 on the Second Market the day before, which made him a paper millionaire.

The listing absorbed the whole market's attention for the first 10 minutes of Kia Ora's listed career. Huge audiences packed the public galleries watching the frantic jostling and shouting of the operators. The fully paid shares hit $5.50 before closing at $4.40 and the contribs sold to $5 before coming back to $4. Australia had another paper millionaire.

The profits were the highest yet taken by stags in a listing, and the glamor rubbed off on other stocks listing that day. Eastern Copper Mines 10 cent shares opened at a dollar and were sold down to 69 cents, still a good price for a company that would spend its poor but honest future trying to breathe life into the marginal Wee Gregor copper mine. Also listed was Eastmet Minerals, whose 20 cent shares changed hands at 44 to 55 cents.

Some bloom even rubbed off onto the all ordinaries index. If industrial shares are rarely mentioned in this history it is because very little was happening in that area of the market. The floor opposite the industrial boards was often empty while a howling pack of brokers surged around the mining posts. Nevertheless, the industrials were now having a small rally of their own. The all ordinaries index in Sydney rose nearly five points to 602 — a 10 percent rise above the 1969 low point which had occurred two months before.

Another feature of trading that day was the 30 cent rise in Quest Mining to close at $1.35. The Age said: "Already the rumor-mongers are working overtime with their customary impeccable timing. The word filtering around the stock exchanges at present is that Quest has encountered extremely interesting nickel values over an impressive width of 67 ft. And the optimists are convinced Quest shares have a fair way to go yet before they reach their peak." Quest promptly denied the rumor.

While the trading floor was running wild, the committee of the Melbourne Stock Exchange were busy disciplining the firm of John F. Robertson Thompson & Co. The firm was fined $5000 and one of its partners, Alan Thompson, was severely reprimanded for conduct unbecoming a member. It was one of the heaviest penalties inflicted by any exchange on a member during the boom.

Robertson Thompson's sin had been to allow a journalist to watch them at work so that he could write a story about how the boom looked from the inside. The journalist was John Sorell of the Melbourne Herald, whose On The Spot column on page two was highly popular because it was topical and well written. Sorell wanted a color story, and the firm had obliged by letting him sit in for a few hours one morning. By having the firm's name mentioned in the press, Robertson Thompson were breaching the exchange rules on advertising, which are very similar to those restricting doctors and lawyers. The firm certainly knew the story would breach the rule and may even have decided that any fine could be written off as publicity costs, but the fine was out of proportion to the offence.[3]

Robertson Thompson might console themselves with the thought that their $5000 was a small legacy to history, because Sorell's article was one of the best pieces of descriptive writing to emerge from the boom. He described the big stainless steel urns of black coffee (already on the boil at 8 am), the banks of chattering teleprinters drowning out the piped music, the operations room reminiscent of RAF HQ in the Battle of Britain and the taut morning conference as the partners and operators try to assess the market.

Two of the operators were munching butter-menthols because they had lost their voices the day before. The operations manager Bryan Frost had received 10 calls from London overnight and had got only three hours sleep. "I live on Ovaltine and sugar," he told Sorell. "There's no time for meals." He had four telephones in his home. The switchboard had been jammed with calls for most of the week. The telephonist said: "It's the most hectic period I've had since I worked at the Southern Cross and the Beatles were staying there." As the morning meeting closed, the first call came through on Poseidon, up $3.50 on the previous day's close. The market was still all right.

The article was a marvellous piece of public relations for the whole broking fraternity. The exchange should have paid Robertson Thompson.

As the week closed there was evidence that the $500 limit and its attendant publicity had helped to reduce turnover in oil stocks. Mining-share turnover had also eased but was still running at well over four million units a day. Friday saw a grim statement from Endurance, whose new board announced that a major orebody did not exist at Attunga.

From London, however, the view was still rosy. The Economist said the ferment in Australian mining shares was not "financial fireworks or stock market jiggery-pokery". "In the northern hemisphere it is still only dimly realised that the whole basis of the Australian economy is being revolutionised — within a few years," the magazine said.

The Australian's financial editor Cedric James, visiting the UK, wrote that more brokers were dealing in Australian shares. Whereas there were normally only six or seven London brokers trading in Australian stock there were now 30 or 40 and most of them were prepared to open their doors at 8.30 am (a huge concession in the City) and to take overnight calls from Australia. Charts were very much in vogue, James reported. The charts of one leading jobber were predicting exceptionally strong projections for Western Mining of up to $21.40. The shares were then $13.55.

Chartists had a degree of influence on the boom. Basically a chartist is an analyst who draws a graph of the price history of a stock and then makes predictions on where it will go next. If enough investors believe these predictions, they can become self-fulfilling. The occult mysteries of the charts are surrounded by a hedge of nomenclature as the chartists discuss head-and-shoulder formations, double bottoms, islands, reversals, traps, accumulation stages and other phenomena. The Rural Bank of NSW surveyed 100 Sydney and Melbourne brokers in mid-1968 and discovered that roughly half the brokers kept charts and about one in eight kept a lot of charts. This survey did not take into account the many chartists outside the broking industry. Some were employed by institutions and others ran private market-analysis services and were sometimes coarsely referred to as tipsters. The most prominent were Cowan Investment Survey and Australian Investment Counsellors, both of whose newsletters contained charts.

Thousands of investors bought stocks during the boom without ever seeing a chart and perhaps without even knowing they existed. But at the heart of the market where the professionals were punting money in large licks every day, charts had a following. Probably every stock that ran in the boom was tipped by a chartist, but like all tipsters they tended to remember their winners and forget their losers. Very few of them tipped the end of the boom, though by February, 1970, it was becoming clear that the game was over.

There is no evidence that Australian investors are peculiarly addicted to chart-watching. The boom chartists never achieved the near-omniscient status accorded to Wall Street chartists during the mid-1960s boom there. I am very sceptical of charts, but if enough people believe in them, they have to be taken into calculation.

Back on the market, the next week opened with Poseidon jumping to $54.50 on the strength of its expected report. The rest of the market moved up with it; Allstate and Newmetal being two to regain ground.

Tuesday saw Kia Ora, in its fourth day as a listed company, reach $10.50 in hysterical trading. The fully paid shares closed at $8.30 and the 10 cent contribs at $6, leaving a gap of $2.30 between the two classes of stock although the theoretical difference was only 15 cents. This was a striking example of the way share prices could get out of line during the boom through a combination of hectic trading and scrip scarcity. Sometimes these misalignments provided profitable opportunities for smart traders. Sometimes they were bull traps. If the gap between two classes of share is $2.30 when it should be 15 cents, it is possible the contribs are undervalued by $2.15. But it is also possible that the contribs are overvalued by $5 and the fully paids are overvalued by $7.15. Many a half-smart trader lost his pants that way.

A different problem was worrying one letter-writer to the Financial Review, who was concerned about the stifling conditions in the Melbourne public gallery. "It is nearly impossible to see the main board from behind three or four rows of people," he wrote. "To view the boards one has to arrive early to get a front position and if this position is left for one second it is lost for good." The writer complained about the shortage of public telephones, the fact that the canny exchange was charging seven cents for calls compared to the standard five cents. Also, the three phones in the gallery were completely in the open, making privacy impossible. The writer finished constructively by suggesting tiered seating in the gallery and private viewing boxes which could be rented out for $50 a week. It was all rather harsh. Only a few months earlier the gallery had been empty. In a few months' time it would be again.

Melbourne at least had a wide public gallery with easily visible boards, although binoculars were usually necessary. Sydney had a tiny gallery with pillars intervening between the public and the boards, and half the boards could not be seen at all.

The tensely awaited report from Poseidon arrived early in the afternoon of November 19 and, characteristically, confused the market. The report confirmed the strike length and width previously announced but did not give assays from the drill-holes. The Melbourne Stock Exchange immediately asked for more detail. It did not get everything it requested but it did get the assays, which were the most important omission from the announcement.

Putting the two statements together, Poseidon had confirmed a minimum strike length of 1000 ft, a width of 65 ft and revealed that the orebody was open at depth of 400 ft. The average grade from 20 percussion holes had been 2.01 percent nickel and 0.28 percent copper in the oxidised zone and 2.52 percent nickel and 0.31 percent copper in the sulphide zone, which included disseminated and massive sulphides. The report also indicated the presence of platinum and palladium and that there was another ore zone separate from the Shirley. Despite the confusion and the query, the report was greeted optimistically and Poseidon shares were chalked up to $55. In London that night the opening sales were at $57.50. North Deborah, Apollo, Newmetal and Target also rose.

The Poseidon announcement and the subsequent query raised yet again the problems surrounding mining company reports. If a company is working continuously on a prospect, when and what should it report? If every company were required to report all preliminary AAS assays, as the Rae Committee thought Poseidon should have done on September 25, there would be masses of meaningless data pouring into stock exchanges and investors would be little wiser. The ideal would be for a company to report whenever something happened to change its position materially. If it drilled 20 dry holes then struck good nickel in the twenty-first, the exchange should be notified immediately. But the key words in this definition are subjective. It is hard to define "good nickel values" and even harder to define "material". Also circumstances differ widely between companies. An average drill-hole by Western Mining at Kambalda would not be worth reporting by itself, but would start a strong run in a speculative miner.

The dilemma was well summarised by the Financial Editor of the Sydney Morning Herald on November 19 following the Poseidon announcements. "One can see the problems of the Poseidon directors," he wrote. "If the angel Gabriel took over as chairman of the company and filled the board with colleagues of his they would still run into trouble over the issuing of public statements. Holes are being drilled day by day. Some provisional guesses as to their meaning may be made by people at or around Windarra. Shares are bought, rumors get around. Some of them may be inspired for profit-taking purposes. But exchanges are bound to ask the directors if they know any reasons for the share movement. Well, do the directors know, or don't they? What exactly do they know? We must strive for perfect solutions but we are not likely to achieve them. The stock exchanges were right yesterday to seek an illumination of the glaring vaguenesses in Poseidon's first statement. The general tenor of their statement was certainly bullish in a general way and the board may have been trying to convey the mood of the company people on the spot at Windarra, to forestall some people getting a special lead on the mood. But when your shares are already standing at Poseidon's level, it is better to impart the maximum precision to what you say from the boardroom."

The more immediate problem arising from the statement was for investors to try to put a value on Poseidon's shares. This became an advanced arithmetical exercise as the boom progressed. The calculation had a nice balance of knowns, unknowns and predictions. This gave it instant charm to high-class game-players, together with the allure that anyone making the right calculation could also make a lot of money. And every now and then Poseidon would come out with a new set of figures which meant you could make the calculations all over again.

It was a fascinating but worrying game. Analysts who had made widely varying assumptions used to agonise over the differences when they met. London analysts made their own set of assumptions.

The name of the game was, "What is a Poseidon share worth?" The arithmetic went like this: the strike length was 1000 ft, the width was 65 ft and the depth was 400 ft plus. This means the total orebody is 26 million cubic ft. But as the orebody is open at both ends and at depth, let's say it's 1500 ft long and 500 ft deep. That would make 48.75 million cubic ft. Seems high? Let's plump for a figure in between, say 40 million cubic ft. This calculation assumes that the orebody is a neat rectangular block. It's not, but we are using average widths and depths, which ought to put us in the ball park.

The ore averages 13 cubic ft to the ton (this is perhaps the most reliable figure in the whole calculation) which means we have three million tons of ore. On an average grade of two to 2.5 percent nickel, the orebody contains about 67,500 tons of nickel metal. Say 70,000.

What is it worth? The current producers' price for nickel is $A2100 an ton, but there is a strong rumor it will rise to $A2600 a ton soon. In fact only two weeks ago the Deputy Prime Minister, John McEwen, and the Minister for Customs, Senator Malcolm Scott, announced that Australia had imposed a limit on scrap nickel exports because there were growing fears of a shortage in the home market. Let's say Poseidon sells part of its output on the free market, where the price is now around $A14,000 a ton. It might average a price of $A5000 a ton. That makes the orebody worth $350 million.

What will it cost to mine? This depends on a number of factors, such as the cost of labor and equipment, the finance costs, the rate of extraction and whether it can be open cut or has to be mined underground. (Eventually Windarra was mined by an inclined spiral shaft and the ore was trucked out.) There are also infrastructure costs to consider such as townsite, railways and associated works. Inflation, by the way, was not a big concern in those days, the rate being around 3 percent. Assuming that operating costs were $200 million, this left $150 million net worth for the company.

If the mine had a life of 15 years, this would represent earnings of $10 million a year. Costs would be proportionately higher in the early years as the mine was established, which would mean lower profits in those years and higher profits later when the mine reached full production. Let's assume Poseidon would reach optimum profits of $10 million a year by the third year of operation. Tax would be no problem because of exploration write-offs. On a capital of 2.5 million shares this would represent a profit of $4 per share. Putting a current price-earnings multiple of 15 times on these future earnings meant that Poseidon was worth $60 per share.

In this abbreviated calculation several factors have been left out. Future earnings have not been discounted back to reach present value. No value has been placed on the associated copper and platinum.

Broadly speaking, all the assumptions built into the above

simplified calculation would have been accepted as fair in November 1969. In a far more detailed calculation, the London brokers, Panmure Gordon & Co, arrived at a value of $380 per share for Poseidon in January, 1970. Yet the ultimate value of Poseidon shares was 0. The trouble, of course, lies in the assumptions. The orebody turned out to be substantially greater, but the grade slipped below the 2 percent mark. Costs rose astronomically while the free-market price for nickel slumped.

In a calculation containing so many assumptions, some are bound to be wrong. The ore may pose metallurgical problems, leading to only partial extraction of the metal. Freight costs may rise, consumers may switch to substitute metals and a host of other unforeseen events may throw calculations out of kilter.

In any coming minerals boom, similar calculations will be made about similar stocks. Because they contain so many figures and appear so learned, they will carry the stamp of authority. In fact, they should be treated with extreme caution. The moral, which should be written on brokers' circulars in letters of fire, is BEWARE THE ASSUMPTIONS. And as soon as one or more of the assumptions turn sour, dump the stock.

In the case of Poseidon, a key factor was the price of nickel. The Inco strike had ended the week before Poseidon's announcement. Inco had lost 150 million lb of nickel output and 140 million lb of copper during the strike and would take four months to return to its former production levels. There was still a nickel shortage, but it was going to begin easing.

Meanwhile, John Gilmour of the Financial Review had been searching the Poseidon register. He noted that Burrill and Associates Pty Ltd appeared to have made paper profits of more than $1 million in Poseidon, which at that stage was $57, having touched $62 during trading on November 20. He also noted the purchases in September. This story was to have far-reaching ramifications, because it was upon reading this article that Dr Anderson in Perth began clamoring for an inquiry into Poseidon.

However, at the end of November, 1969, Dr Anderson was a small, distant ripple on a smooth sea. The free-market price of nickel had touched £6700 a ton, and 40 more mineral floats were reported to be in the pipeline. In Perth Great Boulder announced that it was closing its gold mines in Kalgoorlie after 75 years at Fimiston and was moving entirely into nickel mining. The Fimiston plant was already successfully treating ore from the Great Boulder-North Kalgurli mine at Scotia and would soon be modified to treat nickel ore only. It was symbolic triumph for nickel over gold.

And best of all, in London the Times had dubbed Poseidon "the share of the year — if not of all time as far as stock market performance is concerned". And Poseidon still had a long way to go.

---

[1] General Theory, Ch. 12. If only the rest of Keynes were as readable as his comments on the stock market in this chapter!

[2] After the old market adage: "When the bellhops get in, it's time to get out".

[3] During the boom there must have been a hundred or more stories in which broking firms were named, usually quoting their circulars recommending stocks to clients. None suffered penalties approaching that imposed on Robertson Thompson.

CHAPTER TEN

A Wow of a Christmas

The Melbourne Stock Exchange's corporate section was now illegally taping incoming telephone calls. This was because of the disturbing number of bogus calls purporting to be announcements from nickel companies. A caller would ring up, introduce himself as the chairman or secretary of a company and announce a nickel strike. The exchange had developed a routine of taping each call then immediately ringing back the company concerned to check the announcement. They succeeded in stopping any bogus announcements of this kind becoming public, but they never managed to identify any voices.

London was easier to manipulate and the rewards were just as great. An interesting case study of one share-pusher was detailed in the Financial Review of May 1, 1968, by Trevor Kennedy.[1] The pusher was trying to start a rumor about Longreach Oil. For some reason this stock seems to have had a fascination for share pushers. In 1963 Longreach Oil shares had run briefly, after the market received a fake telegram reporting an oil strike. These attempts to manipulate the stock must have been embarrassing to the company's chairman, Dr Malcolm Mackay, who was one of Sydney's most prominent Presbyterian ministers.

The 1968 attempt to push Longreach Oil began at about 10 pm in London (7 am Australian time) on April 17. The night-duty reporter for the Sydney Morning Herald received a telephone call from an excited young man. He claimed to be a junior reporter on a large London daily newspaper and said he had just overheard a conversation between the manager of the newspaper and a "governor" of Longreach Oil.

The young man said he had heard Longreach had definitely made a nickel strike. Grades and thicknesses of significant proportions had been mentioned. He left his name and address, indicating he wanted to sell the story to the Sydney Morning

Herald. The SMH reporter referred the caller to Kennedy, who was the Financial Review's London correspondent. The caller rang Kennedy next morning and Kennedy sent a memo to Sydney saying a wild rumor was flying about and asking whether Longreach had made any announcement. The reply was that there had been no announcement and the stock was steady. Longreach was then 7/- in London and the equivalent price in Australia.

When the caller rang back, Kennedy told him the story could not be confirmed in Australia, but the man said he intended to buy Longreach and advised Kennedy to do the same. He sounded young and naive and, curiously, did not appear to know how to buy shares.

Later in the day two London brokers rang Kennedy to ask if anything was known about a nickel strike by Longreach. By then the shares had gone to 8/-. Kennedy told them he knew nothing about it and decided to ignore the "story". Longreach shares eased back to 7/6 by the close of the week. That Sunday the Sunday Times (not the paper the caller had claimed to be from) carried a story saying:

"In a market thick with rumors, those surrounding the Australian oil and nickel prospecting companies, magnified by traditional hopes and a distance, need to be taken with more salt and associated sauces than most. But as readers of this column will have discovered, those prepared to take a gamble on the more positive finds can make a handsome return. Longreach Oil has a checkered record in this feverish atmosphere, the shares rising at one point this year to 12/9 on an expected oil well only to fall back despite a fairly wide market to their current 7/6 — a mere 1/6 off the low. But what should really move Longreach now is not oil, for which the company is drilling with a variety of partners, nor its operation in the Mount Isa region, but a plain old nickel strike which, I hear from Melbourne, has a good metal content. Not up to the 9 percent recorded in recent Western Mining digs but highly workable and strong enough to make the current share price irrelevant."

In Sydney, Dr Mackay categorically denied the story. In London the anonymous caller rang the Sydney Morning Herald again and spoke to the same night reporter he had first contacted. The caller gave a different name and address, but the reporter was sure it was the same man. In this call, the pusher drew the reporter's attention to the Sunday Times story. In a social conversation later with a friend from the Melbourne Herald's London office (in those days the two offices were just across Fleet Street from each other), the SMH reporter learned that the same caller had rung the Herald.

From this it did not take long to discover that the caller had canvassed the London offices of all the Australian papers. Also, the caller did not exist on the London newspaper where he claimed to work. As Longreach shares had gone to 11/6 in early trading after the Sunday Times story, the pusher had the chance to make a small killing.

As a small point in favor of the Antipodean press, it might be noted that not one Australian newspaper took the bait, but a London newspaper did. Australian financial journalists were not blameless of reprinting rumors, but they usually identified them as such. Mining share pushers had been working for a long time in Australia and the local press had mostly become hardened to them. London tended to be less critical.

By December 1969, the volume of telephone calls pouring into the financial departments of newspapers was reaching plague proportions. The callers could be roughly divided into the pathetic and the venal. They were all newcomers to a boom market. Hardened market veterans, who had learned years ago to live with unstable share prices, went home at night and slept. But the tyros, sitting home at night biting their nails and worrying about Newmetal or whatever, sought solace by phoning newspapers.

Leslie Carlyon, then finance editor of the Age, told the story of one chap who would ring every night. The caller would ask: "Has Scamander issued a report yet? They haven't...Oh, thanks very much." They all felt very sorry for him.

On the Melbourne Sun, where I was deputy finance editor, the calls became a serious impediment to the job of getting the finance pages out each night. As a popular paper with a large circulation, we probably had every tyro in Victoria as a reader. Some weren't asking for information, they were giving it. "I don't care what the directors said — Pursuit's definitely got oil". And "I won't give my name, but just as a tip, get into Allstate first thing in the morning". And "Acmex has found uranium". And so on.

None of them knew much about mining. A caller would insist on knowing exactly what a company had reported, so a reporter would spend 10 minutes at a crucial time of night reading him a detailed drilling statement. Then the caller would ask: "What does it mean?" As the boom wore on so did my patience until anonymous callers were receiving very short shrift indeed. Carlyon noticed the curious fact that most callers seeking information did not care much what the company's report said. Instead they wanted to know what other people thought of it. Their standard query was: "As a financial expert, do you think this'll put the shares up?"

But at least most punters were still making money, because December was the month when Poseidon completely took over the stock market. Two Melbourne broking firms had been studying the company and released bullish circulars to clients in the first week of December. Guest and Bell, after a series of hypothetical projections on reserves and earnings, estimated that Poseidon would have earnings per share of $12.50 in the year to June, 1974.[2] This meant that on its price of $50 when Guest and Bell did the survey, Poseidon was selling at only four times 1974 earnings. "Australian mining companies with bright prospects rarely seem to be traded at less than a price-earnings ratio of 20 times," the survey said. "Therefore Poseidon shareholders may look forward to a minimum price of $250 a share in the next few years."

Edward Ward & Co was more cautious, but still bullish. Both brokers had discussed their figures with Shierlaw before publishing their circulars. Shierlaw later said he had seen the circulars but had not commented on them.

The circulars triggered another bout of furious trading in Poseidon. The stock, which had been $48 for the first two trading days in December and had moved to $54.50 on the third, hit $70 in Melbourne trading on December 4 before retreating slightly to $67.50 on Friday, December 5. As the board closed for the weekend, traders were reflecting that Poseidon's annual meeting in Adelaide was only a fortnight away. The question on everyone's mind was: "What will Poseidon announce?"

After considering this over the weekend, the traders decided to buy. On Monday, December 8, Poseidon went to $79 during trading, closing at $75. Overnight the stock went over $80 in London. Tuesday saw the market really boil as Poseidon roared to $94, closing at $93. In Perth a single trade of one share was registered at $100. Someone there had been patriotically determined that the Perth exchange should be the first to put Poseidon to the ton.

But the overnight market was weaker in London and Poseidon's price drifted back to $85 for the rest of the week. These were peculiar days as Poseidon hovered in the mid-eighties amid a widening conviction that it would soon see $100 again.

The surge in Poseidon shares had not rubbed off onto the

speculative stocks. The reason was clear. On December 3 the Leonora Mining Warden's Court had begun sitting to straighten out at least some of the massive tangle which had developed over nickel claims in the area. The court had 1700 applications to consider for mineral claims, of which 215 involved formal objections and a further 34 were contested. To the market, the most crucial cases were those of the "Windarra Club" of companies who had pegged areas around Poseidon's borders.

Some of the cases were quite complicated. In one dispute, Amalgamated Petroleum NL had pegged a claim, Apollo International Minerals NL had pegged one next to it and Timor Oil Ltd had pegged the next one again. Objections were lodged to the first two claims by the Perth-based prospecting company Minech Pty Ltd, to all three by Campbell Cooper and to all three again by North Deborah Mining Co NL. In another case, one area was being claimed by four companies — Kia Ora Gold, Conwest (Australia) Pty Ltd, Anglo-American Corporation (Australia) Ltd and Cam Cooper.

Share traders cautiously marked down stock in the Windarra Club, pending settlement of their claims. As the warden's court hearings continued, the bulls became increasingly jittery.

Light relief was provided by Cam Cooper. After the first day's hearing, Cam went out in the middle of the night and pegged another nickel claim. This one covered 6.7 acres and included the court house, the office of the local mining registrar and most of the Leonora primary school. Under WA mining law, a prospector could charge with trespass anyone who entered a pegged claim without a miner's right. By pegging his midnight claim, Cooper now had the right to bar anyone from the court house who did not have a miner's right. This included the lawyers of nearly all his opponents. And as the registrar's office was included in the claim, they could not get to the office to get a miner's right. Cooper magnanimously waived his legal rights, having pegged the claim just for fun, but he had made a point about the state of the mining laws.

The mining laws were not in much more confusion than the stock exchanges. The Melbourne Stock Exchange committee had boldly banned smoking in the public gallery, and stationed a guard there to ensure its edict was obeyed. The ban was unpopular but necessary, because the traders dragging nervously at their cigarettes were overstraining the airconditioning. In future anyone who wanted a smoke to relieve the tension had to sacrifice his position and go have a drag in the corridor.

Then there was the question of Christmas. The Sydney Stock Exchange decided, by a narrow majority, to close during the Christmas-New Year break. The brokers defended their decision by saying they were badly in need of a rest; but there were powerful dissenters. Patrick and Co announced that they would stay open and provide a market for their clients, dealing in Melbourne and if necessary London. The Financial Review sniffed: "From their position of great strength the exchanges cannot act as private groups on some occasions (when they want extra holidays) and as publicly minded utilities on others (when they don't want a regulatory authority above them) and still expect people to believe they are really great private institutions with the public interest at heart."

After Sydney's decision to close, a movement started among the Melbourne brokers to close that exchange too. Simultaneously, a group of Sydney brokers were agitating to have their exchange reopened. Eventually, they both decided to stay open, having lost their holiday and gained nothing but criticism.

One factor influencing the brokers' decision to stay open was the state of the market. It was showing enormous strength, particularly in new listings. Stags were having a literally golden run.

The first listings of the month both had remote connections with Poseidon. Northern Mining Corporation had been jointly underwritten by Shierlaw and Associates and Guest and Bell. (Poseidon obviously hadn't been the only topic of conversation between the two firms.) Northern Mining was one of the better floats of the boom, looking medium solid from day one. It had 4.4 million 20-cent shares, 2.2 million options and a promising iron-ore deposit at Weld Range in Western Australia. Eight years later it would become one of the hottest stocks in the mining list on its diamond interests in the Kimberleys. Its 20 cent shares opened at 90 cents on the first day's trading, closing at 70 cents.

The other listing that day was Amber Gold NL, which included E. Broughton Jensen among its sponsors. The 20 cent shares opened at 60 cents before sellers marked them back to 45 cents. The Jensen magic did not work with Amber Gold, which was never in danger of becoming another Poseidon.

The following Thursday, December 11, saw six new miners list. Star of the show was Laverton Nickel, which established a new listing record by giving stags a 5400 percent return at one stage during trading. The 10 cent paid shares opened at $3 but later drifted to $2.15. Along with Kia Ora Gold and Samin, Laverton was one of the boom's greatest floats. Laverton's main claim to fame, apart from some hot nickel leases, was its modern assay laboratory in the northern Sydney suburb of Artarmon. Laverton was very proud of its laboratory, at one stage announcing that it was being equipped with advanced sophisticated equipment for the analysis of platinum group metals. The equipment was stated to be able to assay to 10 parts per million. This announcement drew some flak from Roy Jones of Burrill and Associates when he was giving evidence before the Rae Committee. Jones said 10 parts per million was the same as 5.9 pennyweight per ton "which would mean that this equipment is totally valueless for the analysis of platinum." Modern equipment, Jones added, could assay accurately to 0.1 parts per million, or 0.06 dwt per ton.

None of this inhibited the bulls of December 1969, and Laverton shares were traded strongly around the $2.50 mark — 25 times their paid value. Other listings on December 11 were Abaleen Minerals NL, whose 25 cent paid shares closed at 55 cents; Beaver Exploration NL 10 cent paid shares sold at 21 cents; Magnum 10 cent paid were 96 cents; Triako 15 cent paids were 90 cents; and Australian Antimony 20 cent paids were $1.20. One vendor to Australian Antimony was a company with the appropriate name of Instant Mining Pty Ltd. Instant Mining's address was care of F.W. Millar in Australia Square. This was Frederick Millar who later became chairman of Thomas Nationwide Transport Ltd. Australian Antimony was underwritten by Patrick & Co and one of its senior partners, Frank Robertson, was on the board.

A broker would have had considerable trouble finding time to attend board meetings in those days, because brokers' offices were under a heavier workload than ever. The man in the street had now become involved in the boom and brokers had suddenly found themselves swamped by thousands of new customers, most of whom had only the vaguest knowledge of the share market. A.C. Goode & Co had just opened a Perth office. Because of the time gap of three hours between Perth and its head office in Melbourne, the WA office used to open a little early at 8.30 am. "They would be queueing up outside the office when we opened the doors," a Goode partner recalls. "They came around physically because they knew there'd be no hope of getting through the switchboard."

It was the same in Sydney. "They used to walk in at lunchtime

— cleaners with buckets in their hands," said an Ords partner. "Once I was working flat out when a client rang. I said: 'We're all in turmoil'. He said: 'Buy me some too'. Perhaps we should have floated a stock called Turm Oil No Liability."

The small punters' instincts were sound because a groundswell of support was beginning to emerge as Poseidon's annual meeting neared. On Tuesday, December 16, the stock closed at $84.50. The next day saw a sudden, awesome surge as the shares soared through the $100 mark. During trading Poseidon went to $110 in Sydney and $115 in Perth. At the close in Sydney the last sale was at $105 for a mammoth rise of $20.50 on the day. As trading closed operators were still bidding $110 but the sellers had been wiped off the board.

Poseidon again dominated trading on Thursday, December 18. The stock opened at $110, ran up to $121 then retreated to its opening price. New listings gave stags another feast. Australian Consolidated Minerals 20 cent paid shares closed at 78 cents. Halls Peak Australia, Overland and Valiant, whose shares were all paid to 10 cents, closed at 50 cents, 23 cents and 29 cents respectively.

Friday, December 19, was the tensely awaited annual meeting of Poseidon. Share traders had been gathering in Adelaide for days. Two stockbrokers had flown in from London. Eastern stockbrokers, shareholders and investment managers were booked into motels. Incredibly, the board had considered banning the press from the meeting. However, the eastern financial press had already arrived equipped with proxies and would have been difficult to remove. (It is almost an automatic precaution for a financial journalist to equip himself with proxies whenever he is travelling interstate to attend a company meeting. Otherwise he could be locked out and have nothing to show the editor for his expenses.)

At 9 am the crowd began pouring into the Building Centre on Adelaide's South Terrace. Long queues formed outside the centre as officials checked the credentials of people trying to get in. The huge turnover in Poseidon's shares had created confusion as to exactly who its shareholders were. About 50 people were turned away as ineligible. Another 500 packed into the Building Centre. The seats were soon taken and about 200 of the audience had to stand at the back of the hall, including several prominent Adelaide stockbrokers. On the platform was Norm Shierlaw, slouched back in his chair, rugged and happy. The rest of the board were seated alongside him, together with Basil Lewis, chairman of Samin.

The meeting opened slowly with the reading of the minutes of the previous meeting. This may have been a deliberate touch of irony, because in 1968 there had been insufficient shareholders to form a quorum. The annual report was then adopted and directors re-elected. Basil Lewis was elected to the board to fill the place vacated by Ganke's nominee, Scott.

Proceedings went mechanically until the special resolution was put to approve the share placements. It was seconded by Robert Holmes, a London stockbroker who said he was attending the meeting on behalf of the Pan-Australian group of trusts. A few rows from the front on the opposite side of the floor sat the slight, bearded figure of Boris Ganke. He now stepped forward and addressed the meeting. Battle was joined.

Ganke moved an amendment to reduce the placements by 90 percent to 50,000 shares and to place the other 450,000 shares at $5 each with shareholders. He was well applauded for this suggestion. In the two-hour debate that followed, Ganke had the debating points but not the numbers. Poseidon was obviously going to need capital to develop Windarra. The easiest way of raising such capital was through a share placement, using Poseidon's huge premium. The shares were now $100 on market but were being given away for $5 and $6 to the placement recipients. Ganke's proposed issue to shareholders would not have raised any more money, but it would have been more equitable.

Ganke argued that Poseidon was wasting an immense asset by placing its shares so cheaply. It would have been different if the shares were being placed with strong financial institutions who could provide loan funds for the mine. But none of the companies were financially strong, with the possible exception of Conwest. The main assets of the Adelaide-based companies were the Poseidon shares which they had been given so cheaply. The placements were diluting the equity of existing shareholders without offering any compensatory benefits.

As Ganke and his associates slogged on in a long debate with the chairman, Tom Hutton, it became obvious that the mood of the meeting was against them. They were heckled from the floor and there was some evidence that the Adelaide Establishment was closing ranks behind the Poseidon board. Adelaide brokers spoke in support of the placements.

Mr A.K. Sangster, QC, spoke on behalf of four of the companies that had received shares. He said the placements had been made following discussions on Monday, September 29, and Tuesday, September 30. At the close of trading on the 30th the shares had been $6.80 and directors had been faced with an opportunity to raise money for development (presuming they had a mine). Under the circumstances the placement price had been very fair. What had been done was regular, for the benefit of the shareholders and no one ought to challenge it.

Sangster added that no assays had been available when the placements were discussed. This was inaccurate, because the Rae Committee later discovered that Burrill had phoned Shierlaw early on Saturday, September 27, with the news that the AAS tests had shown 30 samples from PH2 had readings of greater than one percent nickel. However, none of this mattered much because the meeting was overwhelmingly on the side of the board.

The president of the Adelaide Stock Exchange, Don McArthur, said he and his brothers had been interested in the company from the start. He had been very surprised on learning of the allotments. "However, when I found out the circumstances of the allotments I was more than happy that the directors had done the right thing," he said.

A retired Adelaide stockbroker, Gordon Sunter, was cheered loudly when he said he had not come to the meeting to hear legal quibbles but to hear a report on the company's progress. He said the courtroom, not the meeting, was the place for legal discussions. Hutton, who had allowed the debate to run for two hours, seized the opportunity to ask if he could interpret Sunter's remarks as a motion that the question be put. Sunter agreed. The amendment by Ganke was defeated and the motion to approve the placements passed on a show of hands. Ganke was heckled again when he asked for a poll, but Hutton pointed out that this was the right of any shareholder. The placements were carried by 776,138 votes to 100,030.

The meeting had made the wrong decision. Poseidon had a mine but no funds. The placements had diluted equity without any compensating advantage, except that they helped the existing board retain control. The links between Poseidon and four of the recipients were close. North Flinders, which received 100,000 shares, had been floated in June, 1969, with Shierlaw and Associates as underwriter in combination with Guest and Bell. Three Poseidon directors, Shierlaw, Hutton and C.F. Wegener, were on the North Flinders board and were substantial shareholders of that company. Poseidon was a large option-holder in North Flinders. Nobelex, which received 50,000, had been underwritten by Shierlaw and Associates when it floated in May, 1969. Shierlaw, Hutton and Wegener were substantial shareholders in Nobelex also. Australian Development, which received 50,000, had Shierlaw as a director and substantial shareholder. His firm had underwritten a new issue of shares in March, 1969, for Australian Development and Australian Development had been a sponsor of the Nobelex float. We have already seen, in chapter seven, the links between Poseidon and Samin.

The placements had two side-effects which would ultimately prove fatal to Poseidon. The first was that the threat of litigation was not extinguished by the shareholders' vote at the meeting. Ganke, a determined litigant, had taken out a writ in the Supreme Court of South Australia challenging the placement. Despite the shareholders' ratification, Ganke pursued his lawsuit for another two years. During all this time it was an inhibiting factor in Poseidon's financial negotiations, making it impossible for the board to make a general issue to shareholders or to give any firm guarantee that a further placement might not be upset.

The second, and lesser, side-effect flowed from the placement of shares to Samin. Samin was to prove a weak financial reed. Its other mining operations were uneconomic and as its share price sagged its Poseidon holding became a hostage to fortune. When Samin floundered financially the Poseidon board was forced to come to its rescue with a takeover bid to recapture the 200,000 shares. As a defensive exercise this was temporarily successful but as a financial exercise it was disastrous because it cost Poseidon a net $25 million worth of equity which could have been used to finance the bulk of the Windarra development.

Poseidon, in turn, was doomed to be taken over, Windarra would pass ultimately into a joint venture between Western Mining and Shell, and Poseidon scrip would become worthless.

All this was well into the future on December 19, 1969, as Tom Hutton, the placements having been ratified, rose to address the meeting again. "We now move on to what we consider to be the most important section of the meeting — that of an up-to-date technical report on the Windarra deposit," he said. "It is now my pleasure to announce that a major mining operation will be possible on our Windarra prospect."

He then introduced Roy Jones, who gave details of the geology and drilling progress as Windarra. He said the longitudinal section around the zone in which drilling had positively indicated ore contained four million tons. "This ore is only a small proportion, that is, 1300 ft in length and some 600 ft in depth of a long and probably much deeper zone in which drilling must ultimately be undertaken," he said. "It is not therefore possible to give any indication of the total ore potential of the Shirley zone. It is expected that the overall ore zone will plunge steeply, hence ore should continue for a considerable depth." Within the part of the Shirley zone which had been drilled the average of a total of 943 ft in ore was 2.4 percent nickel and 0.3 percent copper.

After Jones had finished speaking, Hutton said: "I see no reason why any shareholder of Poseidon should not have a wow of a Christmas." The excited shareholders, who had been waiting through the long debate about the placements to hear the good news, cheered wildly. Then one shareholder called for three cheers for Ken Shirley. The response was deafening.

The shareholders flooded out of the meeting, trying to calculate four-million tons of 2.4 percent nickel, dividing it into the number of shares and then adding a subjective "blue-sky" figure to cover the extent of the rest of the orebody. The London brokers who had flown to Adelaide were looking tired but satisfied. One was Holmes, the other was Simon Hunt of Panmure Gordon. Hunt said: "It was well worth coming 12,000 miles for — it looks very good to me."[3] It was Simon Hunt who later produced the Panmure Gordon circular putting a value of $380 a share on Poseidon.

Outside the Building Centre, every public telephone booth for blocks was occupied, invariably by an excited man wearing a suit and tie.

While the long wrangle and the technical discussion had been proceeding, the stock market had been jittery. The stock had opened at $112 and run to $128 but as the meeting dragged on to lunchtime, Poseidon shares sagged back to $110. As reports of the remarks of Jones began filtering through the stock regained strength and closed at $130.

On Monday, December 23, Poseidon staged one of its most breathtaking performances. The stock rose from $130 in Sydney to a peak of $185 before closing at $175. It was the most spectacular day's rise yet in Poseidon, and obviously the product of a lot of arithmetic by the traders over the weekend. The market's fever was heightened by the fact that Western Mining was intersecting good nickel grades at Mount Clifford. Western Mining shares put on $3.55 to $18.50. In terms of market capitalisation its rise was much higher than Poseidon's. A 'Mount Clifford Club' of nickel stocks formed as speculators turned to other companies in the area, pushing up the prices of Australian Consolidated Minerals, Hill 50 Gold Mines and others. BHP was having a run of its own, rising 75 cents to $15.50 on rumors of an oil discovery offshore of New Zealand.

On December 24 stock exchanges adhered to their traditional practice and traded only for the morning session. Volume was enormous. More than 15.5 million shares changed hands in Sydney, compared to only 2 million on the previous Christmas Eve. Poseidon closed $7 higher at $182. Hutton's prediction was correct. Poseidon shareholders were having a wow of a Christmas. The stock had risen by $100 in six trading days.

Bears in London would not have shared the jubilation, because some of them had been savagely mauled in North Flinders. As a Poseidon shareholder, North Flinders had seen its own shares rise rapidly. On December 17 North Flinders had been $2.35. In succeeding days it strengthened to $2.80 and $3.10. After Poseidon's annual meeting North Flinders jumped to $4.30 and on Monday, December 23, the shares reached $6. Then a rumor developed that, apart from holding a block of Poseidon which was soaring in value, North Flinders had struck rich nickel and copper in South Australia. This was alarming news to some London bears who had been selling the stock short.

Early on the morning of the 24th, long before London opened, North Flinders denied the rumors, but in London the bears panicked. They scrambled to cover and North Flinders' already handsome price shot up vertically.

The London Stock Exchange on the morning of December 24 was letting its hair down in traditional schoolboy fashion. Operators were throwing paper darts at each other, playing indoor cricket, practical joking and generally larking about so that the floor of the exchange looked more like an office party than a place of business. The exception was the Australian section, where the action was grim and hectic. North Flinders opened at $9 and soared to $22 as bears bid frantically for cover. In the short morning session Poseidon rose by more than £20, displaying fresh strength to astonish stockbrokers. It seemed as though the whole of Australia was bursting with mineral riches. Shorting Australian stocks seemed suicidal. The Evening Standard's lead finance story began: "The City of London is completely amazed by the Poseidon phenomenon."

One voice of sanity was provided by Dr John Rose, the senior research fellow in the Institute of Applied Economic and Social Research. In the Institute's Australian Economic Review, Rose produced some sobering statistics from Canada which indicated very low rates of success for exploration companies. He quoted the Ontario Mining Association which said that 420,000 mining claims had been staked in that province between 1907 and 1953. About 6600 mining companies had been formed in the same time. Of these companies, 348 reached production stage, but 294 failed to show a taxable profit, leaving only 54 companies which had paid dividends. This was a success rate of less than one company in 100.

Rose was in a considerable minority. More typical was the Age reader quoted in the City Talk column. The day before the Age had run a story on mining claims and to illustrate it had reproduced a sample mining claim form made out to "Fickle Nickel No Liability". The reader rang to ask where he could buy shares in Fickle Nickel.

On the market there had been a flurry in United Motels, a small, sad Sydney company with a history of losses and 1.3 million issued shares. They had been languishing unwanted at about 16 cents in the market until the end of November. Then they rose rapidly to 70 cents in four days' trading with nearly half a million shares being bought. Directors were mystified until a Port Lincoln lawyer rang them on the fifth day and offered the company some nickel leases in return for a placement of 195,000 shares. If the parties seeking the placement had also been the buyers of the stock (which they were) they would have wound up with 51 percent of United Motels and United Motels would have turned overnight from a motel operator into a nickel prospector. It was a typical backdoor listing exercise, except that mining promoters usually reached a private agreement with their target company first.

But this was a sideshow. The main action was in Poseidon and the established nickel miners. The three days trading in the Christmas-New Year week was a huge success, with brisk market action and a packed gallery every day. Turnover in Sydney averaged more than $10 million per day in oil and mining stocks. Poseidon retreated a little to $170 in the first two days. Then on December 31, Poseidon ended the year in typical fashion, rising $40 and smashing through the $200 mark. Spectators in the public galleries cheered wildly as it hit the double ton. It went to $214 before closing on $210 to give shareholders a magnificent New Year's Eve. The strength of its market surges was awesome.

As 1969 ended, it was apparent that while the towering rise of Poseidon had been beneficial to mining stocks, it had been a poorish year for industrials. BHP actually fell during the year from $19.45 to $17. CSR was up from $5.92 to $8.40 but mainly on its mineral interests. ACI was down from $2.62 to $2.23. The Sydney all-ordinaries index had managed only a nine percent gain to 654.03 for the year. By contrast the non-ferrous metals index was up 55 percent to 5737.40. This was a particularly strong performance considering that 1968 had closed on a boom note for miners. At the year's end Poseidon had a share-market capitalisation of $455 million — not far behind CSR at $491 million.

The value of mining and oil stocks traded on the Sydney Stock Exchange had exceeded $1000 million for the first year ever. By contrast, industrial turnover had barely exceeded $500 million. There were signs that the depression in industrials had been partly caused by a diversion of investible funds into the go-go mining shares.

The need for cash to subscribe to mining floats would certainly have diverted some money out of slow-moving industrial stocks.

In London the Economist ran a forecast of Australia's outlook under the headline: "Australia; a Beaut of a Boom". The article said: "Australia more than any other country is being carried along by Japan's explosive economic growth. The spectacular history of the Poseidon shares rather exaggerates the overall mining boom in Australia. But there is little doubt that historians will see the late 1960s as comparable with the legendary Australian gold rushes of the 1850s. Australia is expected to become the world's largest exporter of a whole range of minerals — certainly iron ore, coal and alumina, possibly copper, nickel, pig iron and aluminium."

DIGRESSION III: For Whom The Bell Tolls

Did you hear a bell ring just now?

No?

Then join the rest of the human race because you have proved yet another old market adage which says that nobody rings a bell at the top. If there was a bell it would have rung just three days ago, because that was when the market peaked.

It was December 29, the first trading day after Christmas. The market was looking great. The gallery was packed with speculators, binoculars glued to the boards and leftover-turkey sandwiches in their briefcases. Poseidon was steady on $170, like some colossus gathering strength for yet another huge upward thrust. Market action was brisk as the Sydney floor operators traded $10 million worth of mining and oil stocks on a day of the year when they would normally have been ogling bikinis on Bondi beach.

That was the top. December 29 was the day the Sydney non-ferrous metals index hit 5870.10 — its highest level ever. From here on the heavyweight miners were headed down. The speculative mining stocks still had a final hysterical burst left in them. Patrick's Speculatives Index would peak on January 28 at a market capitalisation of $175.22 million for 30 selected speculative stocks; but they could not fly on their own.

All booms must end at some point, and the reasons are usually pretty much the same. Optimism for a particular commodity — in this case minerals — drives prices to ridiculously high levels. Eventually the market cracks under its own weight. Some of the sharpest rises will have been engineered by smart operators, including a few rises engineered by outright fraud. The more manipulation there is, the more shares there are whose prices are unjustifiable. More of the market becomes fragile.

Australia is not only well endowed with minerals. It is also generously endowed with share manipulators. In 1969 manipulation became a sizable industry. Some examples will be seen within these pages. The veil that normally shrouds their practices was partially lifted by Tom Nestel, managing director of Minsec, in evidence given in camera to the Rae Committee and later printed as Chapter Eight of the Rae Report on "Runs, Pools and Rumors". As the man running the biggest share-trading operation in Australia, Nestel was constantly being offered a piece of the action in pools and runs. Minsec usually rejected such deals, having its own mode of operation, but Nestel had a wide knowledge of what was happening. Nestel was one of the shrewdest operators of his era, with a near-photographic memory. Few men knew more than he did.

His evidence on manipulation is therefore particularly valuable, although the parts of it published by the Rae Committee avoided naming offenders. One thing Nestel confirmed was the widespread suspicion that there had been a lot of churning to hoodwink the chartists. Churning occurs where a block of shares goes back and forth between the same parties in rapid succession, creating the illusion of large volume. If Joe Bloggs sells a block of half a million shares to his wife, she sells it to his investment company and his investment company sells it back to him, the total turnover recorded on change will be one and a half million shares in the company concerned, although Joe still holds the stock. These transactions are also called wash sales.

SYDNEY NON-FERROUS METALS INDEX

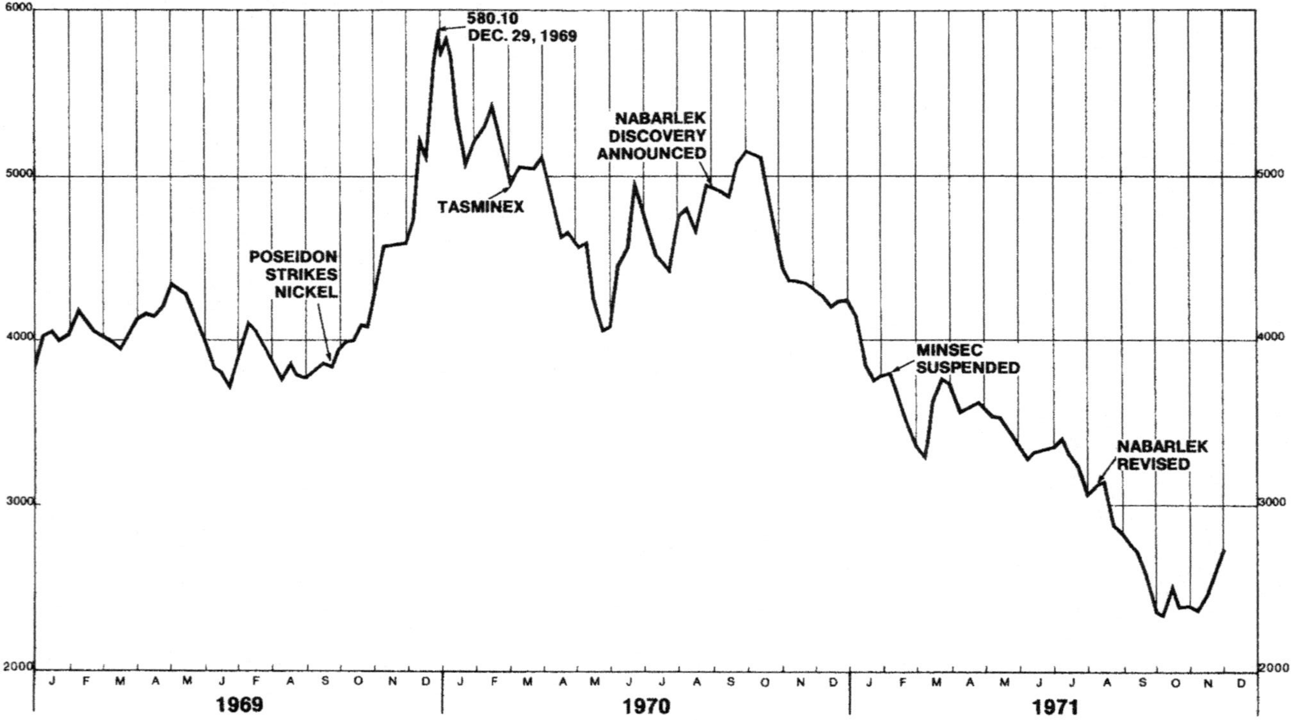

This is Patricks' Speculative Index, probably the best single chart of the boom. It was based on the market capitalisation of 30 stocks, which were changed from time to time but never included Poseidon or Tasminex. Peak capitalisation was $715.22 million on January 28, 1970. — Tasminex's big day. Dividend-paying miners were excluded. It peaks slightly later than the Sydney Stock Exchange's non-ferrous metals index, also shown here.

Churning could be used with great effect to beguile chartists. Their interpretation of charts is particularly sensitive to rising volume and a deft manipulator can churn shares to create a pretty pattern on the charts. Then all the chartists begin recommending the stock and in a bull market where charts have considerable credibility, the stock will really begin moving under the buying pressure.

Nestel said churning often happened because there had been so much avid interpretation of charts. Giving evidence in May, 1971, he said: "A recent one was a case like Sedimentary Uranium where turnover came in, shares went up and people joined in. Everything looked excellent but it turned out that there had been no drilling for uranium for weeks because South Australia was under water."

Another form of manipulation was a pool operation. In a pool, a group of operators band together with, say, $25,000 each. If there are four of them they will agree to put $100,000 into a stock in the hope that this volume of buying would attract other speculators into the stock, pushing it higher still so that the original pool members could sell at a profit. Normally the pool would multiply its buying power by operating on margin, which might mean the members could buy up to $1 million worth of stock on their capital of $100,000. As an inevitable accompaniment, the pool members would begin spreading bullish rumors about the stock as their order was completed.

Pools are very old, probably as old as markets. They have been condemned by every committee that ever inquired into market malpractices, but they are difficult to outlaw. After all, at what stage does a perfectly legitimate share trading syndicate become an undesirable pool? There were many pools formed by wealthy traders during the boom in exactly the way set out above. Arguably, there is no reason why they should be declared illegal as such. After all, the traders are buying on the open market and if they have misjudged they can be left high and dry with the stock.

Smart manipulators knew that they could obtain greater impact, and great freedom from surveillance, if they operated through the London market rather than the Australian stock exchanges. While the Australian exchanges saw plenty of churning during the boom, London was generally an easier market to rig. The mechanics were described in chapter five. Although investors other than brokers had to obtain Reserve Bank approval before they dealt directly in London in Australian shares, in practice large Australian share-traders found no difficulty in evading this requirement. As continual buyers and sellers, they could contrive to operate so that they did not have to make a transfer of funds out of Australia, thus bringing their operations to the Reserve Bank's attention.

The Rae Report said that witnesses testified that wealthy Australian investors wanting to obtain overseas currency in exchange for Australian funds could do so through a well-organised system. Once the funds were offshore, the share-trading profits escaped taxation, of course.

When one broker with a large Australian and international business was testifying in camera before the committee he was asked if he had heard of some investors using "a very expensive communication system throughout Australia which indicates at the beginning of the day what is going to run that day?" He answered "Yes", and explained that he had obtained this information from clients.

When asked if he had ever heard that a particular share was going to run at a certain time and had subsequently noticed that it had, in fact, run, he said: "Yes, this is going on all the time". He also said: "I think London is very bad, the way in which it has the special connections with people out here and arrangements to pay for information. Sometimes I think there are other arrangements, funds not coming to Australia."

Nestel said there were a lot of occasions when a broker had told him that a particular share was going to run. He said that not only were there a large number of organised runs but they were known to a large number of brokers, among whom were WA brokers. From Nestel's description, these brokers and traders appeared to have formed pools and to have relied on their judgment of drilling information and charts. Whenever they knew a company was about to drill and the charts indicated that it was ripe for a run, they would buy, then sell as the stock neared the level they expected it to reach on the charts.

Nestel said other organised runs were based on leakage of assay information. He did not know of speculators or others having access to official assay results before they were released to the companies concerned, but he did know of unofficial assaying being carried out.

As Minsec was in the business of share trading, it was always interested in making a profit, whether the company whose shares were running actually had anything or not. Nestel always used to do a little research on any stock that was supposed to be running before he decided whether to buy some. "One of the worst things you could do in the hysterical days was to talk to a geologist about the share," he said. "He would tell you: 'It is worthless'."

Looking at the various forms of skulduggery that were practised in 1969-70, the small speculator may be tempted to assume that he has little chance of staying alive in such shark-infested waters. The answer is that there is a lot of money to be made in booms and the share market is big enough to accommodate the small speculator. Small speculators can certainly make profits, although, of course, the insiders and big traders are in a much better position.

The most elementary rule is to play with cash you can afford to lose. Brokers are generally reluctant to handle orders under $500, so that should be regarded as the minimum amount needed. Preferably, a small speculator should operate on a minimum of $1500-$2000 so that he can hold three or four stocks, thereby spreading his[4] risks and increasing his chances.

In a speculative market, where stocks can double in a week or even overnight, there is always a chance of catching a flyer. If a speculator decides he can afford to take $2000 out of the nine percent account at the building society and put it into the mining market for a punt, the worst he can lose in a wipeout is $2000 and his gains may run to anything. The whole idea is to limit downside risk while maximising upside opportunity.

Professor Geoffrey Blainey makes the point in The Rush That Never Ended that the speculators who get hurt in a boom are the ones who come in late. The ones who are in early have a much better chance. A speculator who gets set in a market too early, however, can find his money left there derelict and drifting for a year or so. It will be earning nothing because the stocks we are talking about here are not the ones which pay dividends. Unfortunately, nobody on earth can tell you the right time to get set in the market.

The ideal is to get set while stocks are still fairly cheap — around the 30 cent mark, say — and to take three or four companies which are active explorers, or about to be. Where you sell is up to you, but remember it's easier to sell a stock when it is rising. "Leave the last 10 percent for someone else," your broker will say, and he's right. Avoid stocks with extremely tight issued-capital and thin turnover because it is often difficult to unload them.

A good broker is a great help. Small speculators are generally a little better off using smallish to medium-sized firms. If you're trading in mining stocks there are advantages in dealing with a firm that has some mining expertise. Broadly speaking, all

Australian brokers know a fair bit about mining, but some concentrate on it more strongly. The following firms in each city have strong mining associations: Sydney, Clinton Jones & Co, Ord Minnett, Hattersley and Maxwell; Melbourne, Potter Partners, Wallace H. Smith & Co, May and Mellor; Brisbane, Wilson & Co; Adelaide, N.C. Shierlaw & Associates; Perth, D.J. Carmichael & Co.

Avoid companies with empty prospectuses. They are asking you to subscribe money on faith. Why should you show so much faith in someone you've never met? If the 1969-70 boom is any guide, you would also be well advised to avoid mining investment companies, and mining oriented mutual funds, unless they have a long and respectable track record. Invest in companies which have real prospects or are about to acquire them. Don't duplicate by investing in the same situation twice. If you like the prospects of Nabarlek, buy Kathleen Investments or Queensland Mines, but not both. Buy into companies where the management has a record of legitimate exploration, whether successful or not. They may get lucky and they probably know what they're doing. Try to check what insiders are doing by watching turnover in low-priced stocks. Big turnover in tightly capitalised stocks can indicate insiders getting set. It can also be a trap, with insiders churning the stock over to lure speculators in. You pays your money and you takes your choice.

Having done all this, you will wake up one fine day to find the market humming along beautifully. Stocks are up across the board, the indices look like a symphony, the chartists are bullish and each mail is bringing a stack of brokers' buying recommendations. You've run $2000 into $12,000 and after just one or two more trades you can buy a Mercedes.

The only thing you've missed is a faint, distant tintinnabulation. It's time to sell everything and go lie on a beach until it all blows over.

---

1 Now publisher of The Bulletin.

2 In fact, Poseidon make an operating loss of $800,000 in 1973-74.

3 Interview with author at Poseidon meeting.

4 Or her. Many women have a flair for speculation; the masculine gender is used here only for convenience.

CHAPTER ELEVEN

JANUARY 1970: Running Wild

The main news to greet readers of the finance pages on New Years Day 1970 was the announcement that Peko had pulled out of Attunga. "The agreement is at an end," Peko directors said crisply. It was an omen for the rest of the mining market, but it was not readily perceived. Melbourne and Sydney were now turning over a combined total averaging 40 million shares a day, compared to an average of 10 million on Wall Street. There were now 145 listed mining stocks in Sydney, compared with 86 at the start of 1969. The index for mining stocks was standing at near-record levels. Heavyweight stocks such as CRA, Hamersley, Peko, Western Mining and Consolidated Gold Fields of Australia had shown strong gains during the year and were widely tipped for further improvement.

One source of concern was the number of new floats in the pipeline. There were reputed to be as many as 100 more mining companies queueing up to float and eventually list. The subscription money to them would represent a heavy cash drain. Subscribers to these floats would raise money either by selling presently listed industrials or miners, or by diverting new money from them. Either way, listed companies appeared likely to suffer and there was the spectre that the whole market would be swamped by the new floats.

In this scenario, the title of Public Enemy Number One could have been awarded to 33-year-old David Paxton, a former employee in the Brisbane office of Hungerford Kirkhope and Spooner. Paxton announced in January that he was going to float a stable of 11 new companies at the rate of one every three weeks. First cab off the rank would be Barewa Oil and Mining NL. Barewa had a prospectus on issue but was not being underwritten because this would save money. It turned out that the money thus saved would be largely diverted to the promoters, but that would be discovered later.

The remaining 10 companies were to be named from Greek mythology. They were Perseus, Icarus, Cerberus, Hesperus, Orpheus, Pindarus, Pegasus, Patroclus, Proteus and Lapithus Mining NL. They would each raise $500,000 or $1 million, depending on the market, and would issue 50 cent shares in three classes — fully paid, 10 cent paid and 1 cent paid. By the time he had finished, Paxton expected to have a group in which each company concentrated on one particular mineral and the combined stable had a total calling power of $100 million.

Speculators with the benefit of a classical education may have winced at the proposed names. Icarus was a youth who escaped from Crete with his father Daedalus, using wings made of feathers held together by wax. Exhilarated by the sensation of flight, Icarus ignored his father's warnings and flew too high. The sun melted the wax and he plunged to his death.

Patroclus was a friend of Achilles who was slain in the Trojan War. Cerberus was the three-headed dog which guarded the gate of Hades. The name Hesperus had even worse literary connotations.

Even before the Paxton master-plan was unveiled, the Financial Review was expressing concern. In an editorial on January 7 the newspaper said: "The rising flurry of new flotations has always been one of the signs of the closing stages of a particular boom market. It happened in 1960 in finance and property and it happened last year in oil."

This foreboding was reinforced by the mood of London. A lot of English fingers had been burned and they weren't enjoying it any more. Also there were legitimate complaints about scrip delivery. A Mr Cox wrote to the Investors Chronicle complaining that he had invested in three Australian industrial shares a year before and had not received scrip for any of them yet. He was having great difficulty establishing his entitlement to dividends. Mining share-registries were in even greater disarray although dividends were not a problem for most of them, nor ever likely to be.

But the market was still high and there was still money coming in. An abnormal fall in savings bank deposits in November had been blamed on depositors withdrawing funds to play the market. Professional traders and insiders could still make money in an overpriced market. The professional traders had no more faith than the Financial Review in the staying power of the market. What they had was faith in their ability to move in and out quickly when special situations occurred. The name of the game was musical chairs and it was played with complete cynicism.

The non-ferrous metals index climbed back to 5845.13 on January 7, only 15 points below the peak of December 29. Then it shed a staggering 800 points in the next 10 trading days. The heavyweights were in full decline. It was only a matter of time before the speculatives followed them down.

Also on January 7 three stocks were queried by the Sydney Stock Exchange. They were Westralian Nickel Exploration NL, Gold and Minerals Exploration NL and North Deborah. Westralian was the most spectacular of the three, having jumped $1.90 to $6.70 on January 6 and then soaring to $14.50 before closing at $13 on the 7th. After being prodded by the exchanges, Westralian said it was negotiating for leases in WA.

Some recipients of placements had already made a paper fortune out of Westralian. On January 5 a placement had been made of 110,000 20 cent paid shares at a premium of $2.10 a share. This was about as low as Westralian shares had been for a month. The chairman, J.L.Hands, said he was surprised by the soaring share price. He said the placement had been made through a Perth broker and he did not know the recipients. Whoever took the placements was showing a gain of around $11 a share, or 400 percent, in two trading days. The only faster way of making money would be to print it yourself. Hands said: "There was a first-class reason for the placement which will stand scrutiny." He did not give it, however.

A former general manager of Westralian, P.A. Englebrecht, said he thought some of the company's areas were very good. He had left the company shortly before owing to differences between himself and the board. "Some brokers were getting information about the company before anyone else," he said. "I did not agree with this." Hands did not think that information had leaked from the company, but could not give any other reason for the rise in the shares.

Gold and Minerals Exploration, which somehow always managed to get good publicity in the Age, jumped from $3 to $8.50 after touching $9. GME had an indirect interest in a huge lateritic nickel deposit on the WA-NT-SA border at Wingellina. North Deborah, which had been $2.50 on New Years Eve, soared to $7 on rumors of a nickel strike. Strong London buying was reported in the shares, but the same was said about every stock that ran. After directors denied the rumor the shares retreated to $5. This was when North Deborah began selling unissued shares on market through King and Yuill, in its second and third placements.

On the following day, Norseman Gold Corporation leapt by $12 to $31 while its Higginsville partner, Ravensthorpe, rose by $2 to $6. Again there were rumors of a nickel strike and again directors said they knew of no reason for the share rise. On the Melbourne Sun we toyed with the notion of having a standing column entitled "No Reason" and beginning "Directors of the following companies yesterday said they knew of no reason for the rise in their shares..."

By this time Westralian Nickel had run another $3 to $17.70. The large jumps by all these shares, and the speed with which the rumors circulated, gave rise to grave suspicions. It was thought at the time by brokers, stock exchange officials and journalists that professional operators were ramping these and other stocks. Sometimes the directors of the companies concerned were quite bemused by their sudden popularity. "I don't know what the rise is for," said M.R. Lodge, chairman of Norseman. "But perhaps somebody knows more than I do."

The run in Westralian Nickel turned out to be based on one of the strangest deals to be done during the boom. On January 14, a week after the shares' big jump to $14.70, directors announced that they were acquiring prospects in the Murphy Hills and Yalgoo goldfields areas of WA for a cash-and-paper consideration of $19.6 million.

Two transactions were involved. In the first, Westralian would buy 14 claims at Murphy Hills from the Biggs Prospecting Syndicate for $220,000 and 600,000 fully paid shares. The fully paids were not listed at this stage but the 20 cent paid shares had stood at $13.80 the day before the announcement, putting a theoretical value of $14.10 on the fully paid shares and a value of $8.5 million on the deal. The Biggs syndicate was headed by Keith Biggs of Leonora, whom we met before in the Poseidon chapter.

In the second transaction Westralian Nickel bought a 90 percent interest in all the areas held by Magnet Metals Exploration Pty Ltd in the Wyalgee-Narndee area of the Yalgoo for a consideration of $50,000 and 800,000 fully paids. This worked out at $11.1 million on current market prices for Westralian shares. Magnet Metals was owned mainly by pastoralists.

None of these prospects had been drilled. Only surface sampling had been carried out, together with geochemical tests. In return for these leases, the vendors had acquired more than 30 percent of Westralian Nickel, in addition to some tidy cash payments. The vendors were restricted from selling their shares until October 16, 1970. The published version of Nestel's in-camera evidence to the Rae Committee includes one tantalisingly brief snatch in which Nestel says: "There was Westralian Nickel at about the time of Poseidon. Westralian Nickel was magnificently organised and we were buying vendor shares and any shares, presumably. As you know there is nothing there...". The evidence cuts off at that point, which means we may not have the full context of Nestel's remarks.

The shareholders of Westralian Nickel could have complained that their equity in the company was being diluted by the size of the placements. In practice, they should have been selling the shares as fast as they could because there was no way that a group of unproven leases could justify the inflated price of the stock. Westralian Nickel was now capitalised, including the placements, at $45 million. The nickel claims were almost its sole asset. It had run so low on cash, having been floated with a heavy content of vendor shares only two months before, that it was now planning a one-for-four rights issue at a premium of $1.70 to raise working capital.

In the case of North Deborah, whose shares were also gyrating at this time, what the shareholders did not know about their company eventually took 18 pages of the Rae Report to describe. When queried by the Sydney Stock Exchange about the sudden jump in share price, the company said it knew of no reason. Later, North Deborah said that a drilling report was "due shortly". The word "shortly" turned out to mean Friday, March 13, when the company announced it had struck "encouraging nickel values" at Ora Banda. (Ora Banda was a ghost gold-rush town near Kalgoorlie, a romantic relic of the 1890s which had returned to vogue as a potential nickel area.) On Monday, March 16, North Deborah issued another statement saying the mineralisation was "of no commercial value". While the market had been kept in a state of expectancy over Ora Banda for six weeks, holders and buyers of North Deborah scrip did not know that large quantities of unissued stock was being sold on the market.

The punters in North Deborah, Westralian Nickel and their ilk were in a particularly perilous position because the rest of the mining market was falling heavily. On January 13 there were 191 falls on the Sydney mining and oil boards compared to only 39 rises. The only stocks holding the market together at all were Great Boulder and North Kalgurli, which had announced further nickel at Scotia and Carr Boyd.

The market still had some life, however. Samin Ltd entered the lists on January 15 with one of the greatest debuts of the boom. On the first bid the 50 cent shares sold at a phenomenal $30. This price was artificially high and soon tumbled to $21.50.[1] Samin's main asset was its big holding of Poseidon, which were then standing at $215. This alone was worth $12 per share to Samin.

Another memorable float was also in the offing as Alexander Barton announced details of Palgrave Corporation Ltd, whose main asset would be a gorgeous huge opalised rock called the Desert Flame of Andamooka. The prospectus contained several beautiful shots of the opal and was easily the prettiest prospectus of the era.

The float reflected some credit on Barton as entitlements were being offered to shareholders in the collapsed Landmark Corporation Ltd, a Barton company which had sunk during bitter fighting between Barton and a Member of the NSW Legislative Council called Alexander Armstrong. Landmark shareholders would now receive shares in the newly-formed Palgrave. Of course, Alexander Barton and his son Thomas were large shareholders in Landmark, and Palgrave would acquire Associated Gem Corporation from another Barton company, Brins Securities Ltd, but it was still an attractive offer to the small shareholders of a destitute company.

Elsewhere on the good news front, the managing director of Clarence Oil and Minerals Co NL, John Taylor, released a preliminary announcement on his company's areas offshore of northern NSW. Taylor said a seismic survey had been completed and "indications are that the company may have found a vast oil field." What actually had been found was a succession of deep structures. No oil has yet been discovered in them, but Taylor was always one of the mining sector's most irrepressible optimists. In the 1960s, when drilling for oil in the centre of the Grafton racecourse, he reported to the stock exchange that "petroliferous odors" had been detected. The racetrack, however, never became an oilfield. And in January 1971 another of his companies, North West Oil and Minerals Co NL, claimed "the first significant platinum discovery in Australia". On the basis of four diamond drill-holes, Taylor said platinum reserves of 33.1 million ounces could be estimated and that the discovery area at Rudall River in northern WA might "shortly be proven to be a major new mineral province". The platinum mine never developed either.

Even more hopeful was the prospector in Melbourne who pegged the front lawn of the newly-built Ringwood Civic Centre. It covered the main workings of an old antimony mine.

Another index of optimism was the support for the birth of the Paxton empire. On January 20 applications opened for the float of Barewa and closed heavily oversubscribed. The lucky ones were those who missed out.

There was also at this time a rash of backdoor listings. These were causing flutters all over the market, but particularly in small, neglected industrial stocks which might make convenient vehicles for mining promoters.

Sunhill Corporation Australia suddenly ran from 20 cents to $1.20. Savoy Corporation, which had been in receivership from 1965 to 1968, leapt from 48 cents to 90 cents after touching $1.20. Ralta Ltd, which was still in receivership and expected to emerge soon with assets of six cents per share, ran from 15 to 35 cents. Indoor Bowling Australia Ltd, a relic of the bowling boom of the early sixties, was taken over by Colortone Holdings Ltd in December 1969 and became the proud owner of mineral leases a month later. One batch of claims it acquired were the Auguste-Duffy claims bordering Poseidon at Windarra.

While the wheeler-dealers were bubbling with activity, the brokers were showing signs of concern. In Sydney, Valder & Co, whose principal John Valder was later to become chairman of the Sydney Stock Exchange, issued a circular advising speculators and investors to sell at least part of their holdings. "We think all sections of the market could come under some downward pressure in the February-April period," the circular said. "Although the speculative market must be the most vulnerable, all sections will probably suffer. This is because just as the weak go up with the strong so the strong come down with the weak."

The Sydney firm of Malcolm Hazell Steel and Hill surveyed 35 mining stocks listed between June 30 and December 31, 1969, and came to the conclusion that 80 percent of them would never have contemplated listing in any but "the current irrational market". They classified the 35 as shown opposite.

"We draw attention to the fact that the nature and quality of the prospectus inclusions can be taken as a guide to managerial approach," the firm said. "That is, if management regard as significant a certain lease listed in a prospectus on which little work has been done and which received non-committal comment from the consulting geologist, future statements and evaluations by management can be analysed very critically."

PRODUCERS

(a) Present profitable production	Nil
(b) Feasibility study completed, sufficient proven reserves to gain economic production	1
(c) Small marginal operation in existence, further possibility of increasing production	3

NON-PRODUCERS

(a) Drilling within a reasonably defined grade has established a significant deposit needing further evaluation	2
(b) Limited diamond or percussion drilling, or sufficient surface work to justify drilling	4
(c) Surface testing and geophysical results suggesting possible mineralisation	13
(d) Scant evidence of mineralisation but lease in proximity to known find	12

Some brokers were concerned at the way they were financing their clients. A client may have started the boom with only a few hundred dollars and by successful trading had built up a credit balance of several thousand dollars with his broker. Scrip processing delays meant that very little cash had been received although it would come in eventually, provided of course that the buyers did not default and their broker collapse. Meanwhile, the brokers had been allowing clients to trade on their credit balances. This was good public relations, but the brokers had to find more liquid funds to meet the stock exchange requirement that they should hold five percent of customer credit in cash. Some brokers were now demanding that contract notes be paid promptly, regardless of how much was in the client's credit balance. This break in the established pattern of trade aroused resentment and at the time did little to curb the market. A client could easily transfer his business to a more accommodating broker. Also, such restrictions were not easily imposed on large clients.

Another premonition of doom came from Melbourne, where A. Holst & Co released a short, hard-hitting circular predicting that in a few years' time the majority of present "gambling stocks" would be bitter memories to those who continued to hold them.

That the punters were not yet deterred was demonstrated by the Melbourne Stock Exchange's decision to limit viewing time in the public gallery. Henceforth, no more than 70 people would be allowed in the gallery at once, and they would not be allowed to stay more than 20 minutes. Then they would be ushered out to make room for another 70. Commissionaires were appointed to enforce the rule. One bright spark tried to talk his way into the gallery by flashing a pass and saying he was from the CIA.

To make the gallery even less attractive the exchange ripped out the public telephones. The counter to this was attempted two days later when a pair of speculators came in carrying cardboard cartons. They opened them, revealing two brand new walkie-talkies. They tested them along the corridor outside the gallery, then one took up a post by the rails while the other went out into the street to a public phone box. Contact with the brokers had been restored.

Over the water, London had been making some pained noises. At a UK seminar, R. Davie of George Hill & Co, a jobbing firm, said that "A lot of Australian stocks, to put it mildly, are highly suspect". He said there was a "grain of truth" in the accusation that there was a rigged market in some stocks.

His plaints met with scant compassion in Australia. The general reaction from brokers was that British investors who bought doubtful Australian stocks were entering a speculative market with their eyes open; and more stocks were being rigged by London jobbers than anyone else. Charles Rich, chairman of Combined Resources Ltd, which had earned a cash and paper profit of $1 million from the market in six months, described the Australian market as "very much controlled". He said rigging was done by people with "a lot of knowhow, a lot of brains and access to a lot of money". The riggers included both Australian and British operators.

So as January neared its end the market was in an uneasy state. It was now running more on greed than faith, an important difference. The professionals had stopped believing and were cynically playing games. The more responsible elements were muttering doom and distrust. Fresh speculators were still entering the market, but their meagre resources would not hold it up if anything happened to really undermine the already-ebbing confidence. The shadow of Gotterdammerung was looming.

Then came Tasminex.

---

[1] Sometimes a stag would enter a very high opening bid for a small parcel of stock then sell his holding as the market came back. The small loss on the market purchase was amply offset by the high prices for the parcel he had bought at par in the float.

CHAPTER TWELVE

TASMINEX: The Last Straw

As January wore on, it became evident that the market was being sustained by hysteria more than by any tangible factors. An indication of this was the shortening lifespan of market favorites. Western Mining's discovery of nickel at Kambalda had buoyed the market for four years. Poseidon's Windarra find had buoyed it for four months, but at a much higher plane. Now popular stocks were staying in vogue only for a few weeks, or even a few days. The market was in the condition of a junkie needing increasingly big fixes. By January it needed another Poseidon to sustain it, or a collapse would begin.

It was in this nervous market that Tasminex NL suddenly ran from $3.30 to $18.50 on January 27 before closing at $16.80. It was one of the largest runs any stock made in a single day during the boom, either in percentage or dollar terms. Naturally, the rumors were saying that Tasminex had nickel.

January 27 was the Tuesday following the Australia Day long weekend. Tasminex, which had floated in October 1969, was a small, tightly capitalised company with a little over two million shares on issue, including vendor shares. Its most promising nickel prospect was Mount Venn in WA, where the company was known to be exploring.

So much would have been known to any assiduous reader of the financial pages. What was not known was that in the week before Australia Day, a geologist and two of Tasminex's directors had been visiting the site. Rumors of a nickel strike began emerging on Friday, January 23, when the turnover of Tasminex shares increased and their price on the Melbourne Stock Exchange rose from $2.80 to $3.30.

Speculation must have reinforced bullish rumors over the long weekend (usually at this time the weekend break had the opposite effect of a market favorite) and on Tuesday the stock hit $18.50. Shares in companies with claims alongside Tasminex at

Mount Venn also rocketed. Amad NL rose by $1.10 to $5.10, Eastmet Minerals NL by $1.05 to $2, Paringa Mining & Exploration Ltd by $1.50 to $8.50 and Westmoreland Minerals Ltd by 22 cents to $1.10.

At that period I was deputy finance editor of the Melbourne Sun News-Pictorial. Several runs had been occurring in mining stocks over the past few weeks, but this was the largest. I was determined to find out why. I spent the afternoon fruitlessly trying to contact chairman Bill Singline in Burnie and then at the Mount Venn field. Then around five o'clock a message arrived from the Burnie Advocate asking for an interview with Singline, who was a well-known local businessman. The Advocate supplied his home telephone number in Melbourne. I was surprised to find he had a house in North Balwyn. I had been ringing far-flung parts of Australia and here he was in my backyard, so to speak. I rang the North Balwyn number, little knowing that I was about to become involved in the wildest stock run of the whole boom.

Singline told me he had just flown in from Mount Venn. He said the company had made a statement to the exchange during the day and he had nothing further to say.

The statement to the Sydney Stock Exchange had been as follows: "The secretary of Tasminex NL confirmed that exploratory drilling had commenced on the joint venture prospect with Tin Creek Mining Pty Ltd at Mount Venn, Western Australia. The company's chairman and manager are in the area. No information is yet available and the secretary is unable to account for the fluctuation in its share price."

As the statement contained nothing to justify such a large rise in share price, I felt justified in pressing Singline further for an explanation. He confirmed rumors that Tasminex had struck massive sulphides and said it could be bigger and better than Poseidon. This was the story, together with a photograph of Mr and Mrs Singline drinking sparkling wine, which appeared on the front page of the Sun next morning and, perhaps more importantly, was transmitted to London in the normal course of news distribution.

I doubt whether any interview by a financial reporter ever produced a more dramatic effect. I had spoken to Singline after the Australian markets had closed, but the interview reached London before the market opened there. The shares hit $96 in wild early trading. I knew the story would put them up, but I had been expecting somewhere around the $40 mark, given the market at the time. I was shocked when I heard the opening London price.

Next day in Australia Tasminex sold as high as $75 early but closed at $40. By the end of the week they had drifted back to $21 and slowly eased to $14 by the end of February. There was quite a lot of trading in the stock below the $40 level.

On March 26 the Tasmanian Attorney-General appointed J.W. Wilson, a Launceston lawyer, as an inspector to investigate and report on share transactions in Tasminex between November 7, 1969, and March 18, 1970. His report, delivered in August 1970, disclosed that the Singlines had realised a little over $1 million selling shares between January 27 and 30, when the market was at its peak and mostly after Singline's statements to me.

In fairness, it should be recorded that Singline denied several aspects of the interview. Specifically, he denied using the word "nickel". I say he did use the word "nickel"; even if he didn't, I cannot imagine what else we were supposed to be talking about. In his report, Wilson concluded: "I am satisfied that the newspaper report was a fair account of what Mr Singline said...I am also satisfied that the remarks made by Mr Singline to Mr Sykes...were unjustified and irresponsible."

I have pondered the episode a number of times, and in hindsight I still think I was right to interview Singline and publish what he said. If a reporter is not going to try to find out what is happening, he is in the wrong occupation. If a reporter discovers new and valuable information (or apparently valuable information) he has a duty to pass it on to the public. Indeed, if he keeps it to himself, he becomes an insider. But when I wrote the story there were still a lot of things that neither I nor the general public knew about the situation and which only came to light when the Wilson report was published. It is a valuable report because it is one of the few official investigations into share trading during the boom, and as such it sheds some revealing light on how companies were formed and rumors circulated during the nation's most cyclonic bout of speculation ever. The following outline of the Tasminex saga has been drawn largely from that report.

Tasminex was incorporated in Tasmania on August 4, 1969, and floated publicly on October 22, 1969. The directors were Singline (chairman), Johnson, Maxwell Lamb and Richard Morris, who was also secretary of the company during 1969-70.

At the time of the float the company had an authorised capital of $1.5 million, comprising six million 25 cent shares. The prospectus was for the issue of 1,995,000 shares. Of these 464,750 were allotted to the Singlines and three of their companies, Singline Developments Pty Ltd, Singline Transport Pty Ltd and Singline Constructions Pty Ltd. Singline was a self-made man who had left school after reaching grade four in a Tasmanian country primary school. Despite a limited education he had considerable practical ability. He had risen to run his own transport and construction business and he owned one of the largest supermarkets in Burnie. He was rightly proud of his success.

Of the rest of the shares issued in the prospectus, a total of 53,750 were taken up on allotment by Morris and his family, 97,850 by Johnson and 34,000 by Lamb. Thus a total of 650,300 shares, or nearly one-third of the issued capital, was held by the four directors at the time of the listing.

The Wilson report provides an illuminating example of how large parcels of shares were "bedded down" with associates of the board in mining floats of the time. In the Tasminex float, 50,000 shares were allotted to a company named Neerim Pty Ltd, in which Hall and Relph held an interest. These were the partners of Hall Relph & Associates, consulting geologists to Tasminex.

Employees of Hall Relph were allotted a further 14,000 shares, and 37,000 were taken up by partners of the underwriting broker, Clinton & Co. A total of 30,000 shares was allotted to Ian Blake, the manager of Singline Developments, who also managed the Singlines' own transactions in Tasminex.

Others included in the float were Brudenell-Woods (2000 shares); Roland Ralph, a Launceston stockbroker (20,000); and Stephen Quon, one of his clerks (5000). Another 100,000 shares were allotted to a company named Neny Pty Ltd, which operated on behalf of Murray Maguire & Co, who were retained by Tasminex as consultants. Murray Maguire were largely instrumental in arranging the whole float.

Adding these figures together, we find a further 258,000 shares were allotted to interests close to the company, giving a total of just under 900,000 shares. As a total of 856 individuals and companies received shares in the float the average allotment must have been around 1400 shares to those who were not included in the above list of directors, underwriters and associates. In addition, options to take up 250,000 shares at par within five years were granted to Singline and options for 125,000 shares each were granted to Johnson and Lamb.

Tasminex's issued capital was two million shares, which was

tight enough capitalisation. Actually the large holdings by a few groups made the capitalisation tighter still. Tasminex shares would be most responsive to leverage. The allotments from the prospectus may explain to boom veterans why so few shares in floats appeared to be available to the public. As a finance journalist who participated in several floats, my experience was that if I had an account with a broker or a friend in the company or the underwriting firm I could usually get somewhere between 100 and 500 shares. If I had no such connection I could get none. Sometimes my allocation was not enough to constitute a marketable parcel.

Many of the names mentioned above evoke memories of Endurance. Hall, Relph were geological consultants to both companies. Murray and Maguire were retained by Tasminex to assist in the incorporation, promotion and flotation of the company. Brudenell-Woods, another former Endurance director (he was also a disbarred solicitor), received shares from Clintons, who were also associated with Endurance. Soon we will meet some other familiar names.

The prospectus stated that the company would initially concentrate its exploration efforts on iron ore deposits south-east of Burnie. As Singline was a prominent Burnie businessman and iron-ore deposits had been opened recently at Savage River nearby, there is no reason to doubt this was the original intention of the company. But nickel was the name of the game at the time so it is hardly surprising that Tasminex, like many others, was soon lured by the prospect of a West Australian nickel claim.

The shares were listed on the Hobart, Melbourne and Sydney Stock Exchanges on November 13, 1969, smack in the middle of the boom. The listing attracted no special attention, although as a tightly capitalised company Tasminex's shares always commanded a premium on the market. Quon, the Launceston broker's clerk, had sold all his 5000 shares by November 19 at prices ranging from 50 to 95 cents. He said: "They looked like pretty good prices at that stage". At that point Tasminex had no nickel claims.

Tin Creek was a public unlisted company — like Attunga — incorporated in WA on March 14, 1969. It had an issued capital of 350,000 shares held between 50 people. Brudenell-Woods held 36,000 shares and was managing director. Lou Harmanis, the prospector who had been his fellow-director on Attunga and Endurance, held 34,000. A further 32,000 shares were held by members of the Harmanis family and 20,000 by the Brudenell-

Woods family. Messrs Hall and Relph held 8000 and 36,000 were held by three partners of Clinton & Co, George Jones, Graham Hurst and Andrew Clinton.

Harmanis and Brudenell-Woods had originally explored the Mount Venn area about 100 miles north-east of Windarra. They pegged it because of interesting outcrops. Reports of early exploration of the site conflict, but there appears to have been little significant evaluation made.

Negotiations between Tin Creek and Tasminex appear to have begun, through the medium of Clintons, in late November 1969. Originally Tin Creek wanted a million Tasminex shares for Mount Venn plus (in the event that it actually turned out to be a mine) a further 1.5 million shares and a 25 percent interest. In the event, Tin Creek sold Tasminex a 60 percent interest in Mount Venn for $20,000 cash, 250,000 Tasminex shares at par and a loan of $100,000. The deal was settled quickly. Robert Goninon, a Hall Relph geologist, visited the field on November 26 and 27 and described it as a "first-class second-rate prospect." He said it warranted further exploration, "having some potential for the presence of economic copper-nickel sulphides".

On December 23 Tasminex announced the terms of the agreement. They had decided to go into Mount Venn despite some warnings about the people behind Tin Creek. Paul Murray advised Johnson not to have anything to do with Tin Creek and referred to the involvement of Brudenell-Woods in Endurance. The Launceston broker Ralph, who had been secretary of Endurance, specifically warned Johnson about Brudenell-Woods.[1]

Nevertheless, a deal was agreed and an operating committee was established to direct the prospecting program. This committee consisted of Morris and Lamb from Tasminex and Milton Lees and John Sanford sr. of Tin Creek. This committee does not appear to have worked satisfactorily. Hall said: "There were too many chiefs and no Indians. You never knew who was in charge." Morris said he and Sanford had a number of serious disagreements.

Brudenell-Woods and Sanford wanted a drilling program to proceed, despite the opposition of Hall Relph and Johnson, who was manager of the project. In mid-January a percussion drill began drilling for water, simultaneously taking geological samples.

There had been some surface samples taken in June 1969. During the inquiry, Brudenell-Woods produced assays of these samples to Wilson. When Wilson, during his investigation, showed these samples to Goninon, the geologist said the values would warrant an exploration program to determine their source but he considered that the copper values were too high for a nickel prospect. Barry McCammon, another Hall Relph geologist, thought the samples were carefully selected ones and not representative of samples generally in the area. These assays had never been made available by Tin Creek to Tasminex.

The drill quickly found water, as well as producing pulverised samples of dark rock. On Friday, January 23, Johnson, who was visiting the site for the first time, decided to pan a few samples produced by the drill. Panning must be the oldest known method of gravity-separation of heavy metals. By swirling crushed rock or gravel around with water in a shallow dish, an adept panner can make the heavier minerals sink to the bottom. It is still used by prospectors searching for alluvial minerals such as gold and tin, or gemstones such as sapphires. But it will not give you a reliable assay figure for a hard-rock mine because, by definition, you have separated a mineral from the waste material surrounding it.

Johnson's panning produced "a residual heavy metal, or heavy product" in the pan. He showed this to McCammon, the visiting geologist. Exactly what the two men said to each other was later disputed — as was every conversation from here up to $96 a share. Johnson said the geologist identified the sample as sulphides which were anomalous with nickel. McCammon agreed he identified them as sulphides and had been enthusiastic because they were the first sulphides struck at the prospect, but maintained that he had never identified them as nickel. Wilson accepted McCammon's version of the conversation. In doing so he may have been influenced by a third party who was present for some of the conversation. This was Donald Wells, the charter pilot who was flying the parties to and from Mount Venn. Wells, like all WA bush pilots at the time, was interested in picking up any information which might be profitable on the market. He did not hear the word "nickel" used in the conversation between Johnson and McCammon. Wells said: "If someone had said there was nickel in the ground I am quite certain I would have bought shares." McCammon said the sulphides were disseminated rather than massive and the way they had been panned made them a concentrated sample. McCammon took the samples, together with several of his own, back to Sydney for assaying.

McCammon also said Johnson had claimed that sulphides

had been discovered in rock which had been excavated while digging a cellar on the site. McCammon did not inspect it.

It seems incredible that the events outlined above were enough to send Tasminex to $96, but that is what happened. A few panned samples and a hazy identification of some excavated rock were successively exaggerated into a report of a major nickel find.

On January 24, Singline arrived at Mount Venn, but although Johnson appears to have been excited about his sulphides he did not say much to Singline at that point. Explaining this to Wilson, Johnson said: "I said very little to Mr Singline. Very little for obvious reasons...Mr Singline tends to get things distorted and out of place."

Johnson himself had been described by his fellow-director Morris as being garrulous and a super-optimist. Given the opinions they held of each other, the board of Tasminex must have had some problems communicating.

On Sunday, January 25, Brudenell-Woods arrived at Mount Venn, at the head of a small party. He spoke to Johnson. The evidence of both men about their conversation differs sharply. Johnson said Brudenell-Woods had rushed up to him on arrival at the camp and asked what was happening because Tasminex was very firm at $5. As there were a number of people about, Johnson asked Brudenell-Woods to accompany him into the cellar. There he said: "Well, at least we are looking in the right place, Ken. We have got sulphides and McCammon says that we have probably got pyrrhotite with it, and it is anomalous". Brudenell-Woods said Johnson had told him: "You have got it. We have got nickel sulphides". He said Johnson added: "There are sulphides even all over the walls of this cellar". Brudenell-Woods said he and Johnson had then shook hands, put their arms around each other and danced around. The only point of the meeting about which the two men agree is that Brudenell-Woods kissed Johnson on the forehead.

However their excitement was generated, they quickly communicated it to other members of the group. Some cans of beer were opened and a number of toasts were drunk, including one to "nickel in the Cosmo". Cosmo Newberry was the name of the native reserve which covered the Mount Venn prospect.

Brudenell-Woods told the three drillers he would buy them a thousand Tasminex each, and Mr Johnson tipped them $20 each. Brudenell-Woods returned to Perth by air and held a celebration party there that night which attracted the attention of at least one sharebroker's spy.

It is worth pausing a moment to consider the character of Brudenell-Woods, who was a central figure in both the Endurance and Tasminex investigations. Wilson concluded: "I believe that Mr Brudenell-Woods was only too willing to convert the discovery into a major one and that much of the excitement that occurred during and after his visit to Mount Venn was generated by him". Lockhart and Mealey, investigating Endurance, concluded: "He (Brudenell-Woods) represented that the copper mine was viable when he knew that of the four consulting geologists who visited the site up to June 1968 to report on the copper, each one of them had reported adversely, claiming very limited ore reserves. Brudenell-Woods on more than one occasion engaged in the spot sampling of ore and arranged for it to be assayed. This practice of his, together with the statements to the press in July 1968 by Louis Harmanis, a director of Attunga, that Attunga did not need a geologist to tell it that the ore was there, would be ludicrous had not the venture finally returned to these and other people nearly $2 million net profit."

In the Tasminex report, Wilson said: "I am satisfied that during this period (January 24 to 27) the discovery of sulphides, which were then unidentified, became exaggerated into a discovery of massive nickel sulphides. I have summarised...the development of the exaggeration of the nature and extent of the discovery and I am satisfied that Messrs Johnson, Brudenell-Woods and Singline were each partly responsible for that development."

The rumor spread with fantastic speed. On January 23, the day Johnson panned the sample, there had been a strong market in Tasminex shares. This may well have been speculation (any stock with a tight capital in the early stages of an exploration program was liable to such bursts of activity). Johnson does not appear to have spoken to anyone except the geologist at that stage and Singline and Brudenell-Woods were not yet aware of his sampling activities.

Sitting in an office above Menzies Hotel in Sydney that fateful Friday was Surrey Bogg of Mineral Securities Australia Ltd. Bogg was one of the biggest and best share traders of the boom. Operating on behalf of Minsec, which bought and sold hundreds of thousands of shares a day at its peak, he had to be. On that Friday he heard rumors in Sydney that two employees of Tasminex had been drunk in the Laverton Hotel on the previous night and had said their drill-hole had intersected nickel sulphides.

Little more than a hundred yards further up Hunter St, in Patrick & Co's office at the top of Hunter Street, Alan Mann heard the same rumor. Bogg and Mann both bought heavily into the stock. They knew of each other's activities, Bogg's wife being one of Mann's discretionary clients.

On all the other evidence in the Wilson report, the rumors could not have been correct. No Tasminex employee, drunk or sober, could have been talking about nickel sulphides on the Thursday night because Johnson did not pan his sample until Friday. Johnson panned his sample on the Friday morning and dried it in the sun before showing it to McCammon, so they probably had their conversation late Friday morning. Given the 2 hours' time difference between Perth and Sydney, it must have been lunchtime at the earliest in the eastern states. Yet the rumors reached Sydney, apparently via Laverton, before the close of trading that afternoon. The underground network must have been supremely efficient.

On Sunday the 25th, Brudenell-Woods returned to Perth and held a dinner party that night. At about 11 pm eastern time that night, a client phoned Melbourne broker Grant Reed at his home. Reed said the client had heard from a friend in Perth that "the people from Tasminex were celebrating something and the rumors were that they had discovered sulphides while they were drilling for water". Monday the 26th was Australia Day holiday with all the Australian exchanges closed, but that night Reed began buying Tasminex in London. By this time the rumor had also reached London and the shares were very firm.

This is the background to the Tuesday trading when the shares shot to $16.80 after a high of $18.50 Then I rang Singline and interviewed him. The market went to $96 in London before the first edition of the Melbourne Sun hit the streets. I had poured fuel on the flames of an already overheated situation.

The next day's trading on Australian exchanges was the maddest scramble of the boom. While brokers were bidding frantically for stock, they were trying equally frantically to find out what was happening.

In Perth that morning one of Australia's top mineral analysts was about to pay a visit to various mine sites for a client. He had chartered a light aircraft and was all set to go. The night before his client ordered him to change plans and investigate Mount Venn as a matter of top priority. He took off that morning and must have been the first aerial spy over the site. Later the air would become thick with them.

The analyst took off at daybreak and less than two hours later was 400 ft above the drill site. The pilot banked the aircraft steeply, putting it into a slow circle just above stalling speed while the analyst took a series of color pictures, using a telephoto lens. "It was a beautiful prospect," he told me over a Sydney lunch years later. "They had good ultramafics and were drilling in exactly the right spot on the contact. But I could see no sign of nickel. I printed the pictures that afternoon in Perth and blew them up through a projector but all I could see from the hole was black basalt. There was none of the dark green sludge that would have signified nickel or copper sulphides."

The statement made by Singline, and hence my story, were wrong. This should have been evident at least by January 31, when the Tasminex board received the Perth analysis of Johnson's sample. But neither Tasminex nor Singline ever issued a specific rebuttal. The highest reading from Johnson's panned samples had been 1340 parts per million (0.134 percent nickel). This was very disappointing, especially considering it had been a concentrated sample.

Hall Relph made a statement to the Sydney Stock Exchange on January 28 dissociating the firm from Singline's statement. "We have always regarded Cosmo Newberry as a routine exploration prospect in a zone of ultra-basics with associated magnetic anomalies and surface nickel values," they said.

On the same day, Tasminex made a statement saying the company's geologists had reported that samples from the water bore indicated some evidence of sulphides but to what extent was not yet known. "The directors advise caution at this stage," they said.

On February 2, the Tasminex board, under prodding from the Hobart Stock Exchange, issued a longer statement giving fuller technical descriptions of the work being done at Mount Venn. Directors said disseminated sulphides had been encountered in a borehole and stressed that work was in its early stages and assessment of results would take at least two months.

Wilson thought that the board should have repudiated Singline's statements (he had also given a television interview) in clear and unambiguous terms. "A strong and specific repudiation of Mr Singline's reported comments was required in order to remedy the mischief that had been created by those comments," the Wilson report said. "It is certainly beyond my comprehension that the statement did not disclose that the disseminated sulphides had been found in a concentrated sample only."

Meanwhile, a number of people close to the company had

been selling shares at huge profits. The Singlines sold 28,095 shares in the very high market between January 27 and 30 for gross proceeds of $1,028,071. As 25 cent par-value vendor shares, they had been given a value of $7000 when allotted to the Singlines in the float.

Wilson believed Singline should be severely censured for the sales, which had been made despite advice from Morris not to sell. Singline had stopped selling only upon the insistence of his fellow directors. "I do not believe that these sales were due to mere folly or lack of thought on the part of Mr Singline," Mr Wilson said. "I am satisfied that they were made deliberately, and against his fellow directors' advice."

While the Singlines made a million, a small Melbourne broking firm went bankrupt. When Tasminex soared on the market, the firm of Donovan and Stock were ordered by a client to buy heavily. The client was a firm named Stirling Henry Ltd, whose principal director was G.J. Shapowloff, the son of Walter Shapowloff. (Yet another name from the Endurance episode).

After Singline's statement, Donovan and Stock began buying Tasminex but as the share certificates began coming through Stirling Henry discovered that some of the selling had been by Singline interests. On March 4 Stirling Henry sent an urgent telegram to Neil Donovan, the senior partner of the firm, ordering him not to accept delivery or pay for any Tasminex shares in the name of Singline. Donovan took the telegram immediately to the Melbourne Stock Exchange chairman, Sir Cecil Looker.

Donovan said Sir Cecil ruled that the firm was obliged to honor all purchases even if they were made in the name of Singline. Donovan was now in an impossible position, through no fault of his own. If he failed to accept delivery he could be sued by the selling broker and his clients. If he accepted delivery, Stirling Henry would not pay. The only possible solution — of dubious legality — would have been to cover his position by selling. But now Tasminex shares were falling dramatically and the losses would have made this course suicidal. Donovan was faced with the unpalatable choice of three roads to bankruptcy, all immediate.

Eventually Donovan accepted deliveries of Tasminex in the name of Singline to a total of $671,537, but Stirling Henry refused to pay for nearly half a million dollars worth of the scrip. Donovan and Stock were suspended by the Melbourne Stock Exchange. Donovan appointed a liquidator (leading Sydney accountant Ian Ferrier) into Stirling Henry but by that time Donovan himself was bankrupt. Lengthy and complicated litigation followed between Donovan, the Shapowloffs and the NSW Corporate Affairs Commission, and was still continuing in mid-1978. The episode could be summarised by saying that the Shapowloffs paid only part of the money, Donovan was bankrupted and destitute, and the Singlines had a million dollars.

Brudenell-Woods made a profit of $172,000 trading in Tasminex between January 27 and March 3. Wilson said: "I have already criticised Mr Brudenell-Woods for his part in the creation of the myth that massive nickel sulphides had been found at Mount Venn. I regarded him as a most unconvincing and unsatisfactory witness generally. I consider that his trading during the latter part of January and February not only indicates that his sole aim was to make as much profit as possible but it also indicates that he did not believe that any major discovery had been made at Mount Venn. In any event, bearing in mind his position as managing director of Tin Creek and his involvement in the events at Mount Venn on January 25, it is clear that I must condemn his blatant trading in Tasminex shares during the following few weeks when their value was so grossly inflated and when he must have known that he was selling on a false market."

Sanford realised $295,000 on sales of Tasminex between January 27 and February 23. Wilson said: "He must have known that the value of Tasminex shares was grossly inflated...and I consider that it was improper for him to sell so many shares at that time."

Neerim Pty Ltd, the company of Messrs Hall and Relph, sold 30,000 Tasminex for gross proceeds of $835,538 between January 27 and 30. The bulk of both Sanford and Neerim's sales would have been profits, as the shares were originally acquired at much lower figures.

Harmanis made a net profit of $112,781 between January 27 and February 25 selling Tasminex. Tin Creek made a net profit of $382,000 between December 9 and March 3 selling the stock. Johnson, Morris and Lamb sold relatively small parcels during February.

Several of these people continued trading in Tasminex during later weeks. Some of them, particularly Harmanis, made heavy losses buying the stock back during this time.

Wilson found the Hall and Relph sales improper, criticised the sales by Harmanis and Tin Creek, and said Lamb, Morris and Johnson should not have sold. No prosecutions were ever launched against any of the parties involved, but the episode marked the end of the boom's most hysterical phase. The market began sliding, as all boom markets eventually do, and bulls began taking losses. The coming months would see flurries in individual stocks, but the trend was unmistakably downhill.

Tasminex was the culminating frenzy of the boom and its demise marked the end of the boom's highest period, but the boom was due to end anyway. Tasminex terminated it with a bang instead of a whimper.

In the process, Louis Harmanis might have provided a fitting epitaph for the whole speculative era. Harmanis, a frank and colorful prospector who saw his share of action in the 1969-70 boom, was asked by Wilson why he had kept selling Tasminex once he had covered his original stake. Harmanis said: "For the purpose of making a profit of course...We were all out to make a quick dollar; let us be truthful about it."

"I was grateful for his candor," Wilson remarked.

---

[1] Wilson report on Tasminex, p. 15, para. 67.

CHAPTER 13

The View from the Molonglo

While all this had been happening, Canberra had been looking the other way. In the whole of 1969 the stock market had not been mentioned in Federal Parliament, apart from a couple of routine questions about the desirability of establishing a national securities and exchange commission. From October, when the boom was raging, Canberra had been preoccupied by the forthcoming election.

The Liberal-Country Party Government had little desire to intervene in the stock market. Its members tended to accept the mining boom uncritically. A few of them were on the boards of mining companies. The Labor Opposition knew little about the workings of the market, and some sections of the party were ideologically opposed to its very existence. Admittedly, the Commonwealth's hegemony in this area was debatable. There had been sporadic argument since Federation in 1900 as to whether control of corporations ought to be within the domain of the Commonwealth or the State Governments.

Now the boiling market had become so prominent, even penetrating a television prime spot with a show called The Money Game, that the excesses were bound to become a public issue. The trigger was a letter from the aggrieved Dr Anderson of Perth to the Leader of the Opposition in the Senate, Senator Lionel Murphy.[1]

Anderson had sold 4200 Poseidon shares at $1.50 each on brokers' advice on September 26. He had done so after reading the statement released by Poseidon on September 25, which read as follows: "The drilling commenced on September 15, 1969, but due to technical difficulties only a limited footage has been completed. No assay results are available. Therefore the board is unable to explain the sharp increase in the price of the shares."

As Poseidon shares roared off, Anderson became increasingly angry. By Christmas 1969 his holding would have been worth

$800,000. On February 5, 1970, Poseidon shares touched $280, their peak price. If Anderson could have sold his holding anywhere above $250 he would have been a millionaire, although he would probably have had to pay huge tax. This was galling enough, but he was more aggravated by the thought that he had sold while insiders were buying. He became concerned about this after reading in the Financial Review in November that Burrill and Associates had been buying Poseidon before the nickel strike was announced.

Anderson wrote to the chairman of the Perth Stock Exchange, Geoffrey Hynam, asking him to take action on behalf of Poseidon shareholders. In his letter the doctor said: "It would appear to me that either the directors realised the potential of the area being tested at Windarra — even before actual assay figures were available — and failed to inform their shareholders of their knowledge, or, alternatively, that the consulting geologists appreciated this potential but did not inform the directors." This letter was written on November 23. By December 7, Anderson had received no reply so he wrote again, asking whether Hynam believed there were grounds for litigation against the Poseidon directors and geologists.

Hynam phone Anderson some time in the next two days and told him to take up the matter with his own broker and the Adelaide Stock Exchange, which was the home exchange for Poseidon. Hynam volunteered to Anderson that he had been informed that the September 25 statement was "completely factual" at the time. Hynam also said he was satisfied that "the directors and consulting geologists had acted most honorably."

What Anderson did not know at the time was just how close Hynam had been to Poseidon. Between April 18 and May 15, 1969, Mrs Hynam had become the holder of nine percent of the issued capital of Burrill Investments Pty Ltd. Hynam was a close friend of Roy Jones of Burrill and Associates and acted as his broker. Burrill Investments had bought 15,900 Poseidon between 70 and 90 cents during mid-June 1969. Burrill Investments bought a further 4100 Poseidon in mid-July, when Mrs Hynam bought 2000 shares at her husband's instigation. On September 25, the day Poseidon said it was unable to explain the sharp increase in the price of its shares, Hynam bought 2100 shares for himself.

In late November, when the Poseidon directors were facing the challenge from Boris Ganke at their forthcoming annual meeting, Hynam was seeking proxies from Poseidon shareholders in the directors' favor. Burrill Investments began selling its Poseidon holding on New Year's Eve. Hynam's firm, S.G. Brearley & Co, handled part of the orders. Burrill Investments was one of the firm's largest clients.

The Rae Committee concluded: "We have not been satisfied that Mr Hynam's continued resistance to the call for an investigation was not largely to be explained by his unwillingness to risk compromising his personal and broking associations with Mr Jones and Mr Shierlaw; associations which had brought him, through a series of timely share purchases, such an enormous increase in his own and his family's wealth, and which were providing him with a large and continuing flow of broking business."

Although Anderson did not know this at the time, he was determined to pursue his quest. Dissatisfied with the telephone reply from Hynam, he eventually, after repeated requests, obtained a letter from the Perth chairman.

Anderson also wrote to Don McArthur, chairman of the Adelaide Stock Exchange. McArthur's reply, in part, read: "I unhesitatingly say that it is not for me to say whether or not statements that were made were a true indication of the knowledge directors possessed. When asked for a report the directors offered comments and these were accepted by the exchange as all that was necessary at that time and this placed an obligation on the directors' shoulders. It is easy to look back now and suggest what could have been done. As individuals we may all have different ideas looking back, but the decisions at that time are the important ones. Those buying shares at that time took a risk that their interest in them was justified."

Having failed to obtain satisfaction through the much-vaunted self-regulatory system of the stock exchanges, Anderson decided to take his grievance into the political forum. He wrote to Senator Murphy.

He could not have chosen a better advocate. Murphy was a QC with some forensic talent. He also had a grassroots grasp of political realities. He must have sensed that the boiling stock market, the scandalous, unexplained market runs, and the incipient collapse could be exploited politically. Unlike many of his colleagues, Murphy had a good understanding of the workings of the market. There is even some evidence that he may have played the market in a small way.

Murphy and the then Opposition Leader Gough Whitlam had both advocated the establishment of a securities and exchange commission in Australia along the same lines as the United States SEC. The irregularities of the mining boom provided grounds for arguing that such a body was necessary. Also, as the stock exchanges contained few socialists and a host of Liberal supporters, any inquiry into its scandals could damage only friends of the Government. The Senate was the ideal forum from which to launch such an inquiry. At that stage the Senate was emerging as a political force in its own right. The party numbers were close there and an independently-minded Liberal was capable of supporting Labor on individual issues.

Murphy made his first move on January 30, with a public statement asserting that the present voluntary system of control operated by the exchanges was inadequate. Speaking at the end of "Tasminex week", he said: "Most companies are legitimate and honest traders, but there are many others whose activities could only be called bushranging. Millions of dollars have been lost with no redress. Respectable companies have been robbed. It is disgraceful that the present Federal and State Governments are sitting back and doing nothing while people are being cheated."

The chairman of the Sydney Stock Exchange, Jim Cooper, leapt to the defence of the status quo. Cooper said no need existed for an SEC. A great number of the matters referred to by Murphy could be brought under more adequate control by strengthening the Companies Act, he said. The stock exchanges had always been willing to cooperate.

Murphy was unrepentant. He said he had information on stock exchange scandals and that he would reveal it as soon as the Federal Parliament resumed in March. This threat induced considerable apprehension in financial circles, particularly among the shadier entrepreneurs. Parliament was due to resume on March 19, which meant the market had six weeks to brood over Murphy's time-bomb.

One piece of paper which began circulating was an alleged photostat of Murphy's own share dealings. Judging by the extract of the alleged ledger card, which only showed a few entries, he was a modest punter dealing in parcels with total values of a few thousand dollars a time. The Senator, of course, had the same right to play the market as anyone. If he had been doing so, it was a point in his favor, because at least he might know more about the market than some of his colleagues. It turned out he knew quite a lot.

Murphy had hit the right issue at the right time, because the spectacular rise and fall of Tasminex was focusing attention on the growing public discontent with boom companies. Tasminex was at this time on the way down, and so was the rest of the market. Poseidon reached $280 fleetingly on February 5 then retreated. At $280 per share, with 2.5 million shares issued, including the disputed placements, Poseidon had a market capitalisation of $700 million. This was approaching three times the capitalisation of the Bank of NSW, which had 40 million shares selling at $6.40, making $256 million. Poseidon had been capitalised by the market at one-third the value of the mighty BHP which then had a capitalisation of $2150 million.

This was a patently absurd position, but one broking firm was prepared to put a far higher figure on Poseidon yet. During January the London stockbroking firm of Panmure Gordon, which specialised in mining stocks, had prepared a long, detailed survey of Poseidon. The survey appears to have been released late in January to institutions and clients, but on February 10 a summary of it appeared in the Financial Review.

It was a comprehensive attempt to evaluate Poseidon, containing lengthy calculations and four diagrams of the ore deposit. The survey concluded as follows:

"Net profits, assuming the current producer-price of nickel, should amount to some $22.76 million in 1972-3 and $48.805 million in 1973-4, placing the shares on a prospective earnings multiple of 24 and 11 respectively. However, at least some sales of nickel concentrates should be made at higher prices and assuming a nickel price of $5000 per ton net profits would amount to $62.285 million in 1972-3 and $127.915 million in 1973-4. On a present value basis discounting at a rate of 8 percent per year, the shares are estimated to be conservatively worth $300 and more optimistically, $382.

"Despite the rapid rise in the share price we consider there will be a strong rise by the middle of the year as results of diamond drilling at depth and percussion drilling on the Windarra Well zone are issued in quarterly reports. Some consolidation should result when development commences toward the end of the year. The shares are accordingly recommended for purchase."

The survey was signed by Simon Hunt, the broker who had attended the Poseidon annual meeting. From the amount of the detail in the survey he had acquired considerable information during his stay. The survey contained more technical information than had yet been released in Australia.

Panmure Gordon's survey and recommendation were naturally well publicised in Australia, where Poseidon was bubbling around the $210 mark at the time. A few days earlier it had run to $280 during a day when it had opened at $220 and closed at $250. One broker said: "No one can accuse a person paying $280 of being mad, but we think it will be possible to get it at a discount on that particular price." Local reaction to the figure of $382 was faintly incredulous.

Copies of the survey were scarce in Australia. One organisation which began searching eagerly for the document was Patrick & Co. The research staff of Patricks were already sceptical of the value the market was placing on Poseidon. On February 4, the day before Poseidon hit its peak, Patricks' staff geologist Lionel Milligan told a partners' meeting that on what had been found already the stock was "not worth $50". The market was punting that Zone C at Windarra would be bigger than the Shirley Shoot, on which most of the exploration work had been done to date.

Reinforcement for this view came from an authoritative and unexpected source. On February 26, Geoffrey Burrill arrived in Sydney to give a routine briefing to his clients, who included Patrick & Co. No one in Patrick knew it at the time, but Burrill had just been replaced as geological consultant to Poseidon and Samin had been appointed. But this was unimportant compared to the startling news which Burrill brought.

Burrill told Patrick's partners he thought the stock was grossly over-valued. Poseidon was facing very serious water problems. One ton of fresh water would be needed to produce every ton of ore. Water supply would not only be costly, but would delay the start of the mine.

Nor was there any joy in the market. By the end of February the non-ferrous metals index had dropped below 5000. Among the speculatives, A1 Consolidated was $3.90 compared to a top of $5, Allstate was down from $2.80 to $1.52, Carr Boyd from $9 to $6 and Quest from $1.60 to 58 cents. Almost every other speculative on the board was also down from its 1970 peak and some had lost nearly half their value. It was not so much a downtrend as a crash. Poseidon was not holding too badly at $212.

On March 2 an aerial spy over Windarra noticed a drill rig operating six miles south of the Shirley Shoot. The color of the sludge indicated that sulphides were being intersected. Poseidon shares rose to $245.

The possibility of another orebody being discovered gave Patricks' pause, but not for long. On March 11, the research staff received the Panmure Gordon report. Australia's largest mining broker had at last received information which had been known to a London broker since around Christmas.

Patricks' research department began analysing the technical data, including detail of cross-sections and indicated widths not previously known, and came to radically different conclusions to the London firm. They thought the deposit was not as wide as previously thought and that there would be mining difficulties.

Patricks immediately began advising clients to sell the stock, and sold all holdings in discretionary accounts (a discretionary account is exactly what it sounds like: a client deposits money with a broker to invest at his discretion.) The firm telephoned its Brussels office with the same advice and telexed selling orders to its London agents.[2]

Poseidon was around $250 at the time. Considering that the market had already begun to drift, it stood the selling wave from Australia's biggest broker remarkably well. The main selling was from March 12 to 15. It is hard to put a total figure on the volume, but Patricks were probably responsible for selling $3 million worth of Poseidon in 24 hours, most of it in London. Their selling was matched by some desperate buying by London brokers. Panmure Gordon, committed by their own survey, were buying and several other London brokers who had put clients in at high prices bought in an attempt to hold the price. London brokers were defending an Australian stock from Australian selling. Most of Patricks' clients sold around the $210 mark. London was left with the scrip.

The research department's decision to advise selling aroused no small heat inside Patricks because the two senior partners in Sydney, Ric Dowling and Frank Robertson, did not know what had happened until afterward. They had not attended the research-department briefing on Poseidon nor read the material. Michael Baume, who was head of research, said: "The next day Frank Robertson stormed in to me and said: 'You have destroyed the boom'."

More embarrassing was the fact that, unknown to Baume, Patricks had about $1 million worth of Poseidon in their house account at the time. This was a holding of 4000 shares which the firm had been forced to buy to protect itself in an option dispute early in the boom. When Poseidon was around $30, Patricks had bought a call option for a client for 4000 shares. The writer of the option defaulted, which meant that the client could have sought recourse against Patricks. Patricks covered their position by buying 4000 shares and putting them in the house account so that they could deliver if necessary.

Patricks kept the holding secret so that the client would be encouraged to try to get a settlement from the defaulting writer. Patricks' actions throughout in this deal were perfectly legitimate and defensible. Now the market was falling and the firm had got nearly all its clients out (only a few clients held on after Patricks advised selling), the logical thing to do was to sell the house account holding also. They missed the top of the market by a long way, but the price was handsome considering what would happen to Poseidon later. From that point, Patricks took a bearish view of Poseidon and sold it all the way down in the market.

The market was becoming increasingly jittery as press interviews by Murphy hinted at revelations in his forthcoming speech. On March 19, after the opening of parliament by the Governor-General, Murphy rose in the Senate and moved that a select committee be appointed to inquire into the desirability and feasibility of establishing a securities and exchange commission in Australia. It was a long motion, detailing the powers which the committee should have and its terms of reference.

Then Murphy launched into a wide-ranging attack on the excesses of the share boom. He condemned Endurance on the grounds that a false market had been created in the company's shares. He asked how much Singline and his associates had made out of transactions in Tasminex. He urged an investigation of Palgrave Corporation Ltd and called for documentation of the $1 million valuation placed on the Desert Flame of Andamooka. He cited the recent failure of Federated Industries Ltd, where shareholders faced serious losses and the press had reported that the company had "dressed up its statements to make the losses appear smaller than orthodox accountancy would permit". He also attacked the chairman of the Sydney Stock Exchange, Jim Cooper, for a statement that investors who lost money in speculative mining stocks were often the victims of their own ignorance and folly.

Murphy's main fire was concentrated on Poseidon. Anderson's correspondence with Murphy, Hynam and McArthur was tabled. Anderson's argument was put cogently in his letter to Murphy, in which he said: "My complaint is that as a direct result of these reports (by Poseidon) I sold my Poseidon shares on the eve of the company's success while people in the know or in the confidence of the directors were able to capitalise on this lack of published information and make considerable profits. I consider the statements were inaccurate and untruthful and that, while true assay figures may not have been available to the

Jim Say, chairman of Gippsland Minerals NL, with his scintillometer at the Harrietville "Hot Rock" discovery site. It turned out to be thorium. *Picture — Herald & Weekly Times Ltd.*

Newspapermen straggle across the Harrietville-Omeo road. At left, Jim Say points his scintillometer at the hot rock for the benefit of a cameraman. *Picture — Herald & Weekly Times.*

The drama of Poseidon's December 1969 annual meeting over, Norm Shierlaw (left) and Boris Ganke have an amicable chat. Shierlaw is holding mounted a sample of Windarra drill core. *Picture — News Ltd.*

Norm Shierlaw (left) and Basil Lewis, director of Poseidon and chairman of Samin Ltd, at the champagne breakfast in London's Great Eastern Hotel. Despite a big publicity build-up before the breakfast, Shierlaw disappointed the brokers and journalists attending because he had no financing agreement to announce. *Picture — News Ltd.*

There's a little drinking done on these pages. Here, Tasminex's chairman Bill Singline pours his wife a sparkling wine (it wasn't champagne) at their North Balwyn home. Earlier he had stated Tasminex could be bigger and better than Poseidon. As Poseidon later went into receivership and Tasminex struck tungsten in Tasmania, he was correct. *Picture — Herald & Weekly Times Ltd.*

Alexander Barton poses with the two halves of the Desert Flame of Andamooka. In this photo the stone has reflected the white flash of the photographer. Under normal lighting the stone was incredibly beautiful, containing all the colors of the rainbow. *Picture — News Ltd.*

David Paxton's house at 36 Viewmount Road, Glen Waverley, stands vacant in September 1970. The founder and chairman of Barewa Oil and Mining NL has never been seen since. The Age ran this photo across four columns under the heading "David Paxton: Where Are You?" *Picture — The Age.*

Brian Cutler, a director of Leopold Minerals NL, waits to give evidence to the Rae committee. He was jailed for making a false statement with intent to deceive after Leopold announced 5.33 percent nickel. *Picture — Australian Consolidated Press.*

Roy Hudson, former chairman of Queensland Mines Ltd and Kathleen Investments Ltd, in a jovial mood. *Picture — News Ltd.*

The Hudson home in Point Piper, Sydney. *Picture — News Ltd.*

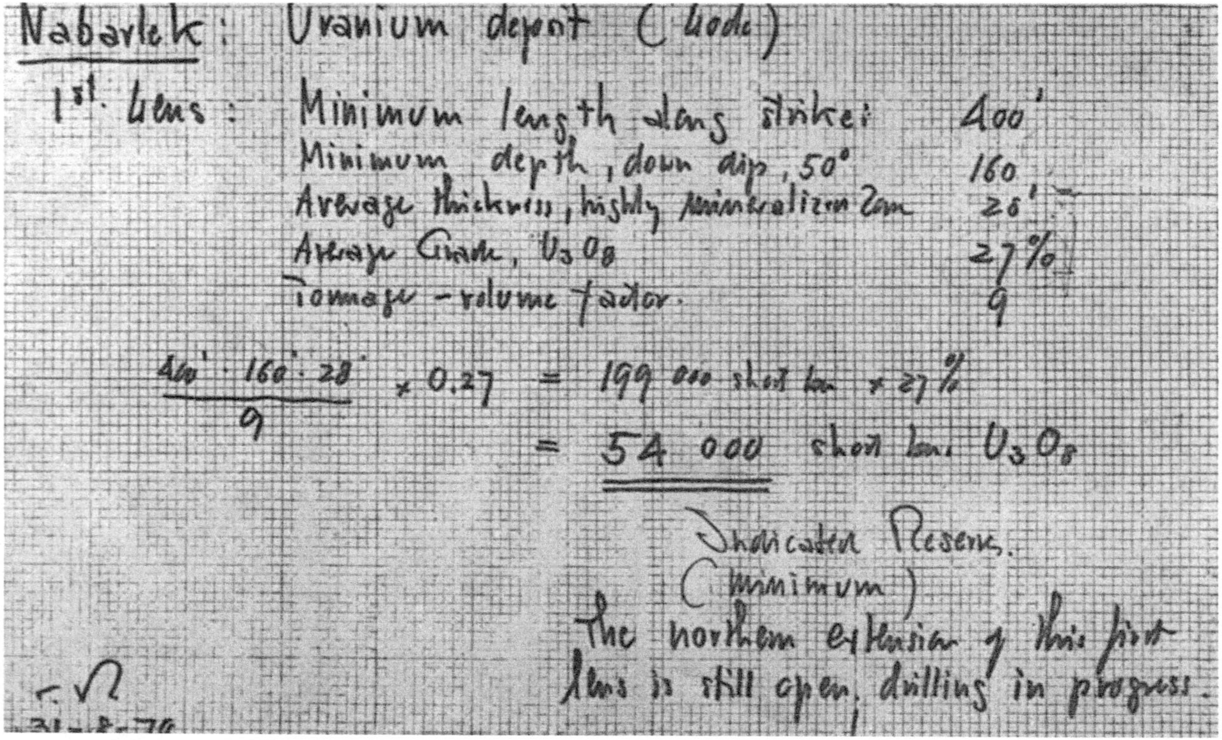

Nabarlek: Uranium deposit (Lode)

1st. Lens: Minimum length along strike: 400'
Minimum depth, down dip, 50° 160'
Average thickness, highly mineralized zone 28'
Average Grade, U_3O_8 27%
Tonnage - volume factor. 9

$\frac{400' \cdot 160' \cdot 28'}{9} \times 0.27$ = 199 000 short tons × 27%
= 54 000 short tons U_3O_8

Indicated Reserves.
(Minimum)

The northern extension of this first lens is still open, drilling in progress.

Dr Emile Rod's calculations of the Nabarlek uranium reserves. Other directors of Queensland Mines Ltd. said that the chairman, Roy Hudson, kept this and another sheet of calculations in a sealed envelope in the office safe. The other directors did not see these documents until Hudson was removed from the chairmanship in a dramatic board meeting on August 12, 1971.

Minsec's managing director Tom Nestel, looking typically efficient. The stand behind him contains a range of mining and technological publications. *Picture — Australian Consolidated Press.*

In typically cheerful mood, Minsec's chairman Ken McMahon has the world at his fingertips. Picture—*Australian Consolidated Press*

McMahon, top left corner of the table, gives evidence in the Senate chamber to the Select Committee on Securities and Exchange. *Picture — Australian Consolidated Press.*

Ric Dowling, senior partner of Patrick & Co, in an apparently persuasive mood. Picture—*Australian Consolidated Press*

Dr Michael Garretty (left) and his wife Elsa and son Peter have a celebratory drink in their Toorak flat. The occasion was a press conference to announce the takeover of Rimibo Resources Ltd. Selected Mining Holdings Ltd, another company in the Garretty group, paid $1250 for the support of more than 100 students who turned up at the Rimibo meeting as proxyholders. Picture—*The Age*

Jim Jamison, liquidator of Mineral Securities Australia Ltd, in his office at the top of Norwich House in O'Connell Street. Picture—*Australian Consolidated Press*

directors, information and evidence of a probably rich nickel find were known but that this information was suppressed and not given to the shareholders." On Burrill's buying, he wrote: "At no time during the months of August and September was I told that the geologists thought so much of the area that they were moved to seek permission from the directors of Poseidon to go ahead and buy up shares in the company...As, in the capacity of a shareholder, I was employing the geologists, I consider I had a perfect right to have been informed of the fact."

Murphy thought Anderson's complaint was reasonable. The Senator said his letter seemed to set out "a classic example of insider trading". By Senate standards the debate on Murphy's motion was not long. Half a dozen Senators spoke. The Labor Senators predictably supported Murphy. The Liberals, Greenwood, Wright and Rae, all agreed with the move to set up a Senate committee, while deploring Murphy's attacks on individual businessmen.

A committee drawn from all parties in the Senate was subsequently set up in April 1970 with Sir Magnus Cormack (Liberal, Victoria) as chairman. He retired from the post upon becoming President of the Senate in August 1971 and chairmanship of the committee was taken over by Liberal Senator Peter Rae from Tasmania. From the start of the committee's proceedings Rae had been showing a keen interest in the inquiry, asking searching questions of witnesses. The committee became known as the Rae Committee after his appointment as chairman.

The committee began with enthusiasm but considerable ignorance. Its members had a large subject to tackle, but no clear idea of how to go about it. Their early hearings were largely an exercise in self-education. Some of the members did not appear to know the fundamental operating principles of broking houses or merchant banks. Senator John Wheeldon (Labor, WA) did not know whether Rio Tinto Zinc Corporation was British or American.

The first witness was Dr John Rose, senior research fellow of the Institute of Applied Economic and Social Research of the University of Melbourne. He was the author of Securities Markets in Australia and the most respected academic specialising in the study of stock markets. Apart from giving evidence, Dr Rose became an informal adviser to Senator Rae. Later he was seconded to the committee as an economic adviser. Later still, Tom Fitzgerald, the highly respected former financial editor of the Sydney Morning Herald, was appointed as another assistant and adviser.

Other early witnesses tended to deal in broad generalities about the securities market. This may have been instructive to the Senators but was not of great use to the rest of the community. However, as the committee began to focus on specific incidents in the market and get down to investigating the details of what happened, they began uncovering a mass of fact which had been hidden from the public eye. Their first detailed investigation was into the alleged insider trading in Poseidon, an inquiry they had to reopen much later after discovering, to their wrath, that important elements of the early evidence had been wrong. From Poseidon, the committee spread its investigation into other incidents. They became increasingly inquisitorial as they uncovered a number of dubious and illegal practices.

Their report of two volumes, and the accompanying three volumes of transcript and one volume of documentation, are still the most valuable single work on the operation of Australian securities markets.

The report took an agonisingly long time to prepare. It was actually written by Rose and Fitzgerald. It was not published until 1974 by which time the boom was well and truly over and Australia was in the grip of a major recession. The report revealed much shady dealing which had hitherto been unknown, and often unsuspected, by the investing public. It established that dubious practices had been endemic in the securities industry during the boom. It was a devastating condemnation of boom-time morality and led to the call for stricter controls on behaviour, including a national companies Act.

DIGRESSION IV: Let's Keep Politics Out of This.

The Rae Committee produced the most wide-ranging survey of corporate activity and securities market misbehavior ever seen in Australia. Its case studies revealed an astonishing diversity of undesirable practices. One of the few areas it did not attempt to cover was the activities of politicans themselves during the boom.

One boom-time phenomenon was the way in which Ministers for Mines in some States associated themselves with company announcements. This is an old political ploy in Australia. Ministers of the Crown like to take as much credit as possible for any good news that is announced. In the mining boom, most company announcements were good news. One long-standing example of this can be seen in Western Australia, where successive Ministers for Mines in governments of different shades have adopted a tradition of announcing the weekly activities of the Wapet consortium. The Wapet consortium (Ampol Exploration, California Asiatic, Texaco and BP) is highly respectable and the practice seems to have dated from the discovery of the Barrow Island field. It is a practice which suits all parties. The Minister concerned enjoys a little favorable publicity and the company has its reputation enhanced by its association with the Minister.

On general principles, however, there are grounds for objection. A Minister may bestow his favors upon a company which later falls into disrepute, and the investing public may have been enticed to buy shares through its association with the Minister. Also, a company with something important to announce should make its statement as soon as possible to the home stock exchange and not route it through a Minister for publicity reasons. Any company which does this regularly puts the Minister and his staff in the position of being insiders in the stock.

One example of this type of statement occurred in March 1970 when the decision of Mineral Securities to take a 40 percent interest in Robe River Ltd was announced not by the companies concerned but by Sir (then Mr) Charles Court, as Minister for Industrial Development in WA. Sir Charles, as an energetic promoter of mining in his State, has perhaps earned the right to get a little publicity from some of the big deals done, although this one later became a source of perilous uncertainty for Minsec's creditors.

One Ministerial statement which must have embarrassed its maker afterward occurred at the opening of the Munga Creek Minerals NL antimony mine near Kempsey on March 2, 1970. The NSW Minister for Decentralisation, John Fuller, speaking at the opening ceremony, was quite optimistic about Munga Creek's prospects.

He said: "Perhaps Munga Creek Minerals will not immediately rival Poseidon in Stock Exchange rating but from reports and expectations the antimony deposit here is rich enough to warrant commercial mining.

"With the price of refined antimony in Sydney at around $7500 a ton, production prospects being what they are, Munga Creek shareholders should be well satisfied with the results of this operation." Later he said: "Here at Munga Creek indications of financial rewards are in the immediate future and it looks as though the operation will have very satisfactory results."

Munga Creek Minerals NL was promoted by Abaleen Minerals NL and Vale Housing Finance Corporation Ltd, two companies associated with George Summers. Summers was also managing director of Munga Creek. Fuller's speech was to open the exploration program, but the company had already put down 10 diamond drill holes, in a 20-hole program. The opening ceremony was held about 10 days after the shares issued in the Munga float had been approved for listing by the Sydney Stock Exchange. Munga Creek listed on April 16, about six weeks after Fuller's speech. The 25 cent fully paid shares came on at 46 cents but soon slid to within a few cents of par.

Any member of the public who invested in the stock looking for a satisfactory immediate return would have been disappointed. In November 1970 directors announced reserves of 585,000 tons of ore. It was never clear how this calculation was made because a later geologist from the Krupp group engaged by Munga said a reserve calculation was extremely difficult.

Munga Creek made losses every year because its costs of production exceeded its revenue from output. At the end of 1973 the company went into voluntary liquidation, never having paid a dividend in its short career. Don Kirkwood, writing in the Financial Review on December 27 1973 said: "If support and interest from politicians alone could ever have ensured the success of a group of public companies, there can be little doubt that Vale Corporation and its associates, Munga Creek Minerals NL and Abaleen Minerals NL, would not be in provisional liquidation as they are today."

At that time the chairman of all three companies was Norman John Mannix, a former Minister for Justice in a NSW Labor Government. His predecessor as chairman was the late Major-General John Stevenson, Clerk of the NSW Parliament.

Munga and Abaleen had had the same board when they had launched. The chairman had been Stevenson and the directors were Sam Cohen and George Somogyvari Summers. Abaleen's first main interest had been in the Torrington district of NSW where tin had once been mined. Its leases there were acquired from International Tin Corporation Pty Ltd. Two shares in International Tin were held by a woman named Iris Mary Cratchley. She sold her International Tin shares to two Summers companies. There is a link here with Hall's Peak

Australia Ltd, because Cratchley and Mannix were both vendors to that float.

Other politicians were directly involved in the mining boom. The most notable was Dr Malcolm Mackay, Liberal MHR for the Sydney suburban seat of Evans and at one time Minister for the Navy. Dr Mackay was chairman of Longreach Oil Ltd and also of its mining spin-off Longreach Metals NL. The Longreach group had close connections with Patrick & Co.

Longreach Metals had an 85 percent owned subsidiary called Longreach Manganese Pty Ltd. Longreach Manganese's biggest deal was to buy two large WA manganese deposits at Woodie Woodie and Ripon Hills from the Mount Sydney Mining Co, in which the Scott family held about a half interest. The most prominent member of the Scott family was Senator Malcolm Scott (Liberal WA). The purchase price for the mines was $2 million, but Longreach was never able to pay the full amount of the debt. Then Japanese customers complained of high silica content in the Woodie Woodie manganese and imposed heavy penalties on Longreach, making it uneconomic to continue production. Mount Sydney resumed the mines.

One boom company which the Rae committee could have investigated without leaving its own chamber was Nickelfields of Australia NL. The directors of this company included Senator Scott and Senator Ian Wood (Liberal Queensland). Nickelfields had one of the shortest lifespans of any boom float.

Nickelfields was floated in April 1970 on a prospectus which raised $1.6 million from the public. Eighteen months later, in October 1971, Nickelfields was put into liquidation. Shareholders got nothing. Out of the proceeds of the float some $650,000 was paid to vendors, and formation expenses, including a $128,000 underwriting commission to Sydney broker Roland Walton & Co. At this stage, Nickelfields had some nickel areas in WA, including claims at Narndee and Mt Clifford, and some wolfram interests in Queensland.

The shares came on fairly in a weakening market. The 10 cent fully paids were first listed in Perth on May 28, 1970 and later ran as high as 80 cents there and to 67 cents in Sydney. They finished 1970 at 33 cents, which was still a handsome price compared with what they would be worth less than a year later.

At June 30, 1970, the company was in a fair cash situation. It had $800,000 worth of mining claims (representing the cost of claims and exploration work), and the same amount in cash and deposits. There were no liabilities, but the formation expenses of $150,000 were written into the balance sheet as an intangible asset. With 18 million issued shares it had 4.4 cents per share cash.

At the company's first annual meeting on December 30, 1970, Senator Scott, as chairman, told shareholders that prospects were good.

He said: "I firmly believe that we may look to the new year with every confidence." The major project was the St George antimony mine in northern Queensland. A crushing plant, ball mills and flotation cells were being built on site. Access roads, an airstrip and a prefabricated township had been established. "The profitability and saleability (of the antimony) appear assured as your directors have been approached by two of the world's largest buyers of this metal who have offered contracts for the continued purchase of the metal," the senator said. Ten months later the company was in liquidation.

The major problems which led to the company's collapse centred around the expense and unprofitability of St George, but shareholders were given no inkling of these developing clouds for some time. In February 1971 Nickelfields negotiated a 75 percent interest in five Indonesian mercury leases. A feasibility study indicated a pre-tax profit of $6 million a year from these leases. In March 1971 Nickelfields announced another Indonesian venture. This was a proposed joint venture with Wattle Gully Gold Mines NL on the Lorentz River gold deposit in West Irian. This deposit was estimated to contain 28.8 million oz of gold valued at nearly $900 million.

In hindsight, Nickelfields was in no shape to embark on new ventures while it was still trying to finance the establishment of the St George mine. It had no significant cash flow yet, and could ill afford to spend even small amounts on new risk ventures. Floods delayed completion of the St George plant, which was proving expensive. The company's recorded exploration expenditure at St George was $338,000 in the March quarter. This sum was approaching half its cash reserves at the previous June 30.

In a major board shuffle, Senator Scott stepped down as chairman, being replaced by Jose M. Safie. Safie, who has since died, was consul in Sydney for the republic of El Salvador and was quoted at the time of his appointment as saying that he represented Canadian mining and European financial interests. He also spoke of the possibility of injecting more cash into Nickelfields.

On July 5, 1971, Safie announced that the company had sold the next two years production from St George. Three weeks later he said the new board of directors felt "most confident that the present position of the company will show an improvement and the company will be placed on a sound financial basis."

Despite the bullish tenor of these statements, some of the smarter shareholders were tiptoeing for the door. The shares had drifted from a peak of 40 cents early in 1971 to 7½ cents at the end of July. As late as September 10, 1971, Safie issued a statement appealing for creditors to be patient and saying that the management was extremely confident of the profitability of the antimony operation. On October 6, Mr Justice Andrews in Brisbane ordered the company to be wound up for failure to pay a debt of $2583.51.

Unsecured creditors were not paid in full and shareholders got nothing back. It might have been difficult to discover who the shareholders were because the liquidators described the share register as "hopelessly in arrears".

The immediate cause of the failure of Nickelfields appears to be that it became insolvent because it spent all its money on the St George mine. It would seem no lender was prepared to advance bridging finance for the project, despite the eminence of the board. This indicates that St George could not have been as attractive a proposition to lenders as it was in directors' statements.

Though the collapse was speedy there does not appear anything sinister about it. But Nickelfields proved that even Senators are not infallible when it comes to mining.

---

[1] Now Mr Justice Murphy of the High Court.

[2] London brokers would not act as agents for foreign brokers who had their own offices in London, so Patricks set up an office in Brussels to handle its London business but also dealt directly from Sydney with the London brokers who were its agents.

CHAPTER FOURTEEN

Decline

If there were any lingering doubts that an investigation of the mining boom was necessary, they were removed by two events in April, 1970.

The first was the demise of Barewa Oil and Mining, which had been refused listing by the Melbourne Stock Exchange. The result was two stormy meetings of shareholders; the man under fire was the promoter and chairman, David Paxton. Paxton, it will be remembered, was the accountant who had announced plans in January to revolutionise mining finance by floating companies at the rate of one every three weeks. The pace-setter for this stampede, Barewa, had become uncomfortably bogged.

The company had raised $1.1 million in an issue of five million shares, of which some were fully paid to 50 cents, some paid to 10 cents and some paid to 1 cent. The fully paids sold as high as $1.40 on the Second Market (in a sale arranged by Paxton), but the shares were quoted there only briefly. That was the only fleeting chance the shareholders ever had to get out of the stock.

By April, the shareholders who had subscribed to the company had become deeply worried about Barewa. At a meeting in Melbourne on April 8, shareholders got some information out of Paxton, but not all they sought. They fired questions at Paxton for two hours, but he refused to reveal what profits had been made by the three promoters of the company from dealings with Barewa. The shareholders foreshadowed a move to have a committee of inquiry appointed into the company's dealings.

It took years and an official investigation before shareholders discovered what had happened and the story is still incomplete in some respects. The inspector appointed by the WA Government, Edward Lloyd, QC, was hampered in his investigations by the disappearance of Paxton, who has never been seen since.

Barewa floated with an "empty" prospectus, containing no reference to any property which the company intended to buy. A more alarming point, which was not established until long afterward, was that it did not have any directors. The directors shown in the prospectus were David Paxton, Jim Say and Ian Cornelius. Say was chairman of Gippsland Minerals (the man who had found the "Harrietville Hot Rock"). Cornelius was a director of Pursuit Oil. In December 1970, a WA Supreme Court judge ruled that none of the directors had been validly appointed, so the original three directors will be called promoters in this story.

Barewa raised $1.1 million in an issue of five million shares divided into three classes. There were 50 cent fully paids, and contributing shares paid to 10 cents and 1 cent. The prospectus was dated January 19, 1970, but it did not mention any arrangements made by directors before that date.

In fact, the promoters of Barewa had been active before then. In early December 1969, Paxton and Cornelius entered an agreement with Say over Petroleum Exploration Permit 67. Paxton and Cornelius, purporting to act on behalf of a proposed public company, Barewa Oil NL, agreed with Say that Barewa would buy a 75 percent interest in PEP 67 from Say's company, Victorian Refining and Smelting Co Pty Ltd. They agreed to buy the interest for $75,000 and $50,000 worth of Barewa shares.

They also agreed that Say would become a director of Barewa, and told him the company would also acquire mineral claims. Say told them that, as he was already chairman of a public mining company, he would not be associated with Barewa's mineral activities. He made it clear that his main interest in Barewa would be exploration of the oil permit.

After making this agreement, Paxton, in an effort to avoid the obligation to disclose it in a prospectus, instructed his solicitors to change the name of the company from Barewa Oil NL to Barewa Oil and Mining NL. Under its new name, Barewa was incorporated in WA on January 6, 1970.

Three weeks earlier, on December 17, 1969, Paxton and Cornelius had incorporated in WA a private company named ISIM Mineral Development Pty Ltd. ISIM's only two shareholders were the private companies of Paxton and Cornelius.

Two days after ISIM was formed, Cornelius introduced Paxton to Robert McEwan, a WA company promoter. McEwan offered Paxton several mineral claims at Pardu near Port Hedland. These claims had been earlier examined and discarded by a subsidiary of an American corporation searching for iron ore. During exploration, minute traces of nickel had been discovered, but they were too low to justify a continued search for nickel.

During the boom the claims were re-pegged by a syndicate. The claims came to the attention of McEwan who sought geological advice about them and was advised they were unattractive. Paxton agreed to buy them for $60,000. The Lloyd report says that when the Pardu claims were offered to Paxton he was sent a document which purported to be a reconnaissance report on the claims prepared for a syndicate member. The document was prepared in McEwan's office and was substantially a copy of a report delivered to the American corporation. The original report indicated that supergene enriched nickel concentrations had been checked for but that the results had been too low grade to continue searching. "Someone, apparently not Paxton, had found this part of the original report unpalatable and therefore altered it so that it would read that supergene enriched nickel concentrations were *found* and that this information was reported with a recommendation for further research," Lloyd said.

"The effect of that alteration was to alter a conclusion that the area was regarded as non-prospective for nickel in such a way as to make it read as if the area was regarded as prospective for nickel. But whoever it was that altered the conclusion of the original report was not prepared to go so far in salting that report as to alter the nickel values mentioned in the original.

"Having received the 'preliminary reconnaissance report', Paxton, in order to induce people to invest money in Barewa and to inflate the price of Barewa shares once they were listed, informed several journalists (and thus the world), two stockbrokers and his own public relations consultant that assays of cores from the Pardu claims had shown nickel values ranging between 1.4 and 1.8 percent. The nickel values actually found were less than one-tenth of those figures and were of no commercial significance."

ISIM also acquired other nickel claims on Barewa's behalf. On about January 6, Paxton offered to buy a half-interest in 10 claims at East Coolgardie, and eventually ISIM bought the interest for $121,000. Within days, Barewa was committed to pay a total of $153,500 for them. Lloyd says the man who had pegged the claims thought they were worth about $10,000.

Between December 19 and January 15 Paxton also agreed to buy 44 claims in the Yalgoo area of WA. Disregarding a share allotment involved, ISIM acquired these claims for about $83,000. Barewa was eventually committed to pay $226,000 for them.

ISIM also had 20 claims pegged at Windarra as close as possible to Poseidon. The cost of pegging and registering these claims was $6400 but ISIM sold them to Barewa for $95,000. "They were very probably worthless," said Lloyd.

Barewa's empty prospectus was dated January 19. On January 20, Cornelius announced to the press that the float had opened and closed heavily oversubscribed. Three days later ISIM formally bought the mineral claims on the terms negotiated by Paxton previously.

On January 26, Paxton, Cornelius and Say held a board meeting of Barewa. ISIM, the company owned by Paxton and Cornelius, sold 91 mineral claims to Barewa for $611,250. The claims had cost ISIM $331,000 only a few days before. Lloyd said that Hervey Newton, Barewa's consulting geologist, was horrified when he later heard how much Barewa had paid.

By this time a large proportion of the $1.1 million raised in the public float had found its way into the promoters' pockets. Barewa had paid:

- $611,250 for nickel claims which had cost ISIM only $331,000.
- $75,000 cash plus $50,000 in shares for PEP67. The total cost of this permit to Say's company (according to an affidavit filed by a later director of Barewa in the WA Supreme Court) was about $25,000.
- $25,000 to PAX Pty Ltd as a consultant's fee. Paxton was a shareholder and director of PAX, which was to receive a fee of $25,000 per year for five years for running Barewa's affairs.
- $12,070 in professional fees for accounting, administrative and share registration services to a firm headed by Cornelius.

The option for PEP 67 required Barewa to drill a well on it within 12 months. Even presuming the company had been left with enough money to afford a wildcat oil well, this requirement was obviously likely to be fraught with difficulty. PEP 67 covered the northern half of Port Philip Bay, and included all the main beaches of Melbourne. The proposal that a well should be sunk in the already polluted bay aroused outraged screams from conservationists throughout Victoria. Barewa claimed there was evidence of oil seepage in this part of the bay. Actually they were right about the oil seepage, but it more probably came from the tankers discharging crude at the Altona refinery. It would have required far stronger geological evidence before any company would have had even a faint chance of drilling Melbourne's precious bay, although it is open to question whether an oil well would have significantly increased the pollution of St Kilda Beach.

Adding the above figures together, it can be seen that some $700,000 of the $1.1 million raised from the public had gone to the promoters as vendor payments or fees. On the night of January 26 the meeting of the three promoters also made a share placement to themselves, with 2 million shares going to Say and his associates and more than 4½ million to the private companies of Paxton and Cornelius. This amounted to 25 percent more shares than had been issued in the prospectus only seven days before. The cash for the placement was derived from the payments made by Barewa to the various private companies of the promoters that night. Thus the promoters converted the cash they had taken from the public float into a controlling interest in the company, virtually for nothing. If Barewa shares were listed they stood to make large profits.

However, the Melbourne Stock Exchange would not list the shares. Barewa's first application for listing was made on January 19, but was incomplete. The incompleteness of the application was a handy excuse for the exchange's listing committee because they were reluctant to list Barewa anyway. The unnerving feature of the whole episode is that the Melbourne Stock Exchange knew nothing of the deals outlined above when it rejected listing. The committee appears to have relied mostly on its instincts. They had a general distrust of empty prospectuses and Paxton's announcement that he was going to list a company a month had not only been brash, but pre-emptive.

It was tantamount to a challenge to the listing committee, who by that time were being prodded anyway to take a tougher line with new prospectuses. Among professionals there was an indefinable feeling from the start that something, somehow, was "wrong" in Barewa. Such instincts are not to be dismissed lightly in the commercial world. At least as many deals are clinched on an instinctive attraction for people or situations as are done on cold figures.

The second application for listing was sent by Barewa on January 27. This application concealed the fact that the funds of Barewa had largely been paid to private companies of the promoters the night before. The application asserted that

Barewa had liquid funds of $1 million. Of all the deals that had been done, Barewa disclosed only the acquisition of the Pardu claims, which were the only ones to have been bought for cost. They were also the ones which Paxton had announced to the press as having good nickel values.

One small fact known to the listing committee by now was that Paxton had been trying to set the price of Barewa on the Second Market. The Second Market, because of its small turnover, may have been manipulated in this way more than once, but in Paxton's case it was an extra small mark against him. On January 29 the Age carried a story by Garry Oliver under the heading "Barewa Outlook Bright", saying that the company's 10 cent shares were changing hands on the Second Market at $1.40. When another journalist, Ian Robertson of Newsday, rang the Second Market for a quote on Barewa he was told, "I can't give you a price because Mr Paxton has not rung up with his bid yet." Robertson told Oliver who passed the information on to David Hume, a broker on the listing committee. Paxton's Second Market activities then gave the committee another reason for refusing to list the shares.

Lloyd, investigating the Second Market trading, discovered that the only Barewa shares traded there had been bought by Paxton from an accountant named John Miller. Miller had given the original instructions for the incorporation of Barewa Oil.

On February 3 the listing committee refused listing and on February 9 Paxton made a third application. He still did not mention any of the mineral or oil leases acquired, apart from the Pardu claims. The committee rejected Barewa again. Paxton and Cornelius then decided to disclose all the claims and hastily assembled geological material about them. By this time the exchange appears to have decided not to list Barewa at any price and rejected Paxton a fourth time.

Paxton and Cornelius had another gambit. They used $350,000 advanced by Barewa to buy a controlling interest on behalf of themselves and Miller in a listed public company called Ronaldson Brothers and Tippett.

The aim was to backlist Barewa through Ronaldson and Tippett, which until then had been a small, unprofitable timber miller in Victoria. John Miller was at that time a representative of the majority shareholder on the board of Ronaldson and Tippett. The backlisting attempt was aborted when the exchange threatened to delist Ronaldson and Tippett. The shares acquired by Paxton and Cornelius were sold back again.

While all these machinations had been proceeding, shareholders who had subscribed to Barewa were in the grip of mounting dismay. They did not know how their money was being used (the full facts did not emerge for years) and they could not sell their shares.

On April 8, Barewa held a statutory meeting of shareholders, as is required for newly-formed companies under the Companies Act. By this time Say and Cornelius had resigned from the board and had been replaced by David Saw and Keith Millheim. Saw and Millheim were not involved in any way in the early maneouvres of Barewa, and later tried unsuccessfully to rescue the company. At the April 8 meeting shareholders fired questions at Paxton for two hours, but Paxton refused to reveal what profits had been made by directors from dealings with Barewa. The meeting finished with nothing settled except that there would be an extraordinary meeting of shareholders on April 21.

It was now becoming evident that Barewa would never be listed. Enough had been disclosed to reveal that its affairs had been run in a most unsatisfactory way. The listing committee had had to decide whose interests it was protecting. If it refused to list Barewa it denied a market to the shareholders who had subscribed to Barewa. Of course, this is always a risk taken when subscribing to a float, because no company is guaranteed listing. If the committee had allowed Barewa to list it would have given an imprimatur of legitimacy to the company and its promoters and would have provided a market where other members of the public would have been allowed to buy the stock (most probably at a premium) in ignorance of what had been happening. The listing committee in such circumstances must always look to the protection of the larger, general public as having priority over the claims of shareholders who have subscribed to a particular float.

Barewa's second general meeting made the first look like sweetness and light in comparison. At the start of the April 21 meeting, interim interlocutory injunctions were issued on Saw and Millheim restraining them from casting disputed proxies. By this time Saw and Millheim were contesting Paxton's control of the company.

The meeting lasted for a rowdy 3½ hours, during which there was doubt about who was the company's official solicitor and Paxton's right to occupy the chair was challenged.

The meeting might have been even rowdier if Paxton had not thoughtfully hired five armed security guards "to ensure the proper conduct of the members at the meeting". Paxton had also changed the locks on the company's offices in Little Collins Street. Possibly this was because Paxton and Cornelius, who was still secretary of Barewa, were refusing to make the company's share register, contracts and proxies available to Saw and Millheim. The meeting was conducted in an atmosphere of siege and ended inconclusively. Paxton held a press conference afterward in an attempt to improve the company's worsening image.

Next month the WA Government appointed an inspector to investigate the affairs of Barewa. In September, 1970, Paxton and his wife, Gay, disappeared. Their house at 36 Viewmount Road, Glen Waverley, a suburb of Melbourne, was deserted. The furniture had gone and what was once a front lawn became a two-foot-high tangle of grass stippled with yellow daisies. The Age printed a photo of the house across four columns under the heading: "David Paxton: Where Are You?"

In December Mr Justice Burt of the WA Supreme Court ruled that Barewa had never had any legally appointed directors. By that time the company was in liquidation.

Say was eventually fined $1300 on charges of issuing a prospectus containing an untrue statement and failing to use reasonable diligence as a director. Dr Gerard Gibney, a consulting psychiatrist treating Say, told the Melbourne Magistrates Court that Say was impetuous and unsophisticated, of average intelligence, and went from depressed periods where he drank excessively and which impaired his memory to periods of having grandiose ideas, over-activity and very little sleep. Mr Duffy, SM, said: "He seems to have perhaps been the victim of his own feeling that other people were as honest as he was." Cornelius was convicted on October 29, 1975, of conspiring with Paxton to cheat and defraud Barewa and its shareholders. He was sentenced to three years' jail, with a 12 months' non-parole period. Paxton was never seen again.

The story had a medium happy ending. Late in 1976, the liquidator of Barewa paid all shareholders 47 cents in the dollar.

As if to prove the boom needed scrutiny from top to bottom, the second scandal of April 1970 was the float of Comalco Ltd. The purpose of the float was to give the Australian public a 10 percent stake in Comalco, the rest being divided into 45 percent holdings by Conzinc Riotinto of Australia and Kaiser Alumina of the USA. Comalco's 50 cent fully paid shares were issued at $2.75 each. The underwriters were Ian Potter & Co, with Daysh Renouf & Co acting as associated brokers in New Zealand.

It was the blue-chip float of the period and consequently attracted a lot of attention. Gossip before the prospectus was issued indicated that there would be some interesting names on the float list. The same gossip was circulating in New Zealand, which disturbed the finance editor of the Evening Post there. On April 29, 1970, he wrote that the company had asked the organising brokers to make direct offers of shares to "selected parties". Among those who had been offered fairly substantial allotments were a number of men in important positions in the newspaper industry.

This was the first public airing of what was known to be happening. In May, the then Prime Minister John Gorton was offered shares in the float and rejected them. He told his Ministers not to accept shares in the float either. In New Zealand the Prime Minister, Keith Holyoake, issued the same instruction to his Cabinet.

When the float list became available in June the Melbourne Age inspected it and discovered that thousands of shares had been allotted to high-ranking politicians in Victoria, NSW, WA and in Queensland (where nearly half the Cabinet was involved). Allotments included 1500 to Mrs M.I.Askin, wife of the NSW Premier; 1000 to Tom Lewis, then NSW Minister for Lands and later Premier; 1500 to Queensland Treasurer and Acting Premier, Gordon Chalk, and another 1000 to members of his family; 1516 to Sir David Brand, Premier of WA; and 1500 shares to Charrita Pty Ltd, a private company associated with Charles Court, the subsequent Premier. The Victorian Premier, Sir Henry Bolte, received shares and there were also financial journalists on the float list. According to one report, the Age's researchers were aided by a helpful lass on the Comalco share register who told them the names of the people behind some of the private companies which appeared as allottees.

The issue stirred a considerable scandal. Comalco had obvious reasons for wishing to be on friendly terms with the Queensland Government in particular and the issue was seen as a form of bribery.

There were mitigating factors. For one thing, some of the people on the list were CRA shareholders or clients of Ian Potter and Co (which had a long list of silvertail clients) and were therefore entitled to participate in the issue. For another, Comalco shares were generally believed likely to come on the market at around $4. This would represent a mark-up of $1.25 on a $2.75 share. This was small compared to the multiples of five times, 10 times and higher that stags had enjoyed on many boom floats only a few months earlier and whose float lists had rarely been inspected. From the float list the parcels offered to leading State politicians appeared to range from 500 to 1500. As the expected mark-up was around $1.25 a share, this could hardly be called a large bribe. It looked more like a clumsy attempt to buy goodwill and it seems certain that the company rather than the underwriter was behind it. The feeling at the very top of the company, so the story went, was that it would become the sort of "in" stock that top people would talk about at cocktail parties.

Whatever merit there may have been in the idea, it was atrociously timed, occurring at a period when the market was already in disrepute. After this youthful indiscretion Comalco led a life of corporate rectitude mining bauxite.

While these scandals were having their run, the troubles of Poseidon were expanding. The goodwill generated by the company could not live indefinitely on the heady expectations of Christmas 1969. The market was waiting for Poseidon to announce how the Windarra mine would be funded. Inevitably, as the good news did not materialise those waiting gradually became nervous. By the start of April 1970, the shares were down to $178, $100 below their peak of two months before.

The full story of the decline and fall of Poseidon was a long, agonised affair which falls outside the timespan of this history. As Poseidon gave the boom its name and as its decline illustrates some of the falsities upon which the boom was based, it is necessary to explain why its shares should go from $280 to zero in seven years.

There were three basic reasons underlying Poseidon's downfall:

1. The nickel market stagnated. Nickel demand by western countries reached 437,000 tonnes per year in 1972. In the next two years it rose to 562,000 tonnes, but then it collapsed and by 1976 was back at 440,000 tonnes, showing virtually no growth in four years. This stagnation in demand was accompanied by a huge growth in world nickel production, especially in Third World countries. Nickel output rose heavily in the Dominican Republic, South Africa, the Philippines, Greece, Indonesia and Rhodesia and more capacity is still coming on stream in Guatemala. Communist countries, notably Cuba and Yugoslavia, also became heavy surplus producers.

Meanwhile, the stagnating economy of the world from 1973 onward drastically reduced stainless steel production. Further, the high prices reached by nickel forced steel producers to search for cheaper processes using substitute metals. The result has been that less steel is being produced using less nickel. Half the current nickel production goes into the manufacture of stainless steel. In the USA, the world's largest user, 65 to 70 percent of stainless-steel production is in the form of austenitic (nickel-chromium) steels. These steels usually contain 8 to 12 percent nickel but the high-nickel steels in this category are facing increasing competition from low-nickel steels using only 3.5 to 6 percent nickel and substituting manganese, copper and nitrogen for the balance. The other major category of stainless steels is the ferritic, accounting for 25 percent of production. Ferritic stainless steels previously contained 0.5 percent nickel, but are now being developed in forms containing none.

With demand static and a glut of supply, nickel markets suffered. The producer price rose from US 94 cents per lb in 1968 (annual average) to US$1.29 in 1970, but the European free-market price fell from $3.90 in 1969 to US$2.00. When Western Mining opened Kambalda it had been able to sell all its output on the free market for the first two or three years, giving the company an enormous cash flow. Western Mining then, in a move which suited consumers but was also very sound strategy, switched to supply on long-term contracts at producer prices. Poseidon in mid-1970 was unable to follow the same initial strategy.

2. Inflation. The operating costs of a nickel mine escalated at a far greater rate than could have been foreseen in 1970-1. No one could blame the Poseidon board for failing to foresee this. Australia's inflation rate had been below three percent throughout the sixties until the last three months of the decade, the period in which Poseidon took off. Typically, an analyst projecting future earnings for a company at this period would have estimated 5 percent as a maximum future level of inflation. Few analysts foresaw the apocalyptic impact of the OPEC price rise in 1973.

Capital costs escalated for the same reason. Not only was Poseidon going to be more expensive to operate than originally thought, it was going to be more expensive to construct. This ate heavily into future earnings. Also, the cost of financing soared as interest rates rose sharply.

3. Poseidon's eventual grade was almost half that originally reported. Early studies by analysts of Poseidon relied heavily on the 3.56 percent figure in the statement of October 1, 1969. An ominous trend that went undetected in early statements of Poseidon was a gradual erosion of this figure. Ultimately

Poseidon's average reserve grade was about 1.9 percent which, in combination with other factors, was a marginally uneconomic grade. Shierlaw himself told the Rae Committee: "Anything past two percent is economic grade".

In fact the 3.56 percent assay stated in the October 1 report was wrong. The real assay for that intersection was 3.17 percent. The assay average announced at the annual meeting of December 1969 was 2.4 percent. In retrospect, nothing about Poseidon is more surprising than the way in which this figure was so gaily greeted in the euphoria of the time. In truth, it was an ominous portent.

Analysts' calculations on Poseidon rarely made any allowance for mine dilution or losses in treatment, which effectively reduce the recoverable grade of the ore. Mine dilution occurs as low-grade ore or waste rock (sometimes called country rock) is removed with the sulphide ore. The presence of this poorer material might reduce the average mill grade of contained nickel from 1.9 to 1.3 percent. The published data for the years 1974 to 1976 suggest that richer sections of Windarra were worked initially to provide average mill grade of 1.85 to 1.97 percent. This was before losses in concentrating which approximated 30 percent. In general, for an established sulphide nickel mine, losses in milling approximate 15 percent of the contained nickel and another three percent might be lost in smelting and refining.

Given these circumstances, Poseidon's best hope was to bring Windarra into operation as quickly as possible to catch the high nickel prices while they lasted. There was a clear precedent for this in the rapid establishment of Kambalda by Western Mining. However, Western Mining was a larger company with strong associations in the Collins House group. Poseidon was small by comparison and for all practical purposes had no big brothers. Its task in raising capital would be harder.

Simon Hunt, whose access to Poseidon information appears to have been second to none at this time, estimated that the capital cost to establish a mine and mill to produce one million tons of ore per annum should amount to $28 million, with a smelter costing a further $12 million. The Panmure Gordon circular said: "Finance is no major problem as the company will have received numerous propositions. At this early stage, development will probably be financed mainly out of debt-borrowing, linked with a small equity issue."

Panmure Gordon was wrong. Finance was to prove a major problem for Poseidon, although the concept outlined by

Panmure Gordon always appeared feasible. Let's look at the arithmetic.

How much is needed? On Panmure Gordon's figures, about $40 million. This figure could be lopped by $12 million if Poseidon dropped its plans for a smelter and sent its concentrate to Western Mining for toll smelting. This always looked a better proposition but Poseidon was always very eager to have its own smelter. In the event, the development of Windarra would cost more than $50 million without a smelter and the concentrate would be treated by WMC on a toll basis anyway. But for the purposes of the sort of calculations made in April 1970, let's take a capital figure of $40 million.

How much did Poseidon have? At June 30, 1970, Poseidon had a little over $2 million in cash and liquid assets. Prudence would dictate that this be retained for working capital and any cost overruns on capital items. It therefore needed $40 million.

How would Poseidon raise $40 million? This was the question that was never satisfactorily answered. Ideally, the company would have preferred to raise as much as possible in equity while giving away as little equity as possible. Let us say Poseidon managed to place 10 percent of the equity to a company at $100 a share, it would issue 250,000 shares and raise $25 million. In return for this direct slice of the action the new equity partner would provide a further $15-million loan to the company and the problems of developing Windarra would be solved in one swoop.

Obviously, to get any such deal off the ground would require a partner with considerable funds and would involve much haggling over the terms of the equity issue. No such deal was ever done. Poseidon received several offers, but all of the offerers wanted a controlling interest in Poseidon, i.e. more than 50 percent. The directors, having discovered a nickel mine, wanted to develop it as the base of a new mining house. Their dream was never realised. It is one thing to find a mine but another to develop it, and Poseidon's board were soon bogged down in financial negotiations which never reached a satisfactory conclusion.

Ganke's lawsuit dogged Poseidon. Shierlaw says he would have raised ample finance with a one-for-one share issue at $25 each but for the fight with Ganke, which prevented the board from making any general issue.[1]

This would have been an ideal solution, raising ample funds without incurring external debts. However, control of the company could have been thrown open. To finance his own rights-issue Shierlaw would have had to raise $3.75 million or relinquish some on-market. It would have been feasible if the rights of the placement-recipients to an issue had not been contested.

Instead, Poseidon dealt with offers from international giants. One which was particularly interested was the Anglo-American Corporation of South Africa. Oppenheimer, the chairman of Anglo-American, flew his private jet to Adelaide to have secret talks with the Poseidon board. The visit only became known when the jet was spotted parked in a corner of Adelaide airport. Poseidon also had talks with various overseas nickel producers. Inco and Falconbridge were both very interested. They were all prepared to meet all costs, including a smelter, but they all wanted over 50 percent. "Well over 50 percent," says Shierlaw.

In mid-1970 Shierlaw and Lewis set off on a trip around the world. They examined a flash smelter in Finland, refining operations in Norway and other examples of treatment around the world. They also negotiated with a number of overseas companies.

The purported highlight of the tour was a champagne breakfast in the old Great Eastern Hotel in London's Liverpool Street (a name familiar to Monopoly players). Shierlaw was widely expected to announce not only Poseidon's latest exploration results at this meeting, but also details of the financing. A total of 75 stockbrokers and financial journalists was originally invited to the breakfast, but eventually it was attended by almost everyone who could squeeze into the room and there was precious little champagne. Bruce Dunfield, a Canadian public relations officer whom Shierlaw had acquired en route, told one shareholder: "Just turn up — but you betta wear a suit of armor; it's gonna be a fight to get in."

Poseidon's statement, released simultaneously by the board in Adelaide, was a disappointment. No new figure was put on reserves, although detailed drill assays were given. The only clue to the expected size of the orebody was a statement that the board was considering initial production at a rate of 700,000 to 1.2 million tons annually. As a mine normally operates on a minimum of 10 years reserves, this indicated an orebody of somewhere over 12 million tons. Worst of all, after all his time negotiating in Australia and overseas, Shierlaw had no financing agreement to announce. For an occasion that was hoped to be a celebration this was very flat news.

The London Stock Exchange marked Poseidon back $3 to $118 on the report. It was now nine months since the announcement of the discovery. Poseidon had not yet used its mammoth share premium to finance the mine through an equity placement. Any such placement would certainly have to be made at a discount to market, but the premium would still be handsome. Consider the trading range of Poseidon during 1970.

Month	High	Low
January	$250	$190
February	$280	$212
March	$245	$140
April	$180	$66
May	$91	$52
June	$145	$64
July	$144	$77
August	$87	$67
September	$98.30	$73
October	$93	$65
November	$67	$54
December	$65.25	$39

In this time, particularly in the first six months, the opportunity was missed. The lower figures from August onward are partly the result of the capital being expanded by the takeovers of Samin Ltd and Kalgoorlie Lake View and Star Ltd.

The comparison with Western Mining is revealing. It took WMC only 16 months from its first drill intersection at Kambalda to bring its mine into production. WMC had certain advantages. Its first drill hole was so rich that there was no question about the mine's viability, whereas Poseidon's reserves took time to establish as the grade weakened in subsequent drilling. WMC had powerful and established associates in Collins House, whereas Poseidon had none. WMC financed Kambalda's development entirely from share issues, whereas Poseidon was forced to look for loans.

Eventually, Poseidon took five years to bring Windarra into production. This lethal delay was mostly due to the long negotiations to procure finance and sales contracts. Poseidon's directors wanted a sales contract so they could use it as a collateral for borrowings, thus enabling equity capital to be kept relatively small. Windarra would then be a highly geared mine in that borrowings would be high relative to shareholders' funds. Such a mine is only a viable proposition if its earnings stay high enough to service the interest and repayments on the debt over several years. The borrowings have to be repaid within the life of the mine — a period which will sometimes include times when metal prices are low. A mine financed by equity capital alone can shut down in lean years and incur only care and maintenance costs. Dividends to holders of equity capital can be withheld until profits return. But a mine operating on borrowed money has interest payments to meet whether the mine is open or closed. Such a mine has to be exceptionally rich to cover interest payment through bad years as well as good and Windarra was not a rich mine.

Anglo-American made a loan of $3 million to Poseidon, but its offer was considered to be "not in the best interests of Poseidon shareholders". The loan was repaid in 1971 when Poseidon entered into a joint venture agreement with Union Oil Development Corporation of the USA and its subsidiaries, Homestake Iron Ore Co Ltd and Australian Hanna Ltd. Union-Hanna-Homestake had an orebody at South Windarra of comparable size and similar mineralisation. The concept was that the partners would jointly develop the two mines. As part of the deal, Union-Hanna-Homestake guaranteed a $3 million loan from the ANZ Banking Group to allow repayment of the Anglo-American loan.

However, late in 1972 the Americans pulled out and sold their interests back to Poseidon for $7 million, comprising $3 million in shares and $4 million to be paid as a virtual royalty on the mine. Union-Hanna-Homestake was secured by a first charge over the South Windarra mine which they had previously owned outright. It was now three years since the original discovery and Poseidon still did not have a satisfactory financial agreement or an operating mine.

The delays were looking increasingly dangerous. Poseidon was missing the top of the nickel market. It was also entering a period of rising inflation, although the OPEC price rise was still a year off. With no mine, Poseidon had no income apart from interest earned on borrowed funds; nor did the company have a sales contract.

In September 1972, the chairman of Amax, Ian MacGregor, said Poseidon had lost a chance to sell nickel and have its smelter financed by Amax. MacGregor implied frustration as he spoke of "talking with Poseidon and their numerous partners from time to time". It must be said that Poseidon negotiated with a lot of companies without ever reaching a satisfactory or lasting agreement. Whatever the faults of the foreign mining companies, one is forced to assume that some proportion of blame for the various aborted negotiations must lie with the Poseidon board.

Then in October 1972 — three years after the discovery — Poseidon announced it had a sales contract for five years to Western Mining and Sherritt Gordon Mines Ltd of Canada. The Australian Industry Development Corporation agreed to provide temporary finance to allow the mine development to proceed. On January 3, WMC became a joint venturer at Windarra, taking up a 50 percent interest in the mines and financing further development.

By this time Poseidon had developed other problems. The first was Samin Ltd, which had proved to be a disaster. The company had been formed with the old Burra copper mine and its placement of Poseidon shares as the major assets. In August 1970, less than a year after Samin's float, Poseidon announced a takeover bid for the company. Lewis said the merger would eliminate areas of overlap and duplication between the two companies. What must also have been a consideration was that Samin's shares had slid as low as $8.80 during 1970. It was an increasingly tempting target for a takeover raider, because control of Samin would also bring control of its 200,000 Poseidon shares. These shares, placed for defensive reasons, had become a hostage to fortune. Poseidon's bid for Samin was a share swap, involving the issue of a further 540,000 Poseidon shares in return for the whole of Samin's issued capital. As Poseidon shares were worth around $75 on market at the time, this represented an outlay of $40 million worth of equity. But as Samin held 200,000 Poseidon shares which were returned to the company, this should be offset, making the net cost of the Samin purchase $25 million in Poseidon scrip. An equity issue of this size would have been far better used to raise finance from a major financial institution or mining company.

One justification for the takeover was that Samin expected to earn profits of $2.2 million before tax on its mining operations alone, meaning Burra. In fact, Burra turned out to be a running sore and a diversion of badly needed funds. Samin had already spent $2 million developing the old copper mine, the history of which went back to 1845. The plant was commissioned in 1970, when production began. Production setbacks were caused by the installation of the wrong filters in the plant, but output expanded in 1972. Burra lost $160,000 in 1971-2, $300,000 in 1972-3, $780,000 in 1973-4 and $2.8 million in 1974-5. At least $5 million was spent on this venture and more than $4 million in losses flowed from it.

Poseidon's other major takeover, of Lake View and Star, seemed like a good idea at the time. Poseidon acquired Lake View in late 1970 with the idea of using the old Kalgoorlie miner's plant to treat its nickel and the dam to provide its water. But when the price of gold boomed in 1970-1 Poseidon decided to sell the company's gold mining assets to an associated company. Then gold prices fell again and Poseidon had been saddled with the high cost of a new plant at Windarra.

This takeover cost Poseidon a further 150,000 shares. Thus the Samin and Lake View takeovers had together enlarged Poseidon's share capital by 30 percent.

It was not until September 1974 that production of nickel concentrates began at Windarra, five years after the discovery. Then in addition to its other woes Poseidon struck mine problems. Like many other large sulphide orebodies, Windarra was subject to vertical shearing. The ground was unstable and there was one nasty rock fall. Windarra was mined by sinking an inclined drive around the orebody in a spiral. The miners struck trouble early in the upper levels. The serpentinite was heavily faulted and caved in. The Kalgoorlie miners now working at Windarra were used to dealing with narrower orebodies. As an old Barrier miner, Shierlaw was sanguine about the problems of wide, unstable lodes. "We had mining problems and faulting later on," he said. "But the problems associated with mining were no more than you would associate with any new orebody."

The orebody was 60 to 70 ft wide, compared with an average six ft at Kalgoorlie or 20 ft at Kambalda. Western Mining eventually took over as mine managers.

Poseidon suddenly became aware that it had a problem. Low nickel prices were making Windarra only marginally economic. Poseidon now had large debt commitments and had been late to see the storm signals. It had maintained lavish offices and high exploration expenditure. Poseidon tried to sell its share of the mine to WMC but Western Mining refused to make an offer.

Then Poseidon canvassed other major Australian miners and still received no offer for its once priceless mine. The AIDC, as the major creditor, appointed a receiver. Had it been any company other than Poseidon a receiver would have been appointed much earlier.

Poseidon was delisted from the Australian stock exchanges on December 22, 1976. Seven years earlier it had jumped from $130 to $175 on the same date. The company's debts were such that there was no hope for shareholders. Poseidon's half-interest in Windarra was sold to Shell Australia for a reported $30 million.

But there must have been a curse on the mine because only five months later, in February 1978, Windarra closed. The dream was gone.

---

[1] Interview with author, June 1978.

CHAPTER FIFTEEN

NABARLEK: The Orebody That Shrank

The market continued to sag until the end of May 1970, when the non-ferrous metals index and the Patricks Speculatives Index bottomed. The cause of the June recovery in both heavyweights and speculatives was to some extent a reaction to the long fall in the market. By May 27 the non-ferrous index was down to 3782.23. At some point a rally was bound to occur, and this was the point. The rally was prompted by the news that the WA Government's four-month-old ban of pegging was about to be lifted. This had been imposed to help Mines Department officials clear up an avalanche of paperwork; now the staff had almost cleared the backlog.

On June 5 the ban was lifted and a new pegging rush started. Many field teams had camped overnight on selected sites and began pegging as soon as they heard the news on the radio. There were few clashes between rival teams, and the first claim documents were lodged at Leonora 2½ hours after the ban was lifted. The paperwork was starting again.

The renewal of pegging gave grounds for fresh optimism and lifted the market for a while; but optimism alone could not sustain the market and it was soon drifting again as investors waited for some good news.

The news drought was broken on August 12 when International Mining Corporation NL announced that surface samples from its Trough Wells prospect had averaged 1.5 percent nickel. IMC shares, which had been languishing in the market at 25 cents, having been as low as five cents early in the year, leapt to $1.90 on the news. Trough Wells, near Southern Cross in WA, was a good prospect but surface samples are an unreliable indicator of what lies below the surface. These samples were highly encouraging, and in a market starved for good news there was a sense of relief in the discovery of a new hopeful. IMC, despite its imposing title, was a small listed company with five million shares and five million options. It had the rare distinction of being chaired by a woman, Mrs Millie Phillips. The shares hit $5.90 at the end of September when IMC reported values of up to 3.74 percent in a percussion drill hole at Trough Wells. The share price soared to $7.20 and ran even higher to $9.50 in London. However, shrewder judges were holding off. They were worried by the absence of detailed drill assays because the 3.74 percent figure was meaningless unless the intersection width was known. When details were released on October 6, it was discovered that the 3.74 percent was over three feet. The first four drill holes had intersected relatively narrow widths of sulphides, varying from seven to 10 feet, with nickel values ranging from 0.65 to 3.74 percent. This was IMC's high point, and the shares gradually retreated from here. But by now the speculators had another game to play because IMC had been overshadowed by a spectacular new find.

On September 1, 1970, the mining world was astounded by a statement from the chairman and managing director of Queensland Mines Ltd, Roy Hudson, that his company had discovered 55,000 short tons of uranium oxide at the phenomenal average grade of 540 lb per ton at Nabarlek in the Northern Territory.

This made Nabarlek easily the richest uranium deposit in the world at the time. On one estimate, Hudson's announcement had increased the western world's known reserves by 10 percent. The announcement not only sparked a sudden leap in the price of the shares of Queensland Mines and its parent company Kathleen Investments Ltd, but also excited international attention.

A year later, on August 12, 1971, the mining world was astounded again — this time by an announcement from Queensland Mines downgrading the reserves and grades to about one-sixth of those announced 12 months before. This announcement came as Hudson was forced out of both companies in a bitter fight. The initial statement had been issued at a time when Queensland Mines had not received a single drilling assay. When correct assays had arrived three weeks later the company had made no attempt to correct the report. The original report had prompted thousands of investors to buy shares in these companies at inflated prices. Excessive buying of Queensland Mines and Kathleen Investments shares was the direct cause of the downfall of Mineral Securities Australia Ltd and was at least a major contributor to the collapse of Patrick Partners.

The character of Roy Hudson is central to the Nabarlek debacle. At 64, he was a tough, grey-haired veteran of 30 years in the rough-and-tumble world of Australian mining. He had owned several mines and explored for more in New South Wales, Queensland, the Northern Territory and the Solomon Islands. He had discovered the Savage River iron ore mine in northern Tasmania,[1] and the royalties had made him wealthy, with an annual income well over $100,000.

When Ric Dowling had formed Kathleen Investments in the 1950s in a fight to retain some Australian ownership of the Mary Kathleen mine in Queensland, Hudson had been a lawyer at Broken Hill with some experience in mining deals. Dowling approached him to handle the negotiations and legal work, which Hudson did very ably. Control of Mary Kathleen Uranium Ltd went to Rio Tinto Zinc Corporation of London, but Dowling and Hudson won their fight to retain a 35 percent stake in the mine for Australians. Uranium had welded an alliance between them, and uranium would split it asunder.

A veteran of the boardroom as well as the mining camp, Hudson was a shrewd old fighter and looked it. He was tenacious enough to retain control of both Queensland Mines and Kathleen Investments for a year despite mounting dissension with directors. It was not until he collapsed at a board meeting that other directors could gain control. They took over while he was being attended by a doctor in the next room.

Hudson was tough enough to force the resignation of Dowling from the Kathleen Investments board at a time when Dowling was the most powerful stockbroker in Australia. This happened at a tense, dramatic meeting of 500 shareholders in Sydney's Anzac House. Hudson commanded the stage throughout, adroitly handling any criticism from shareholders until Dowling, emotional almost to the point of tears, unexpectedly announced his resignation.

Hudson's implacable hatred of Dowling, after he learned that Dowling's associated companies had been heavy secret buyers of the stock, was a key factor in the events that followed. Despite Hudson's protestations of concern for the shareholders, he also had a passionately personal attachment to both companies. Trying to take control of them from him was like trying to take a cub from a tigress. The parties who tried to move in on the companies were savagely mauled.

Kathleen Investments was launched on stormy waters in 1958. At this time a company called Australian Oil Exploration Ltd held 35 percent of Mary Kathleen Uranium Ltd. AOE was nearly bankrupt in late 1957 and during a trip to London Dowling had heard the company was about to sell but to RTZ. On his return to Australia, Dowling sought the support of clients and associates to keep a stake in MKU (then a world-class uranium supplier) in Australian hands. Kathleen Investments was floated as the vehicle to buy the 35 percent. Hudson, a friend of three of the partners of Patrick & Co, was retained as legal adviser to the company and given a board seat. Sir John Northcott, invited by Dowling to be the first chairman of Kathleen Investments, died shortly after the company's formation. Hudson had been chief executive at the time and now became chairman also.

Queensland Mines was formed in 1959 with Factors Ltd as a 51 percent shareholder and Stanley Korman as chairman. After Factors failed, Kathleen Investments acquired the Queensland Mines shareholding in 1965. It became a wholly-owned subsidiary of Kathleen Investments, then was floated as a public company in 1967.

Both Kathleen Investments and Queensland Mines had enjoyed fair support during the Poseidon boom, mainly because of their excellent uranium prospects at Westmoreland in western central Queensland.

Kathleen Investments had an issued capital of 7.5 million fully paid 50 cent shares and 1.1 million 30 cent contributing shares. These were $6 and $4 on market respectively as the New Year of 1970 opened. Kathleen Investments at this point had the stake in MKU, some beach sand interests and an investment portfolio, but its share market value was largely predicated on its 51 percent holding in Queensland Mines. Queensland Mines, with only 4.9 million 50 cent fully paid shares and just under a million 20 cent contribs, was more tightly capitalised and its shares were $12 and $8.50 respectively at the start of 1970.[2]

Queensland Mines owned the uranium prospects at Westmoreland and Nabarlek. Westmoreland was later overshadowed by the large rich uranium deposits to be found in the East Alligator Province of the Northern Territory, but at the time was ranked as one of the most promising finds in Australia. Nabarlek, sometimes called Oenpelli after the nearby mission, did not begin to attract attention until April 1970.

In April Queensland Mines began an airborne spectrometer and magnetometer survey of Nabarlek. On May 31 the survey revealed a number of anomalous areas "and a very prominent cluster of strong anomalies" at the area later called Nabarlek. On June 22 Hudson told the board that a quick perusal of the survey records indicated "an outstanding anomaly",[3] where the grade had exceeded the measuring capacity of the instruments. This was welcome news because the companies' shares had been subsiding as the boom ran out of steam. By mid-1970, Kathleen shares were down to $3 and Queensland Mines to $5.50. In six months the market had wiped out half their capitalisation.

On July 3, Queensland Mines made its first public announcement about Nabarlek, saying that assays of surface samples showed a very high uranium content and that it could be in shareholders' interests to wait for further information from the company. Queensland Mines had already begun an intensive campaign of radiometric gridding, trenching and sampling.

The results were spectacular. The 12 foot deep costeans (trenches) were being dug by backhoe and as it sunk through the soil it began to reveal the rich yellow and orange ochres of pitcheblende, sometimes only four feet from the surface. Excitement mounted until the fifth costean, designated N-5, cut through a 2½ foot reef of pitcheblende assaying an incredible 72 percent uranium oxide. Canadian and US mines at the time were working uranium ore grades as low as two pounds to the ton, or 0.1 percent. This reef had assayed the equivalent of 1440 lb to the short ton. This was phenomenally rich. The ore within a distance of 30 ft was highly mineralised, ranging from 10 to 500 lb per ton. An internal managing director's report by Hudson on July 23 said the strike of the lode would appear to be 900 ft with a possible extension under a laterite covering to the south.

On the same day, diamond drilling had begun at Nabarlek. This must have been an intensely exciting time for the field geologists and the directors. If they could prove a sufficient quantity of this amazingly rich ore, they would have a discovery of world significance. At this stage the boards of Queensland Mines and Kathleen Investments were identical. Hudson was chairman and managing director of both. Other directors were Dowling, John Roberts, Harry Ferguson and Adelaide stock-broker C.P.Tilley. Roberts was a director of Castlereagh, the Longreach group and other companies closely associated with Patricks. Tilley was one of the original participants in Kathleen Investments, being the former chairman of AOE. Ferguson was a director of Pioneer Concrete Services Ltd and various companies including Longreach Oil, Cultus Pacific and

Abrolhos Oil. Ferguson was alleged by Hudson to be one of the 'Patrick bloc' on the board. Ferguson was certainly a director of companies with close links to Patricks, but if Hudson's allegation were true Dowling, Roberts and Ferguson would have constituted a majority on the two boards. But Hudson had firm control of the board at this time and did not lose control until three months after he had forced Dowling and Roberts to resign.

Following the exciting field report of July 23 Dowling, Roberts and Ferguson decided to visit Nabarlek. Queensland Mines' senior geologist, Dr Emile Rod, went with them. Three other people on the trip were Tristan (later Sir Tristan) Antico, chairman of Pioneer Concrete, Jim Millner, chairman of Washington H.Soul Pattinson & Co, and Jim Hohnen, a West Australian investor and mining engineer. Antico, Dowling and Millner were all directors of Castlereagh Securities Ltd, Patrick's recently formed investment company.

Hudson had been invited to go, but had declined. He later said he did not like the idea of people other than directors visiting Nabarlek but since the others were being taken by directors he felt he had no right to object.

The party visited Nabarlek on August 27 and were most impressed by what they saw. "Undoubtedly visual reception is a thing that does affect you very much," Dowling said later. "I must admit I was terribly impressed with it." That night the party stayed at Mudginberri, Antico's pastoral property nearby. Antico sent a radio-telephone message to his secretary from Mudginberri ordering 10,000 call options in Queensland Mines. Hohnen subsequently bought 5100 QM shares and 1000 options plus 800 shares and 2000 options in KI. There is no doubt they were all impressed.

There is one other event of August 27 that should be recorded. On that day Hudson's wife died. He was described by close associates as "very down" at the time. It was during this time that crucial events occurred within the company.

August 28 was a Friday. Turnover in Queensland Mines shares rose fivefold. It is hard to attribute this rise in turnover to the touring party. They had not been in direct contact with anyone and were now on their way home. Antico had put through a call option order and Hohnen's buying did not begin until the following Monday. However, word may have been filtering around that a high-powered party had been impressed by a visit to Nabarlek. Hudson himself testified later to the speed with which news spread. He told the Rae Committee he had heard on August 28 that Antico had ordered 10,000 Queensland Mines options the day before. Hudson said he had been informed that a telegram had been sent through the Katherine Post Office. Katherine is the nearest town of any size to Mudginberri. Thus Hudson learned, within a matter of hours, the exact details of a broking order sent from a remote Northern Territory cattle station. If this information had leaked to Hudson, one can be quite sure it had leaked elsewhere as well. Hudson told the committee he did not know who gave him the information.

On Monday August 31 the Sydney Stock Exchange asked Queensland Mines if directors knew of any reason for the rise in share trading volume. Hudson called in Dr Rod, who was now back in Sydney, and asked him to prepare a statement for the exchange by noon the next day. Next afternoon, after some discussion, Hudson dictated a statement with Rod's approval and sent it to the exchange. This statement — wrong and with inadequate foundation — was allowed to stand uncorrected for nearly a year. In view of its devastating results, it is worth reprinting in full:

"The directors of Queensland Mines Ltd refer to their report to the stock exchange dated July 3 last advising the discovery of a uranium anomaly named Nabarlek situated some 18 miles from the Oenpelli Mission and 170 miles east of Darwin.

"The anomaly is in the Archaean-metamorphic rocks with a strike length of 3600 ft and an average width of 300 ft. The uranium occurs within the anomaly in large lenticular pitcheblende lodes dipping at an angle of 50 degrees to the east-north-east.

"A two-foot thick core of the pitcheblende lode, surrounded by a thick crust of bright yellow uranium ochres, was exposed by costeaning at a depth of four feet below the surface and has been drilled to a vertical depth of 130 ft.

"Intersections by diamond drilling were made of massive pitcheblende with a maximum true thickness of 12 ft and of a grade of 1300 lb to the ton of ore. On each side of the pitcheblende core the metamorphics contain patchy pitcheblende giving a high grade mineralised zone of a combined average width of 28 ft.

"Of 10 diamond drill holes, 7 intersected the first lens with a minimum length of 400 ft, one was between the first and second lenses and was barren, and two drill holes intersected a second lens of patchy and disseminated pitcheblende of an average grade of 120 lb uranium oxide per ton of ore.

"The northern end of the first lens is still open. Drilling is continuing to the north and at depth.

"Drilling and costeaning of the first lens gives indicated reserves of 55,000 short tones of uranium oxide of an average grade of 540 lb per ton of ore." Yours faithfully,

Queensland Mines Ltd
E.R.Hudson,
Chairman and managing director."

The Rae Report described this announcement as "grievous misrepresentation". In the first place, the fact that drilling or drill holes are referred to six times gives the impression that the assays were based on drill cores. They were not. The company had not received any assay reports on drill cores at this date. Of the 10 holes referred to, only two had struck ore of 27 percent (540 lb) grade, both times over short intersections. Drill hole Na-1 had hit an astoundingly rich patch between 140 and 150 ft with values ranging from 13.2 to 48.7 percent. Hole Na-4 had struck 26.6 percent ore over 2.4 ft and the next 2.6 ft had assayed 36.7 percent. No other assays from these holes came close to the average grade stated by Queensland Mines. The assays for these 10 holes arrived at Queensland Mines three or four weeks after the statement of September 1. They were not released to the public for another year — until after Hudson lost control of the board.

A second criticism is of the use of the term "indicated reserves". In mining the terms "indicated", "proven", "probable", and "inferred", used in conjunction with reserves, have carefully defined meanings. At the time the definition of indicated reserves from the US Bureau of Mines and the US Geological Survey was: "Ore for which tonnage and grade are computed partly from specific measurements, samples or production data and partly from projection for a reasonable distance on geological evidence. The sites available for inspection, measurement and sampling are too widely or otherwise inappropriately spaced to outline the ore completely or to establish its grade throughout."

The Nabarlek estimate was not made on such data. In evidence to the Rae Committee, Hudson said he probably used the word "indicated" in the sense of "estimated" and went on to argue that the term "indicated" was used in its technical sense only by mineral producing companies, and not by exploration companies. The Rae Report squashed this contention, saying:

"We do not think such an argument...deserves to be taken seriously. The reaction of some experienced and qualified mining investors to the September 1970 announcements is sufficient to refute it, and Mr Hudson must have observed this response at the time. It is equally impossible to accept the proposed distinction whereby the untried and speculative class of mining companies, often possessing less experienced managements, are permitted and expected to use words more loosely and irresponsibly than established, producing companies. Under such a convention, the scope for double-talk would be unlimited."

A third criticism of the September 1 statement is that the reference to the open north-end aroused hopes that the ore reserves could be even greater than stated. These flames of hope must have been fanned next day when, in response to a stock exchange enquiry for more details, Hudson issued a further statement. This said, in part: "The orebody is minable by open-cut methods, is mineralised at the surface by high-grade disseminated pitcheblende mineralisation commencing from 4-16 feet below the surface. Mining costs would be small per ton of ore and are of no significance per lb of uranium.

"The area is accessible by road to Darwin during the dry season, but would not be accessible without further road construction during the wet season, and is well supplied with water. Having regard to the grade of ore, only a small treatment plant would be required and costs and infrastructure should not exceed $3 million.

"In estimating the reserves, the result from drilling of a second lens has not been taken into consideration as only two drill intersections have been made to date. The 7 drill holes which intersected the first lens were over a distance of 400 feet and to a depth of 130 feet. The lens is still open at the northern end and drilling is continuing."

These statements of September 1 and 2, surpassing the wildest hopes of investors, could only have one result. The price of the shares roared. Investors, bloody and battered since Tasminex, regained confidence. This was a world-scale discovery by an established and respected mining group.

Kathleen Investments, which had been $4.80 on August 31, jumped to $8.10 the following day and hit $15.50 by the end of September. Their peak was $17.50 in October. Queensland Mines had been the subject of more speculation. From a low of $7.20 in August, these shares had risen to $11.20 on the eve of September. By the end of September they had vaulted to $43.

Their peak was $46 in early October.

It may or may not be a coincidence that the peak in the share prices followed very closely on the receipt by Queensland Mines of the first drill assays from Nabarlek. The cores from these drills had been analysed by Amdel in South Australia. There is some uncertainty over the date Queensland Mines received the assays, but they seem to have been given to the company by telephone on September 22 and actually delivered to the office on October 2. The shares peaked a few days later. The price stayed high for a long time, drifting downward only gradually; but by the end of December Kathleen Investments were back to $11.80 and Queensland Mines had been pared to $30.

These price movements were of great importance to Australia's two most successful boom-time institutions — Mineral Securities Australia Ltd and Patrick Partners. Australia had not seen such successful entrepreneurs as Minsec in living memory. By December 31 Minsec was locked into a shareholding that would destroy it.

In mid-August Castlereagh Securities had bought 15,000 QM shares. Patrick Corporation Ltd, the other investment company associated with Patrick Partners, bought 14,400 at the same time. This was after the 72 percent costean sample but before the Dowling-Antico visit to Nabarlek. As Patricks had formed Kathleen Investments and Dowling was a director and shareholder, it was natural that Castlereagh and Patrick Corporation should have invested in the Nabarlek companies. Castlereagh sold 5000 of its QM shares on the rising market between August 28 and September 1. After the September 1 announcement Castlereagh decided to sell the remaining 10,000 QM and to begin buying KI shares, aiming at an eventual total of 500,000. On prices of a little over $8 for Kathleen at the time this meant Castlereagh was prepared to commit at least $4 million to acquire about six percent of the company. The switch by Castlereagh from QM to KI was strategic and long-term. QM shares were only trading counters. Ultimate control of the deposit lay through KI.

Minsec also jumped into the ring, starting a buying campaign in QM on September 2. Castlereagh and Minsec quickly became aware that they were bidding against each other in a bull market. On September 15 they agreed on a joint buying effort, in which purchases of Kathleen Investments shares by Minsec and Castlereagh would be coordinated. All buying would be channelled through Patrick Partners and split equally between the two parties.

This joint buying led to the formulation of a concept for a company to be called Power Resources of Australia Ltd, which would hold strategic parcels of major energy stocks. The Kathleen Investments parcel would be put into this company, which would ultimately be floated publicly. Castlereagh and Minsec decided to put a coal stock in too and chose Thiess Holdings Ltd. They also contemplated an oil stock.

In conformity with this policy, Minsec and Castlereagh widened their buying to include Thiess shares. Their aim was to buy 1.8 million Kathleen Investment shares, representing 20 percent of that company, and 1.4 million Thiess, representing 12 percent of the big Queensland coal miner. By December 31, 1970, Minsec held 746,908 Kathleen, 428,635 Queensland Mines and 22,780 QM contributing shares. The cost for these was $11 million for the Kathleens and $16.6 million for the QM, a total of $27.6 million.

This had all been financed by borrowings. Minsec's average costs had worked out at $14.67 for Kathleen, $37.08 for Queensland Mines fully paids and $32.95 for the QM contribs. By New Year's Eve they were taking a paper loss of more than $5 million on this portfolio. Castlereagh's buying of Kathleen had been the same as Minsec's, which meant they were taking a paper loss of $2 million on their portfolio by then. As can be seen from the prices, much of their buying had been at the top of the market. Indeed, their buying was largely responsible for the high prices.

The scale of buying was reflected in market turnover. Queensland Mines shares had turned over at about 70,000 per month in Sydney in July and August. In September, turnover leapt to 582,000 shares, followed by 147,000 in October and 171,000 in November before lapsing to former levels in December. Kathleen Investments also had a turnover around the 70,000 mark in July and August, but climbed to 1.3 million in September, 502,000 in October and 350,000 in November. On these figures, the bulk of Minsec and Castlereagh's shares were bought in September and October in the peak market. Their joint buying accounted for two-thirds of all trading in Kathleen between September and December and Minsec had accounted for nearly half the QM purchases. All these shares were bought on the false premise that Queensland Mines had 55,000 tons of 540 lb to the ton ore.

It was not until August 12, 1971, that the new Queensland Mines board and the public learned that the reserves amounted to only 8932 tons, some of which was at 16 lb per short ton and some at 240 lb. This was only one-sixth of the amount stated as indicated reserves a year before. The news shocked financial circles, despite persistent rumors that the mine might be downgraded. A greater shock was the disclosure that the original announcement had not been based on any drill assays.

The September 1 statement was prepared by two men — Hudson and Dr Rod. Hudson, although he had spent half a lifetime in the mining industry, was not a qualified geologist. Dr Rod had obtained a Ph.D. in Berne, Switzerland in 1937 in geology, mineralogy, petrography and chemistry. He did not have any other degrees, because in Switzerland it is possible to get a Ph.D. after six years study, without acquiring other degrees first.

Having obtained his Ph.D. at the age of 25 Rod joined Royal Dutch Shell in the Hague. In 1938 they sent him to Guatemala for two years and from there he was transferred to the jungles of the upper Amazon in the backlands of Ecuador. He was sent to the Dutch East Indies next but had been in Java only three months when the Japanese occupied the island. After the war he took a six-month rest in Switzerland before Shell sent him to Peru then once more back to Guatemala. In 1948 Shell were going to send him to Borneo, but Rod had had enough of the Dutch East Indies and joined Atlantic Richfield, who sent him to Venezuela.

He worked for that company in Turkey, Bolivia and the Spanish Sahara before being transferred to Australia in 1962 to work first in Gippsland and then in Brisbane. In 1968 he left Atlantic Richfield to join Queensland Mines.

Rod's experience had been mainly in petroleum geology, but like all geologists he was expected to keep an eye out for any minerals that might interest his company, especially in the Spanish Sahara and the upper Amazon. When he joined Queensland Mines he was widely travelled, with 30 years experience as a geologist, but he had no experience in uranium mines.

On September 1, Queensland Mines had not been asked to calculate its reserves. All the Sydney Stock Exchange had asked for was the reason, if any, for the rise in trading in the stock. On receiving the query Hudson asked Rod to make an "assessment" of Nabarlek. Rod took this to mean a calculation of reserves.

On this point the Rae Report said: "The stock exchange had not, as should have been obvious to Mr Hudson, asked

Queensland Mines to pronounce on the size of its estimated ore reserves, let alone to issue a wildly speculative 'guesstimate' ... No outsider was setting a time limit on the company converting the preliminary encouraging announcement of July 3 into definitive terms. If specific information was not available on September 1, the company was entitled, and was required, to say so. It could well have said that it was awaiting drill assay results from the laboratories."

Instead, a reserve calculation was made. The calculation of ore reserves is a technical matter involving geological assumptions and a detailed explanation of it is difficult for a layman to comprehend. But as Dr Rod's assessment of Nabarlek was the subject of so much subsequent criticism it is only fair to give a brief outline of his methodology.

The information he had available was the geological mapping of the area, the assay results from 15 costeans and spectrometer checks on some of the first 12 drill-holes. Split cores of the first 12 drill-holes had been sent to Amdel for analysis and those assays were not yet available. However, it is possible for a geologist to get a reasonably accurate reading of uranium from the cores on site. With high grade ore it is possible to pick the bright ochres of massive pitcheblende by sight.

When Dr Rod visited the site on August 27 he had a light German geiger counter called a Berthold-1200. Once this instrument is calibrated to allow for the effect of other radioactive minerals, a geologist can use it to make reasonable estimates of uranium oxide levels in a core sample up to a level of 50 percent. Dr Rod knew already that the Nabarlek ore contained no thorium and hardly any potassium (two other radioactive minerals often found in association with uranium) so he could calibrate the Berthold to take readings for uranium.

Armed with these spectrometer readings, Dr Rod began his calculations. He made a model dipping at an angle of 45 degrees, which was the slope of the mineralised zone, toward the east. He took an 18 ft zone near the surface averaging 15 percent uranium oxide for 14 ft and 70 percent for four feet. He knew that the average for the mineralised zone in the drill intersections was 38 ft so he averaged the width as 28 ft (halfway between 18 and 38 ft). The ore grade he calculated was the average of 14 ft times 15 percent and four ft times 70 percent, which works out at 27 percent. This is equivalent to 540 lb per ton. Working on a 28 ft average thickness over a strike length of 400 ft, he calculated that the ore contained 55,000 tons of uranium oxide. In fairness to Dr Rod, he had made an attempt at conservatism by taking a 400 ft strike length at a time when mineralisation was known to be extended over some 700 ft.

The weakness in this calculation is the assumption that the amazingly rich 70 percent grade could be taken as typical of grade. However, I do not know enough about geology to say whether Dr Rod's calculation was soundly based. Nor did the Rae Committee, so they hired the Canadian-based firm of consulting geologists, Watts, Griffis and McOuat, to review the data on which the Queensland Mines statements were based and to comment on their validity.

The Watts, Griffis and McOuat report, prepared by Jack McOuat and Terence Willsteed, was presented on October 20, 1971, to the committee. Unfortunately it is too long to reproduce in this book.[4]

The McOuat report was a scathing indictment of the original Queensland Mines statement and the company's subsequent silence. McOuat and Willsteed said Nabarlek should be a "relatively small, high-grade profitable mine" and agreed that with some minor criticisms the field programs seemed to have been conducted normally. The exploration program was described as "workmanlike". Apart from that, it had little good to say about the company.

"The September 1 and 2 statements were not supportable by technical evidence and were merely geological 'guesstimates' at best," the McOuat report said. "They should not have been made available to the public. Any independent geological examination would have disclosed the weakness in the statement and its assumptions.

"...As of September 1, the company had made a significant and high grade surface discovery of uranium. An analysis of surface assays and other information showed possible dimensions of the deposit of 700 ft in length by 60 ft average width on the surface, which averaged 1.2 percent uranium oxide. They also clearly showed the mineralisation was erratic in grade and width."

The 72 percent costean sample, which had been given such an important weighting in Dr Rod's calculations, appeared to have been a selected grab sample and could not be considered representative.

"Geological descriptions of the drill cores from the first 10 holes completed showed that while the zones persisted to at least some 130 ft below the surface, it was still very erratic in both shape and mineral content, and that there could be barren or very low-grade areas within the zone.

"Except as geological 'doodling' there could not have been a tonnage and grade estimate made, and even the estimate that was made and published appeared to ignore the technical data that was available at the time."

McOuat and Willsteed said the calculation of reserves or grades could not have been possible on the data available at September 1 and the statement could not have been founded on fact or even reasonable geological assumptions.

The McOuat report by its nature was an exercise in hindsight. But whatever the merits of Dr Rod's original estimate, they were clearly changed when the Amdel assays of the first drill holes came to hand. The arrival of these assays also meant that the company was now in a better position to calculate its reserves than it had been on September 1. Why did not the company immediately alert the market that the fabulously rich grade would not be attained?

Hudson later defended his action by saying that each drill hole had varied materially and until the end of June 1971 he had accepted the advice of Dr Rod and geologists working under him that the original estimate of tonnage would be maintained. In a statement in August 1975 (made after the section of the Rae Report on Queensland Mines had been released) Hudson said: "On October 7, 1970, I visited Nabarlek without Dr Rod and accompanied by a senior Government geologist. I made all the information available to him and he informed me that in his opinion, the tonnage estimate was correct, but that the grade would be lower than stated. I was not materially concerned with the fall in grade because of the high nature thereof. When doubt arose about the reserves position, Amdel was commissioned to make an independent report, and I assigned members of the staff of Kathleen Investments to report also."

Hudson said he still believed the board made the right decision in all the relevant circumstances, in the best interests of shareholders, to defer any further statement until drilling was completed and a final assessment of reserves could be made.

In evidence before the Rae Committee, Dr Rod said he was disappointed to see the results, which were much lower than he had thought, but he did not consider it necessary for the moment to recalculate the reserves. He thought that before any revision the company should drill many more holes to have a much better picture and do some statistical research to find out the average grade. This led to the following exchange:

Senator Rae: Do you agree that from the time either in late September or early October when the assay results were received from Amdel all the assay results which you received tended to take away from the original estimate of the reserve which you had calculated for the purpose of September 1?

Dr Rod: That is correct. What actually appeared was the grade. As long as you can compensate with more tonnage of ore you are safe. I was watching this closely, and I think it was in May that I made an estimate. I said: "Well we certainly do not have a lot more than 25,000 or 30,000 tons proved." Based on the drilling results I was quite aware of this in May. I knew, though, that unless further drilling proved more tonnage and better grades, and unless we could find more ore to the north or along the southern extension we could not make it. But I was very confident. I had actually no doubts that eventually we would make it. I was still full of hope that the 55,000 tons would be here.

The expression "full of hope" struck Senators as less than scientific. Replying to a further question Dr Rod said: " 'Hope' in this sense, for the geologist, is a strong desire, a strong expectation to find something." Dr Rod's hopeful stance was not shared by the McOuat report which said that no statement of material facts to rectify the original announcement had been made until August 12, 1971, although data demonstrating the need were available, at the latest, in early October 1970.

The report goes on: "Incredibly, no effort seems to have been made by the company to check these guesstimates at any time until mid-1971. No apparent inspection of assay data or drill logs seems to have been made. If such an inspection were made it would have become immediately obvious to a technical person or even a person experienced in arithmetic that the average grade and tonnages publicly stated could not be achieved. This should have been immediately obvious upon an inspection of the assay data available up to hole 10, hole 23 or hole 47.

"In our opinion the geological logs of drill holes Na-1 to 10 showed conclusively, *even before the assay became available,*[5] that the deposit was erratic and could not be reasonably expected to have the uniformity, grade or tonnage stated or implied.

"Even if by wishful thinking it was hoped that the ultimate tonnage could be developed, it had to be geologically obvious that by drill-hole Na-23 it was then at best a remote possibility."

Hole Na-23 was important. Completed on October 6 and sent to Amdel for analysis on October 14, it had been put down only a few feet away from Na-1, where the richest strike had been made. The best intersection in Na-23 was 1.5 ft of 1.3 percent uranium oxide, compared with grades of up to 48.7 percent in Na-1, only 15 ft away. Also when Na-23 was completed it confirmed, together with the rest of the drilling program, the closure of the mineralised zone.

This was equally important. The amount of uranium contained in the Nabarlek orebody was calculated on two factors — the tonnage of ore and the grade. Dr Rod had reached a figure of 55,000 tons of contained uranium oxide on the basis that he had 199,000 tons of ore averaging 27 percent. (Actually this comes to 53,730 tons, but this is a minor discrepancy.) Any drop in grade would have to be matched by an increase in ore to maintain the amount of uranium oxide in the deposit. If the grade dropped to 10 percent, there would have to be 550,000 tons of ore to still make 55,000 tons of contained uranium oxide, for example. This equation prevailed down to the cut-off grade where the ore was too poor to be mined economically.

By itself the rich grade was not important. It made little difference to the economics of the mine whether the grade were 540 lb or 100 lb to the ton. But if any dilution in grade was not matched by an increase in tonnage, reserves would be reduced. As long as the deposit was open in any direction there was room left for hope, but once the mineralised zone was fully delineated hope would be supplanted by arithmetic.

However, all these doubts were in the future as the market euphorically pushed Kathleen Investments shares up in September 1970. The scale of buying was such that on September 17 Hudson hurried to Canberra and persuaded John Gorton that foreign investment in Kathleen Investments and Queensland Mines should be limited to a total of 15 percent, with no more than five percent hold by any single foreign party. Gorton, being a dedicated economic nationalist, agreed and introduced the Nabarlek Ordinance in those terms. He was able to take this action because Kathleen Investments was an ACT-registered company.

By the end of September there were two big secrets about Queensland Mines and Kathleen Investments. The first secret was that Narbalek's reserves were not as high as stated. From the McOuat report, this should have been evident to Hudson and Dr Rod from September 22. But it was not known by the investing public nor even by Hudson's fellow-directors.

Harry Ferguson told the Rae Committee: "Mr Hudson has consistently withheld information from the board, and he did not inform directors that the first assay results indicated a marked downgrading until these results had been in his possession for almost 12 months. Not only did he refuse access to geological reports, but he presented to the board only his own interpretation of them."

The second secret was that Dowling and Roberts, as directors of both companies, were also part of a group which was in process of buying a substantial interest in those companies without informing either Hudson or the other directors. Dowling and Roberts were both directors of Castlereagh, which was collaborating with Minsec in an effort to acquire 20 percent of Kathleen Investments. Originally Minsec had suggested buying 51 percent of Kathleen but this had been ruled out as impractical. Twenty percent would have made Castlereagh (or Power Resources of Australia) the largest stockholder in Kathleen, although its voting clout would have been reduced substantially by the sliding scale of voting under Kathleen's articles of association, unless it split its shares.

The Rae Committee believed Dowling and Roberts should have disclosed details of their buying to the Kathleen Investments board. On this point Dowling said: "So what would the board do with the information? Would they circulate it to shareholders? I do not know what they would want it for."

Ingenuous, to say the least. Any company is interested in a major change in its ownership and it is particularly desirable that substantial changes in directors' shareholdings be known fully and promptly. It would be hard to imagine Ric Dowling taking the same complacent view if he were chairman of a company where other directors were making a similar undercover raid on the shares.

A different pointer to Dowling's attitude comes from the report into the affairs of Mineral Securities by the inspectors appointed by the NSW Government, called the Rath report.[6] Ken McMahon, in evidence to the investigators, said Minsec had been anxious in December 1970 to reveal the joint purchases of shares. McMahon said: "Mr Dowling agreed, but requested Mineral Securities to hold off as Kathleen Investments were to have a board meeting in a few days time and would have to make its regular quarterly or half-yearly report; Mr Dowling said that if Mr Hudson resented the moves which were going on he was quite capable of issuing a report that would be detrimental to our proposals."[7]

Later McMahon said: "Mr Dowling said that if Mr Hudson knew before we had a pretty substantial percentage and disapproved, he would quite possibly throw a spanner in the works."[8]

This had become a high-flying, cat-and-mouse game because Hudson was already suspicious. On his own evidence he had been suspicious since the proposed visit by Dowling's party to Nabarlek on August 27. He told the Rae Committee: "One of the purposes of the visit may have been to try to acquire an interest in Kathleen Investments. I have no basis for my reason but you asked me why I was suspicious. They are the facts as I knew them that made me suspicious."

Hudson later said he became aware of Castlereagh's buying as soon as he saw the price of the shares rising. He made inquiries and found they were being registered in the name of Patrick Nominees. This did not necessarily mean they were being booked for Castlereagh, but Castlereagh had only been floated a few months before and still must have had a lot of cash on hand and four Castlereagh directors had just visited Nabarlek. A far less astute man than Hudson could have come to the same conclusion. On his own evidence, Hudson's suspicions were in existence before September 1, although Castlereagh had only bought a few QM until then.

A detached observer could have derived a certain amusement from the subsequent board meetings. Hudson's suspicions were tantamount to certainty and he was suppressing geological information from the board. Dowling and Roberts were suppressing news of their share purchases from Hudson. One wonders what other subjects there were to talk about.

If Hudson suspected Castlereagh of buying as early as August, it is worth asking why he went to Gorton on September 17 with fears of foreign buying. Perhaps he feared Castlereagh would resell overseas. In fact, the Nabarlek Ordinance became a very large problem for Minsec because it limited the number of buyers to whom they could sell when they ran into trouble and tried to unload their KI and QM holdings. Hudson was never formally told of the PRA buying until McMahon saw him at the end of January 1971, when Minsec was only days away from collapse.

When McMahon saw Hudson, an uneasy suspicion was spreading about the validity of Nabarlek's ore grades. It had been spreading for some time. On March 25, 1971, Minerals Week published an article which said: "Japanese companies such as Mitsui & Co and Mitsubishi Shoji Kaisha had conducted their own surveys of Nabarlek and claimed that the amount and quality of the newly-found ore deposits are only about one-half to one-fifth of those reported by Queensland Mines".

When this article was shown by Roberts to Hudson at a board meeting in early May, Hudson told directors the statement was completely incorrect. Later Hudson told the Rae Committee that a Mitsui representative had seen the orebody on October 19, 1970, but no one from Mitsubishi had seen it. Hudson said Mitsubishi acted for RTZ and were competing with Queensland Mines for sales in Japan.

An interesting point in this piece of testimony[9] is that Hudson said the Mitsui representative saw drill results up to hole Na-10. This was more than the Australian public saw for another 10 months, but on that basis Mitsui apparently came to a far more accurate conclusion about the size of the orebody than Dr Rod. It also provides justification for McOuat's assertion that on the basis of holes 1 to 10 the deposit could not reasonably be expected to have 55,000 tons.

The Japanese are excruciatingly thorough in their investigations of potential ore suppliers. Queensland Mines would never have had a chance of concluding a supply contract to any Japanese uranium buyer without first releasing to them every detail of its orebody.

Others visiting the orebody reached similar conclusions to Mitsui. It has never been previously published, but Noranda and Patrick Partners both sent geologists to inspect Nabarlek in November 1970.

Noranda sent a small team. Patrick Partners were represented by their own geologist, Lionel Milligan, accompanied by an independent consultant, Geoff Hannes.[10] Hannes was a former general manager of Queensland Mines and had been largely responsible for the company acquiring the Nabarlek property.

The Noranda and Patrick teams looked at the drill cores and results and received an unpleasant surprise. By that time the limits of the orebody had been fairly well defined. They found that the geometry of the orebody had been satisfactorily outlined (the McOuat report later agreed on this) but that the

540 lb grade had been extrapolated from limited early knowledge and that there was now abundant evidence to refute it. The more drill intersections you have in an orebody, the less influence any single intersection has on the weighting of averages.

Milligan estimated that the orebody contained 10,000 tons of uranium oxide at worst and 20,000 to 25,000 at best. He thought 10,000 the more likely figure. Hannes thought Milligan's assessment harsh and that it was more probably 25,000.

Noranda's geologists had been to the site a little before the Patrick Partners team. After the Patrick team returned they quietly compared notes. The estimates were almost identical, Noranda having estimated 9000 to 25,000 tons. Thus three visitors to the site — from Mitsui, Patricks and Noranda — had all estimated considerably lower reserves than Queensland Mines had published. Their estimates were based on visits within three months of the September 1 announcement.

On his return Milligan reported his findings to the head of Patricks' research department, Michael Baume. Both understood the gravity of the situation because they knew of the heavy buying in Kathleen Investments. They reported their findings to Dowling.

Dowling took no action on this report, which must have reached him by early December 1970. By that time Castlereagh and Minsec had basically stopped buying and the share price was weakening. Despite his inaction, Dowling could hardly have ignored the report. It was too vital to his own future.

In evidence to the Rae Committee[11] Dowling said he did not think any of the staff of Patricks had visited Nabarlek between September 1970 and May 1971. His memory must have slipped because this visit was in November 1970.

Upon receipt of this information Dowling was also in a position where, if he did not have a conflict of interest, his various responsibilities certainly sat uneasily with one another. As a director of Castlereagh he had a duty toward its shareholders and an interest in seeing the value of its major investment maintained. As a director of Kathleen Investments and Queensland Mines he had a responsibility to see that the companies' public statements were accurate and not misleading. As a stockbroker he was responsible for advising clients on whether to buy, sell or hold the stocks. As a member of the Sydney Stock Exchange committee he had a responsibility, if he suspected a company was seriously misleading the public, to protect the larger public interest by seeking information from the company. It is difficult to see how anyone could have satisfactorily reconciled these divers responsibilities.

As a director of Queensland Mines, Dowling could have pressed for more information. Asked whether he was happy to be kept "in the dark" about Nabarlek, Dowling told the Rae Committee he had been very concerned, "but the board meetings were advised that it was drilling up according to the program, which meant the situation had not altered". Questioned further on this point, he said he did not see any reason why a company should not rely on its managing director and geologists. He said that at one meeting, in about October 1970, Hudson had told the board "that in his view there were 95,000 tons of uranium oxide".

In fairness, it should be recorded that Dowling later said (commenting on the Rae Report) that while he had been involved in different interests, actual conflicts among his own interests had not arisen. He said that he did not advise his clients when he was on the board of a company and that any conflicts of interest were easily resolved.

One point in Dowling's favor is that although he had a sizeable personal holding in KI, he did not sell any stock before or after his resignation, despite the growing doubts about Nabarlek's viability.

By November 1, geological evidence indicated to senior executives of Queensland Mines a definite closing off of the mineralised zone. On November 24 Hudson told the board that, by contrast with the early indications at Nabarlek, "some of the current drilling has shown that the pitcheblende has become disseminated on each side of the massive pitcheblende with consequential lowering of grade". This was the first evidence of Hudson advising the board of a downgrading of the ore.

According to board minutes, he concluded that drilling was not sufficiently advanced to give any real indication of reserves, but it indicated that reserves would still be 55,000 tons although the overall grade would probably be between 150 and 200 lb of uranium oxide per ton. On January 19 a further report from Hudson to the board said the indications were that the grade would be lower than 540 lb but there seemed little doubt of the tonnage of 55,000 being achieved. On February 5, 1971, Queensland Mines made its first public statement since September 1, saying that 47 drill-holes had been completed and affirming reserves of "at least 55,000 tons of uranium oxide". No mention was made of the assay results or the falling grade, except for a statement that "no calculation of proved reserves and grade is possible at this time". The report received scant analysis, because the day before Mineral Securities had been suspended by the Sydney Stock Exchange and the whole Australian share and money markets were in the grip of a major panic.

On March 17, the Minsec holdings were sold to Noranda Australia Ltd and the AMP Society; Terry Rodgers and E.C.Kennon were appointed to the Queensland Mines to represent the respective interests of these new major shareholders. Rodgers was managing director of the Australian subsidiary of Noranda of Canada. More importantly, he was a mining engineer and became the only director of Queensland Mines who was technically qualified to appraise the Nabarlek deposit.

The New Zealand-born Rodgers was a pleasant-mannered executive with greying hair. A former general manager of the Radium Hill project in South Australia in the early 1950s, he was an expert on uranium mines and his company Noranda was currently delineating the Koongarra deposit not far from Nabarlek. His hair was about to become greyer. He began his board appointment on relatively cordial terms with Hudson, but he developed doubts about Nabarlek and began to fight the chairman for fuller disclosure. This soft-spoken technician would eventually succeed, after much trauma, in ousting Hudson. Rodgers, who said he was "emotionally upset" after his calculations first revealed the drop in Nabarlek reserves in mid-July 1971, had a series of increasingly heated clashes with the redoubtable Hudson.

The next few months were tempestuous. On May 6 Hudson told the board there had been a further drop in average ore grade to 120 lb. On May 26 the matter was ventilated publicly when Tom Nestel told the Rae Committee that rumors had been circulating for some time that drilling at Nabarlek had not confirmed either the reserves or the grades claimed on September 1.

Two days later was the Kathleen meeting — one of the stormiest annual meetings in Australia in years. It had been public knowledge for weeks before that Dowling was under pressure to resign because of Castlereagh's secret buying of KI shares. Both Dowling and Hudson had been seeking proxy support in the days before the meeting. Dowling and Roberts, who were both facing reelection, sent a statement to shareholders supporting their actions. In a shock move at the meeting, Dowling announced his resignation after an emotional speech. He appeared close to tears as he spoke. The contrast with his usual image as a calm, self-assured top financier was startling. Roberts also resigned rather than carry the fight to a showdown. Any such fight would have resulted in victory for Hudson on the proxies. At both this meeting and that of Queensland Mines five days later, Hudson parried shareholders' questions about Nabarlek reserves.

For the moment, Hudson had retained dominance of the companies, but in mid-June 1971 Rodgers visited Nabarlek and began having his first doubts about grade levels and reserves. Rodgers was in an unenviable position. Noranda had bought a stake of nearly five percent in Nabarlek and he was its chief representative in Australia. If the reserves at Nabarlek were lower than stated, as he suspected, his company's investment would fall in value. If he were instrumental in forcing such disclosure he would be triggering a fall in the value of his own company's investment.[12]

He had only been on the Queensland Mines board a few weeks and was now on collision course with its formidable chairman. Rodgers is not a naturally aggressive person and must have winced at the prospect. Nevertheless, he went ahead.

He made a series of requests, by letter and at board meetings, for an updated reserves statement. Hudson resisted on the grounds that no new reserves estimate should be published until further exploration work had been completed. Their relationship cooled markedly. One board meeting lasted from 3.30 pm until well after midnight as Rodgers spoke to directors at length. Their letters, previously addressed "Dear Roy" and "Dear Terry", changed to "Dear Mr Rodgers" and "Dear Mr Hudson".

The climactic board meeting came on August 12, when the staff technical committee of the company, comprising Dr Rod and two other geologists, reported that reserves were not expected to exceed 9000 tons of ore. Rodgers pressed for an immediate statement to this effect. According to Rodgers, Hudson resisted this on the grounds that it would destroy confidence in the company and its prospects of negotiating sales contracts.

From this point, the normally dry minutes of a board meeting disclose a volcanic series of confrontations and maneouvres. Rodgers declared that Queensland Mines might be inadequately led by its chairman. Hudson fainted and was taken to another room. Other directors believed his illness may have been feigned. Queensland Mines' secretary Lincoln Madden, a longtime associate and staunch adherent of Hudson, left to get a doctor for his chairman. In his absence Ferguson assumed the chairmanship of the meeting. Rodgers resigned, saying, he considered "his presence on the board would undoubtedly contribute further to the deterioration of Hudson's health". Ferguson foreshadowed a motion appointing Millner to the board.

Then Madden returned and violently attacked Ferguson for resuming the meeting in Hudson's absence and for abrogating a previous agreement in moving the appointment of Millner. He also deplored the resignation of Rodgers at a time when the board needed solidarity. The board voted to appoint Millner despite Madden's strong objections.

On the same day, a Friday, Queensland Mines announced its downgraded reserve figure at the end of trading. On August 24 Hudson was removed as chairman of Queensland Mines and was replaced by Millner. Two weeks later Hudson resigned as managing director of both companies.

Hudson bitterly resented his overthrow and subsequently tried to regain power. He called an extraordinary meeting of Kathleen and tried to unseat the new board with a slate of directors headed by Sir Walter Michelmore. He failed — after yet another long and explosive meeting — but his bitterness was unassuaged.

One consolation was his share-trading profits. Hudson and his family company, Talbot Investments Pty Ltd, had bought 8500 QM shares between February and July 1970 and 5500 KI in April 1970 for a total of $67,000. After the September 1 statement Hudson and Talbot began selling. A total of 4000 Queensland Mines were sold on September 2 and 3. On September 24, two days after Queensland Mines had received telephone advice from Amdel of the first drill assays, 4000 KI were sold and 3000 QM.[13] By that time KI was over $15 and QM was $40. With 2000 more Queensland Mines sold on November 18, these sales grossed $355,000.

It must be emphasised that the trading volume was insignificant compared to Hudson's total holdings in the two companies. By the end of 1970 Hudson and Talbot's combined holdings totalled 95,000 KI shares and 28,000 QM; but his profits had been large. On his trading in the two stocks in 1970

Hudson had made a gross profit — before brokerage and tax — of nearly $290,000.

Hudson told the Rae Committee he had sold shares, after consulting his accountant, to raise money for other commitments, notably a tax bill totalling nearly $250,000. He said the only shares he could sell without incurring further tax were Queensland Mines and Kathleen Investments, as he had held equity in these companies for 13 years.

The Rae Committee concluded: "On each of these selling occasions Mr Hudson was privately aware of developments which widened the glaring discrepancy between the ascertained geological facts and the state of confident belief in the market to which he sold the shares. Each of the selling transactions coincided with an advance in his personal understanding of the discrepancy. Mr Hudson's explanation of the sales does not alter the grave impropriety of the share dealings. This is a case of 'insider trading' with a peculiarly objectionable twist. The person who made profits from his possession of the information that made a mockery of the market's belief in his company's shares was also one of the persons responsible for misleading that market for a period of nearly a year."

The committee had other criticisms too. The report said there was "a continuing vein of speciousness" in Hudson's arguments. On the Dowling-Hudson clash, it said: "This bitter fight left the interests of the shareholders disregarded and virtually irrelevant, and, accordingly, the shareholders suffered the consequences of the power play of the members of the two boards."

These and other criticisms hurt Hudson, but the sufferings of others were greater.

As 1970 closed, Mineral Securities had barely a month to live.

---

[1] When Hudson lost control of Kathleen Investments the new board disputed his right to the Savage River royalties on the grounds that he had been working for the company at the time. The company's claim failed.

[2] The gap between the fully paid and contributing shares of both KI and QM is sympotomatic of the market at the height of the boom. Investors, particularly large institutions and foreign investors, were wary of partly paid shares. Action tended to concentrate on the fully paids. In this chapter the prices quoted are those of the fully paids unless otherwise stated.

[3] Think of the earth as a magnet, with the lines of magnetic force running roughly parallel to the meridians. A magnetic survey plots deviations in these lines of force. Where they gather into a tight cluster they form what is called a magnetic anomaly, which is plotted in a very similar way to the isobars that show a cyclone on a weather chart. Such anomalies warrant further investigation by geologists because they may be caused by concentrations of metal in an orebody.

[4] It can be found on pp.2343-2382 of the transcript of evidence to the Rae Committee, Part II, Volume 2.

[5] Author's emphasis.

[6] The similarity of names with the Rae report is regrettable.

[7] Rath report, p.213.

[8] Rath report, p.214 para 680.

[9] Rae Committee transcript, Part II, Volume 2, p.2188.

[10] Brother of Jack Hannes, founder and chairman of Hanimex Corporation Ltd.

[11] Rae Committee transcripts, Part II, Volume 2, p.2013.

[12] As noted earlier, Noranda geologists had already assessed the orebody. Noranda must have bought its shares on the basis of their most optimistic projection, i.e. 25,000 tons. Noranda's decision also influenced the AMP's decision to buy.

[13] Rae Report, Part I, Volume 3, p.11.

CHAPTER SIXTEEN

The Fall of the House of Minsec

The collapse in Queensland Mines and Kathleen Investments shares had thrown a mighty spanner in the Minsec works.

In only five years of existence Minsec had achieved the most remarkable success of any company in Australia's history. Starting from scratch in 1965 with an issued capital of $300,000, it had acquired assets worth more than $100 million by the end of 1970. The stated philosophy of the company was to make profits from share trading and to channel these funds into long-term mining investment. Its success in both fields was such that by the end of five years it was easily the greatest share trader Australia had ever seen and had achieved international fame as a glamorous new mining group. By the end of 1970 Mineral Securities Australia Ltd controlled more than 20 percent of the world's rutile production and was the second largest tin and wolfram producer in Australia. It controlled Robe River Ltd, which had the largest stake in the Robe River project, which in turn was based on the world's largest iron-ore contract to date.

The rise of Minsec had been regarded by the market with admiration. The boldness of its large-scale share dealings earned it the awe of the whole marketplace. Being a new company it had no legacy of failure. Being a successful share trader it was admired by those who respected a top operator — and this respect was enhanced by the sometimes breathtaking audacity of Minsec's mammoth deals.

After it had been playing the market for three years, Minsec began ploughing its profits into direct mining investments by buying control of companies. The objective of securing Australian ownership of mining was universally applauded. Minsec looked like becoming the first major Australian mining house to be formed since the rise of the Collins House group in the 1900s. This has always been a popular cause in Australia where foreign investors control large sectors of the mining industry.

Sadly, all these shining virtues of Minsec would eventually become tarnished. Despite its reputation as Australia's number one share trader, its principals overstayed the bull market and were losing large sums on trading at the finish. Its largest ever market manoeuvre breached the fundamental commonsense of trading. In the long run, it aided foreign takeover of the mining industry because nearly all the mines it bought had previously been in Australian hands, whereas after Minsec crashed many of them passed into overseas control. It was a tragic end for a company which had excited such high hopes.

In its original stages, Mineral Securities was the brainchild of Ken McMahon. He was born at Coolah, about 250 miles northeast of Sydney, and his father was a storekeeper at nearby Leadville. Going to Scots College in Sydney, he took a French course and scored a dismal 25 percent. So he switched to the alternate course, which was geology. It's a sobering thought, but if McMahon had possessed any gift for languages, Minsec might never have happened.

He passed his Leaving Certificate in 1944 with first-class honors in geology, and joined the Royal Australian Navy. He served in the Philippines, Indonesia and Japan until being demobilised in 1946, but saw no action. ("My father used to say 'Another free cruise on the taxpayers?' after every voyage," McMahon said.[1]) After the war he made a false start in accountancy with Hungerford Spooner & Co but soon returned to his studies in geology. In 1950 he graduated as Bachelor of Science. He once joked he would still be there if he hadn't been coached by his wife, who was a year ahead of him.

His first job was with Colonial Sugar Refining as a mining surveyor. CSR sent him to their blue-asbestos mine in the arid Wittenoom Gorge of WA.

"We went there two days after our marriage," he said. "It was blistering hot — 115 degrees in the shade. The heat radiates from the ironstone in the area and makes the climate even worse. We had to eat our dinner sitting on dynamite boxes — empty of course."

After this romantic honeymoon, he spent two years at Wittenoom, then went to the other extreme working in a gold mine in British Columbia where it snowed for nine months of the year. Then he worked for American Smelting & Refining Co (ASARCO) in Washington, Idaho and Arizona as a shaft miner, shift boss, geologist and engineer. Returning to Australia he joined Commonwealth Mining Investments Australia Ltd in 1958, rising to the position of manager by 1961, when the British-controlled Consolidated Gold Fields of Australia took over the company. McMahon was sacked, apparently for being too independent minded.

McMahon was unemployed but he had accumulated a modest amount of capital. He also had mining experience, some important contacts and an idea. He had been strongly influenced by the concepts of Commonwealth Mining Investments, a Sydney-based investor in mining shares and situations, and wanted to create something similar. Although he subsequently established friendly links with CGFA the episode had given him first-hand experience of a foreign takeover. He would become a strong apostle of Australian ownership.

With the backing of Ken Craig and Reg Hare of the Metals Exploration group he formed a company called Kenneth McMahon and Partners Ltd. The company's name was a misnomer because in reality it was all McMahon. He held 20 percent of the shares while Craig and Hare were sleeping partners with 40 percent each. The company's funds, amounting to $10,000, were provided from overdraft accommodation by Craig and Hare. Eventually McMahon bought them out and owned the company entirely.

McMahon and Partners' business as a consulting geologist grew rapidly, with strong emphasis on South-East Asian countries. The work in these countries was unprofitable overall but McMahon was a strong believer in fostering mining in the area. The profitable end of the business was Australia, where McMahon and Partners worked for local firms and a large number of overseas companies. Its clients included Freeport Sulphur Co, Cerro Corporation, Cominco, US Steel, Union Carbide, Anglo-American and Union Corporation of South Africa. After the crash of Minsec some of these companies would reappear as bidders for parts of the empire. The beach-sands interests were sold to Union Corporation of South Africa and the controlling interest in the Aberfoyle group to Cominco, which is a subsidiary of Canadian Pacific.

In 1965 McMahon formed Mineral Securities Australia Ltd as a share trading and investment company. Its initial issued capital was $300,000, raised with some difficulty by McMahon. Ironically for a company which would become a fervent supporter of Australianisation, it was poorly supported by local financiers and more than half the capital was raised overseas. Originally McMahon had set out with a larger concept, planning in 1962 to form a company with a capital of £700,000. He persuaded one bank and one life-assurance company to put up 10 percent each, provided he could fill the rest of the subscription, but he could not, and the plan was abandoned. Minsec[2] was started with more modest funding aims, but even so there was little enthusiasm for the project in Australia, apart from one or two stockbrokers. Eventually 60 percent of the capital was raised from foreign countries, mostly Canada.

Minsec began as a public unlisted company. It had directors and shareholders but never any employees, because it was always managed by Kenneth McMahon and Partners. It was the employees of McMahon and Partners who bought and sold shares and ran Minsec. Accounting work was done for Minsec by A.H. Dickins and Co.

Minsec's entry into the share market in late 1965 was perfectly timed. It had struck at the start of the boom. BHP had already struck oil and gas in Bass Strait and WMC was about to strike nickel in Kambalda.

Minsec had entered the share market at the right time, in the right area and with the right expertise.

The story is told by Minsec's mushrooming profits.

PROFITS AND DIVIDENDS 1965-6 TO 1969-70 ($000)

	1965-6	1966-7	1967-8	1968-9	1969-70
Share Trading profits	21	161	1887	1946	12,418
Mining Profits (net of outside interests)	-	-	-	-	2400
Interest & Divs. received	6	28	46	124	415
Minorities	-	-	303	523	7220
Tax Provision	21	161	1584	1423	12,707
Dividends Paid	-	33	91	282	770

The jump in 1967 followed the company's decision in that year to list publicly. The reason for this was that the company was being treated as a private company for tax purposes so the directors decided to expand capital and list. By June 30, 1970, the issued capital of Minsec had expanded to $10.25 million, comprising 4.5 million fully paid $1 ordinary shares and 5.75 million fully-paid $1 preference shares on a fixed dividend of 5 percent.

Two points illustrated by the table are the group's low taxation and low dividend payout, resulting in a very high retention of profits. Tax was largely avoided by subscriptions to new mineral share issues. Earnings from share speculation were channelled into floats. At a 47.5 percent company-tax rate, this meant the savings in tax reduced the cost of share subscriptions by almost half. In 1969-70 the group's growing profit meant it had a growing need to subscribe to floats. In that year subscriptions would have had to reach $12 million to avoid tax. This was largely met by the float of Robe River Ltd which Mineral Securities helped to sponsor. The low dividend payouts enhanced Minsec's cash-flow and enabled it to grow at an accelerated pace.

More impressive than any of these figures was the rate of expansion of Minsec's share trading. In 1968-9 Minsec bought $31.4 million worth of shares and sold $16.5 million. In 1969-70 it more than doubled this volume to $73.8 million in purchases and $49.7 million in sales. In the first seven months of 1970-1 (up to its collapse) Minsec's activity leapt again with $75.2 million worth of purchases, including $29 million on Queensland Mines and Kathleen Investments, and $38 million in sales. For the last 18 months of its life, Minsec was the heaviest share trader Australia has known.

To make an adequate comparison, the Rae Committee sought evidence from the AMP Society. The AMP is the largest life office in Australia, with roughly one-third of the total assets of all life offices in this country. It is as large as its next three competitors[3] combined. The AMP's investment manager, Alan Coates, told the committee his society's rate of share purchases at that time was about $40 million a year and its sales about $8 million. The National Mutual was then buying at the rate of about $16 million a year. By contrast, in the calendar year 1970 Minsec bought $107 million worth of shares and sold $47 million (not counting a $6 million transaction on both sides which came to be known as the "Robe shuffle".)

Thus Minsec's total turnover of $154 million was more than three times the volume of the AMP Society. "If figures were available for the share-trading business of all members of the Minsec group in 1970, including the mutual funds, we believe that the total would not be less than four times as much as those for the AMP Society and would exceed the turnovers of all life offices in Australia combined," the Rae Report concluded. The life offices were far bigger, but Minsec traded far more frequently.

The scale of Minsec's trading is also shown by comparison with volume on the Sydney Stock Exchange. In 1969-70, when markets were dominated by the Poseidon boom, the total business recorded on the Sydney Stock Exchange was $1900 million. The turnover of the parent company Mineral Securities Australia Ltd alone in this period was $122 million. Taking into account share trading by its funds and other members, the Minsec group's activity must have accounted for something approaching 10 percent of Sydney's turnover in this huge year. Of course, much of Minsec's actual trading was done on other exchanges but the vast bulk of it would have been reported through Sydney.

As it had expanded, Minsec had begun forming and acquiring other companies. In December 1967 it floated a listed company named Petroleum Securities Australia Ltd to trade in oil shares. In October 1965, shortly after its formation, Minsec had acquired a 50 percent stake in the explorer, Amad No Liability. In later years Minsec floated several more companies, but the controlling interest was generally split between several members of the group. By June 1970, the Minsec group looked like the accompanying diagram (p.242). In addition there were various other wholly owned subsidiaries.

Minsec's biggest asset is not shown on the chart. This was the brain and personality of Tom Nestel, the young Yugoslav who had become Minsec's managing director and was now widely respected as the best and biggest stock-market player in Australia. Nestel ran the investment and trading department, consisting of 10 employees. It was this handful of individuals who were responsible for all of Minsec's huge trading activities, with Nestel the presiding genius.

Nestel's Jewish parents had brought him to Australia from Belgrade at the age of seven. He studied mechanical engineering at the University of NSW, decided to become a mining engineer instead and worked underground for a while in NSW coal mines. After 10 years in mining, much of it at Broken Hill, he joined McMahon at the formation of Minsec. When younger he had gambled on horses and cards: now he was to gamble for far bigger stakes. Tall, slightly stooped, black hair brushed straight back, Nestel always bore an air of dark intensity. He was dedicated to Minsec and the Presbyterian Church. During the week he literally lived Minsec, often getting to the office at 6 am and staying there until late at night. As an elder at the Chatswood Presbyterian Church he had not missed teaching Sunday School for years until the weekend after Minsec crashed.

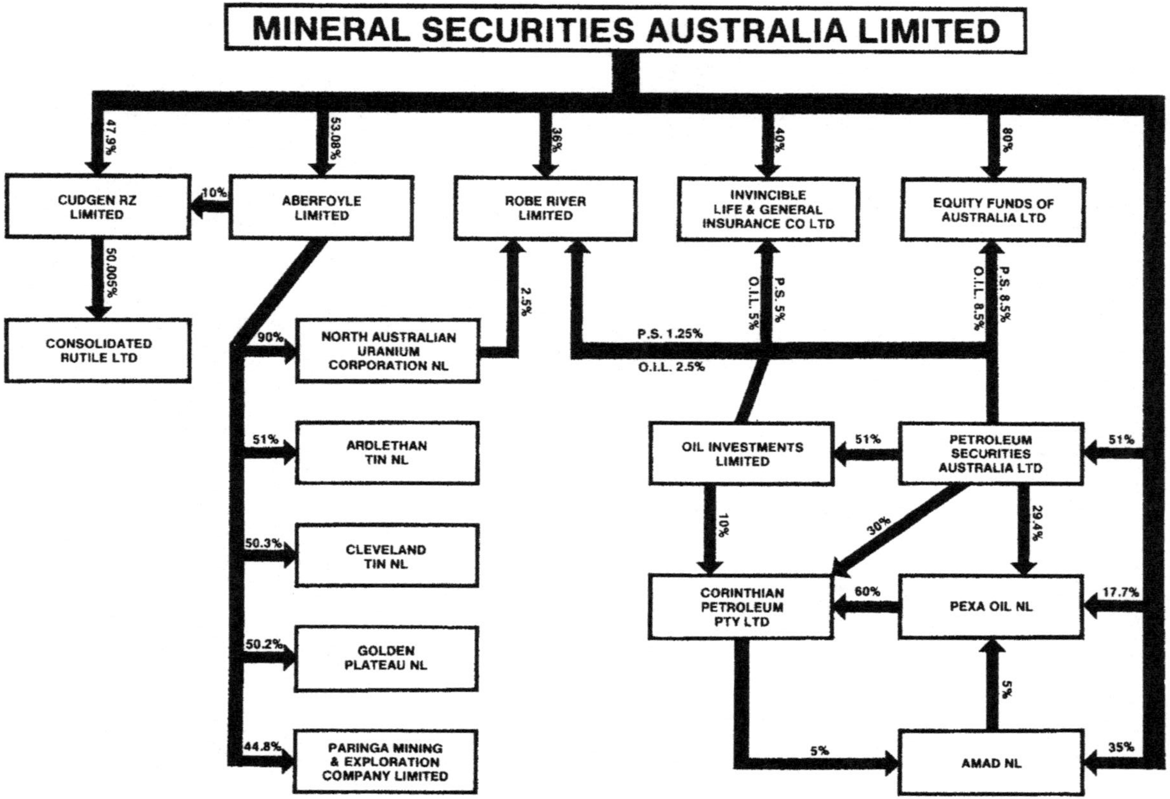
MINERAL SECURITIES AUSTRALIA LIMITED
47.9%
53.08%
36%
40%
80%
CUDGEN RZ LIMITED
10%
ABERFOYLE LIMITED
ROBE RIVER LIMITED
INVINCIBLE LIFE & GENERAL INSURANCE CO LTD
EQUITY FUNDS OF AUSTRALIA LTD
50.005%
2.5%
O.I.L. 5%
P.S. 5%
O.I.L. 8.5%
P.S. 8.5%
CONSOLIDATED RUTILE LTD
90%
NORTH AUSTRALIAN URANIUM CORPORATION NL
P.S. 1.25%
O.I.L. 2.5%
51%
ARDLETHAN TIN NL
OIL INVESTMENTS LIMITED
51%
PETROLEUM SECURITIES AUSTRALIA LTD
51%
10%
30%
29.4%
50.3%
CLEVELAND TIN NL
CORINTHIAN PETROLEUM PTY LTD
60%
PEXA OIL NL
17.7%
50.2%
GOLDEN PLATEAU NL
5%
44.8%
PARINGA MINING & EXPLORATION COMPANY LIMITED
5%
AMAD NL
35%

He could be rude at times to colleagues, but they always respected his abilities. He was taciturn, with little light conversation.

It would be hard to imagine a more complete contrast to the gregarious McMahon, who had a joke business-card reading: Ken McMahon — International Lover — Dancer of Note — Last of the Big Spenders — Known to Take an Occasional Sip of Sherry — Owner Seaforth Hilton — Soldier of Fortune — Wars Fought — Revolutions Started — Tigers Tamed — Bars Emptied — Computers Verified."

This unlikely couple combined brilliantly. Both had immense drive and won deep loyalty from their employees. To an outsider the Minsec offices in the lower floors of the Menzies Hotel had a family spirit about them. The rooms were small and cramped with ugly air-conditioning tubes hanging from the ceilings, but the dingy quarters emphasised the speed of Minsec's expansion. They were planning more palatial offices in Macquarie Street, but Minsec collapsed before they got there.

The office of the silver-haired McMahon reflected his taste and flair. This was unusual in the Australian mining industry, whose leaders are often ruggedly plebeian. McMahon's office did not contain any of the usual rocks and maps of a mining company chairman. Instead there were original paintings from the countries in which Minsec and McMahon and Partners had operated. Behind his desk was Monkeys, Bali, by Donald Friend.

The staff were exuberant yet professional, and worked like slaves for McMahon and Nestel. Minsec inspired a very rare dedication in its employees. Not only were they growing rapidly, inspiring visions of wealth, but they had a missionary sense of purpose. They were welding a great, new Australian mining house. Time and investigations revealed a lot of flaws in their methods, but belief in the vision was deeply held.

For a time Minsec's performance was exciting enough to overcome anyone's doubts. Their market coups were legendary and as the volume of trading grew, so did the size and boldness of their operations. They would buy and sell up to 10 million shares and options a day, and on one day achieved a turnover of 21 million units. In *The Money Game* Adam Smith[4] talks about the great days of Wall Street in the mid-1960s and the funds "where they move so fast that the man at the wall end of the desk sometimes ends up buying what the guy at the window end has been selling". It wasn't quite like that at Minsec, but it must have come close in their wilder moments.

Minsec divided its shareholdings into long-term investments, comprising the companies it controlled or hoped to control, and trading stock, bought for short-term trading. Nestel and his team reviewed the market constantly and changed their position occasionally. Sometimes they might be almost entirely out of the market as a trader, but still in as a buyer of long-term holdings in stocks such as Aberfoyle or Consolidated Rutile.

In March 1969, Minsec took stock of the market and decided that the established mining stocks had risen as far as could be justified on fundamentals. Nestel decided that henceforth this was a trading market. Over the next few months Minsec sold its holdings in heavyweight miners except for Peko, which had recently acquired Mount Morgan, and for stocks such as Aberfoyle and Cudgen where it was acquiring control. It began trading more heavily in the speculative end of the market.

By the middle of 1969 Nestel and his men were increasingly unsure about the direction of the market, which was wallowing. Surrey Bogg, the cheerful young metallurgist who became Nestel's top share trader, recalls that it was a period of "agonising uncertainty". When Poseidon ran at the end of September Minsec was actually short in the stock by about 6000 shares. Nestel was taking a holiday at the time but he returned in the first week of October.

The first weekend of October 1969 was the Labor Day long weekend in Sydney. Nestel, McMahon, Bogg and others spent the three days in conference, arguing which way the market would go. They decided Poseidon was the detonation the market had been waiting for and that Minsec should buy across the board. Bogg recalls that it was a great relief to be committed to a specific policy.

The next three weeks saw a huge buying wave from Minsec. They had already covered the Poseidon short. Now they bought the stock heavily. It was around this time that Minsec had its turnover of 21 million shares in a day. This was the day it bought 15 million shares in Bounty Oil Ltd. Minsec was a heavy trader in the oil stocks because it was an area of the market where a dealer could get huge volume at low prices, representing excellent opportunities for leverage. During the four months of the boom — October 1969 to January 1970 — Minsec made about $12 million profit. Much of it was made before the end of November and much of it was made on Poseidon. The bulk of Minsec's Poseidon holding was picked up between $20 and $30 in October. This $12 million represented almost all of Minsec's share trading profit for the full 1969-70 year, which was $14.5

million before the post-balance-date writedowns.[5]

While Minsec probably made the biggest profit of any boom-time trader, its overall trading record looks less brilliant on closer inspection. The first requirement — indeed almost the only requirement — for a share trader is to be able to pick the swings in the market; when it's topping and when it's bottoming. Starting in 1965, Minsec hit a big boom wave at birth and until the end of 1969 had not seen a really bad wipe-out in the market. Minsec rode the wave spectacularly well, better than anyone else, but it didn't cut out when the first danger signs appeared and wound up riding right on to the rocks. It had a lot of company on the rocks, but there were plenty of smaller players who survived and in investment, survival is the first test of ability.

Minsec's size made it vulnerable. Being far bigger than any other share trader in volume it was also more exposed. As long as a market is rising, big buying will push it up and it is relatively easy to sell large blocks of stock. When the market tops — and sooner or later all markets top — there is a sudden dearth of big block buyers. It is difficult to unload large blocks on the way down without pushing prices lower. If a big trader is caught long at the top of the market and can't sell, the consequences are often fatal.

All this is elementary in share trading, and financial history is littered with the skeletons of those who ignored it. No matter how big or brilliant a share-trading group may be, it cannot repeal the market's laws of gravity.

At first, Minsec monitored the boom pulse well. In early February 1970, just after making $1 million on Tasminex in one killing, Minsec's top men decided the boom was ending. At a meeting on February 18 the board confirmed the general manager's action in not buying any more trading stock, with the exception of new issues and placements below market prices, and support of the rights market. Thereafter the company virtually sold all its trading stock.

This involved some losses. The Rath Report[6] showed that the three major share trading companies in the group — Minerals Securities Australia Ltd, Petroleum Securities Australia Ltd and Oil Investments Ltd — made total losses of $798,000 in April 1970. In May they lost $363,000 and in June a further $445,000. The Rath Report concluded that these losses were not the result of imprudent management. By the end of May, Minsec's share trading portfolio had been reduced to three stocks with a market value of $90,000, which was slightly below cost.

Then Minsec made an expensive mistake. Anticipating a market rally in June, it re-entered the market in a big way. By June 30 its trading portfolio had been built up to $13 million. Ominously, this was $800,000 over the market value of the shares bought. In the same month Minsec decided to commit $17.5 million to a long-term investment in the Robe River venture. The group had thus invested more than $30 million within the space of a calendar month — a breathtaking sum by Australian standards.

The re-entry into the market was an error of judgment. Having decided to quit the market in February, and having achieved that objective despite some losses, Minsec had returned with a vengeance. The biggest single position taken was in Poseidon. Of the $13 million in portfolio at June 30, $4.5 million was represented by Poseidon, all bought that month. As the new financial year started Minsec continued trading Poseidon. By July 10 it was showing a trading loss of $789,129 of which $426,592 resulted from sales of Poseidon. At that date it was still holding 47,566 Poseidon shares which were showing a further market loss, on paper, of $2.1 million. By the end of July Minsec's trading loss stood at slightly more than $3 million, of which $1.9 million had been on Poseidon. The decision to re-enter the market was proving disastrous.

Nestel acted to correct the error. In a memo on current investment policy dated July 24, Nestel referred to "the trauma of the last three weeks" and said the market was expected to drift lower. Selling of trading stock had been completed and liquidity had been fully restored. "From now on the policy is to SIT and WATCH," he wrote.[7] "Shorts will be covered as they become profitable, but averaging down of existing long positions will not be attempted on any worthwhile scale until a much clearer picture emerges both on Wall Street and at home."

More trading stock was sold in August. For that month trading losses totalled a further $1.5 million including $832,766 on Poseidon. (In June when Minsec entered the market, Poseidon raced from $64 to $145 then slipped to $128. By the end of August it was $72. Minsec appears to have bought at the top and was probably responsible for pushing the stock to $145 when otherwise it would have been in a downtrend.)

On August 12 the general manager (Kenneth McMahon and Partners) reported to the board that: "In line with general policies already discussed, the liquidation of all available share trading positions was carried out through July and early August, with the aim of anticipating what is thought to be a major decline in the mining and oil share market".

Minsec had made a bad start to what increasingly looked like a bad year on the share market. One solution was to take some of the losses in 1969-70, when the group had made a healthy profit. In the year to June 30, 1970 Minsec had made a share trading profit of $14.5 million. On August 12, 1970, the company secretary was instructed "to investigate what provision should be made in respect to prospective share losses or variation in market prices as at June 30, 1970, and to report to the board at the next meeting". The matter was canvassed at length with the auditors and eventually the board decided to write down value of the trading stock at June 30 by $2.1 million to take into account post-balance-date losses. Of this figure, Poseidon accounted for nearly $1.3 million. After this writedown, Minsec's trading profit for 1969-70 appeared as $12.4 million.

The Rath and Rae reports differed over whether or not this writedown was adequately explained in the Minsec 1970 accounts. The Rath Report thought disclosure was adequate, but the Rae Report did not. I would say that the accounting treatment of the backdated losses was correct, but the disclosure was inadequate. Note 10 to the accounts did say that the value of stocks had been adjusted to reflect lower prices since balance date, but did not say that this had involved a writedown of $2 million.

The overall loss of $3 million in the market was a bloody nose, but no worse. After all the group had made $14.5 million in 1969-70 share trading and only had to meet low tax provisions and dividend payments. The mid-1970 foray back into the market was Minsec's biggest error in share-trading strategy. What would bring the group apart would be its errors in long-term strategy.

The Minsec group took three major decisions on long-term investments in 1970. These were to take over Aberfoyle and Cudgen, to gain 51 percent of Robe River Ltd and to buy substantial shareholdings in Kathleen Investments, Queensland Mines and Thiess Holdings.

The Cudgen-Aberfoyle takeover was a paper exercise in which Minsec shares would have been swapped for shares in those companies. This involved no cash outlay, although Minsec and its two funds were net large buyers of Aberfoyle and Cudgen shares on market until the end of October.

The acquisitions of Robe, Kathleen, QM and Thiess were to be on-market for cash. These were expensive exercises, involving huge sums by Australian standards. They were undertaken on a market where mining shares were generally falling — a fact already well recognised by Minsec and, specifically, by Nestel.

Minsec was short of cash, raising the elementary problem of how to finance the share buying. It planned to meet part of the costs by share placements and issues. The rest would be financed by borrowings and the Kathleen-Thiess purchases would be paid for by the public when Power Resources of Australia was floated. The first half of this financing plan could not be faulted. It was not only shrewd use of Minsec's substantial share premium, but it also expanded the group's equity base. The second, and larger, part of the financing was dubious strategy and the direct cause of Minsec's ruin.

It is worth pausing for a moment here to reflect on the nature of gearing in companies. Gearing is the ratio of borrowings to shareholders funds. The desirable level of gearing varies considerably depending on the nature of the business. A money-market operator dealing in Commonwealth bonds and fixed-interest paper may finance a high turnover by large borrowings and need only a small issued capital. A manufacturing company would be lower geared.

Historically, mining companies in Australia had operated on little or no gearing. Mining operations would be funded entirely by money subscribed by shareholders and retained profits. This limited the amounts of finance available but also limited the risks. In bad years — which come to all mines — dividend payments can be suspended, but interest payments must still be met. Australian miners kept their exposure to debt low. BH South, for example, had external liabilities of only $17 million at June 30, 1969, which was one-third of shareholders' funds of $51 million.

This pattern was broken by the CRA group, which pioneered the technique of borrowing funds against sales contracts. Hamersley, for example, launched in 1966 with $130 million in borrowings when it had only $1.5 million in subscribed capital. By 1969 this imbalance had been corrected but the ratio of borrowings to shareholders funds was still roughly 1:1.

This technique was adopted by other mining companies, who gradually became higher geared as the boom continued and borrowings were cheap. This led to trouble for many groups — notably Metals Exploration — when the boom collapsed. Minsec was already operating on sizeable borrowings. It had external debts totalling $28 million at June 30, 1970, compared to $39 million in shareholders' funds (excluding minority interests). The difference with Minsec was that it was borrowing money not to develop a mine, but to buy shares.

Borrowing money to buy shares is not per se disastrous. It is a technique which has been used extensively by entrepreneurs for takeovers of industrial stocks. However it is only viable if (a) the income from the shares is sufficient to cover the interest bill on the borrowings, or (b) the company's profits are running strongly enough to cover the interest bill, or (c) the market value of the shares is rising.

In Minsec's case the group was losing money on share trading in the last half of 1970 and none of the shares it bought were yielding dividends except Thiess; the yield on Thiess shares at the end of December 1970 was 1.1 percent. The overall market was falling, one of the few factors holding it up being Minsec's own buying.

The logical alternative was to finance the share purchases by long-term borrowings. The failure to restructure its borrowings from short to long-term ultimately resulted in Minsec's collapse. The final form of financing, the float of PRA, was never firmly arranged. Undaunted by his July-August setback in the market Nestel began planning a far larger share-buying program. At a fateful board meeting on September 9 he proposed an expansion plan which would include buying strategic holdings in Queensland Mines and Kathleen Investments, buying a majority stake in Robe River Ltd and a takeover bid for Cudgen and Aberfoyle.

The buying would be done through Minsec and the group companies. The two Minsec-sponsored mutual funds would buy into the same stocks and also support Minsec's own shares so that they would not weaken before the share swap was offered for Cudgen and Aberfoyle. It was an orchestrated buying scheme.

At this point Minsec was perfectly healthy, having absorbed its mid-year losses. It could have survived as a large successful mining and investment group. The expansion program of early September would bring about its collapse in the spectacularly short span of five months.

The Robe River buying was based on what would now be called inside information. Robe River Ltd had 35 percent of the giant Robe River iron ore project in WA. Robe River Ltd had floated publicly in July 1970. It had been the slowest of the Pilbara deposits to get going because it was based on relatively low-grade ore and the major Australian financial institutions were already committed to the high-grade projects such as

Hamersley and Mount Newman. But in early 1970, with Japanese iron-ore demand booming, it suddenly became feasible to upgrade Robe River's ore by pelletising. Darling and Co, who were underwriting Robe River Ltd, had Daniel K. Ludwig (purportedly the world's second richest man) as a possible participant, but early in the year his group dropped out. Darling and Co approached Minsec and asked if they would like to become the major shareholder. "We said yes in 10 days," said McMahon. "Probably the quickest we have made up our minds about anything."

Minsec took 14.4 million ordinary $1 shares in Robe River and also made a $3.6 million, interest-free convertible loan to the company. When Robe River listed, it was almost a subsidiary of Minsec.[8] McMahon and Nestel were directors of Robe River Ltd and McMahon and Partners was mining consultant to the company. By the standards of the day, Robe River was a "heavy" float. It was in a large long-term project from which no early spectacular profits were expected. Brokers and professional traders were lukewarm about the company. However, there was a strong response from the general public which surprised McMahon, Darlings and most market professionals. When the shares listed on August 16 they sold at $1.60. They had generally been expected to come on around par value of $1.

Later Minsec was suspected of propping the price at this time. Actually, Minsec was a heavy seller, unloading more than 400,000 Robe River on August 20 and 21. If Minsec had not sold these shares, at profits of around 60 percent on the subscription price of $1, its August trading result would have looked even worse. As it was Minsec recorded a net loss of $1.6 million.

The market premium on Robe shares was a result of spontaneous enthusiasm. As the provisional liquidator of Minsec, Jim Jamison, told the Rae Committee: "The stupid public paid $1.60 for them the first day they came on to the market."

The stagging of Robe was profitable and masked the losses as the trading portfolio was sold off, but together with some share placements by Robe reduced Minsec's equity in the company to 40 percent. Minsec's expansion program approved by the board on September 9 envisaged making it a subsidiary again by buying some 10 percent of the capital on-market. From mid-September Minsec's buying began to push up the price of Robe, which went through the $2 mark and finally peaked at $2.55.

As the word seeped around that Minsec was buying,

professional traders were bewildered. Minsec had already secured a huge stake in a company which was not generally regarded as a glamor stock, and now it was buying more at over $2 a share. Their previous holding gave them effective control, so the only logical explanation was that they wanted to consolidate Robe River into the Minsec group results. McMahon and Nestel, as directors of Robe, had learned from Cleveland Cliffs, the operator of the Robe venture, that initial production from the project would be greater and that profits would be greatly enhanced.

During the buying of Robe someone in the Minsec group saw an opportunity to paper over the past losses. The bulk of Minsec's Robe holding had been acquired in the float at $1. If these could be sold on-market to another member of the group at, say, $2.30, Minsec would make a profit of $1.30 per share. So Mineral Securities Australia Ltd began selling Robe River shares to the broking firm, Hattersley and Maxwell, who sold them to Mineral Securities wholly-owned subsidiary, Minsec Pty Ltd, which financed its purchases by an interest-free loan from Mineral Securities.

The profit thus attributable to Mineral Securities was purely a result of playing with numbers and in commonsense terms was not a profit at all to the group. It is as though a man had a handkerchief which cost 50 cents in the right-hand pocket of his trousers and a $2 note in the left hand pocket. If the left pocket buys the handkerchief from the right pocket for $2, the right pocket has made a profit of $1.50 and the left pocket is now holding a handkerchief at market value of $2, but the man is no better off. However, what does not make sense to a layman frequently makes sense to accountants. Nestel consulted the auditors of the group, Bowie, Wilson, Miles & Co, in mid-October a few weeks after the big shuffle began. The auditors were uneasy about it but raised no objections. The share switch went ahead on the grounds that it would result in a valid profit.

In all, 5.5 million shares were shuffled from Mineral Securities through Hattersley and Maxwell over three months. The total value of the sales was $13.5 million, on which Minsec Pty Ltd claimed a profit of $5.5 million. Other group members also sold to Minsec, raising the total number of shares to 6,175,000 in the shuffle and the total "profit" to $6.6 million.

While engaged in this share-shuffling exercise the group had almost doubled the cost of its Robe holding. In July 1970 the group had held 50 percent of Robe River at a cost of $14.6 million. At December 31, 1970, the group held close to the same percentage of Robe at a cost of $28.1 million. After allowing for early sales and the $13 million switch the group appears to have spent a net $7 million expanding its Robe holding. This would not have been fatal by itself. Perhaps the most valid opinion on this is that of Minsec's liquidator Jim Jamison. In evidence to the Rae Committee he said: "If they had spent their $7 million only on Robe River I would not be sitting here now, I would suggest."

The nail in Minsec's coffin was not made of iron, but uranium. Simultaneously with the Robe River purchases, the Minsec group had been huge on-market buyers of shares in Kathleen Investments and Queensland Mines. As part of the same strategy they had been buying Thiess.

In hindsight it is easy to criticise Nestel for his optimism about Nabarlek, but these were still optimistic times. Although a number of obviously speculative mining companies (often called "desperates") had come and gone, Australia could still boast an enviable number of recent, world-scale mineral discoveries. Queensland Mines was an exploration company with a small capital, but it had a highly respectable history and was run by the man who had discovered Savage River. And while it was always realised that an early reserve statement was an approximation (if Nabarlek had turned out to contain 50,000 or even 45,000 tons this would have been regarded as acceptably close to the first estimate) no one imagined that it had been overestimated by a factor of six.

Nestel had been on holiday on September 1 when Queensland Mines made its announcement. (It is ironic that such a dedicated worker should have been away from the office for both the Poseidon and Queensland Mines announcements.) Surrey Bogg bought a few thousand Queensland Mines on-market for trading purposes. When Nestel returned to work a few days later, he learned from his investment department that Patrick Partners had been heavy buyers of the parent company, Kathleen Investments. On September 9 the Minsec board approved the expansion program which included the buying of Kathleen Investments and Queensland Mines. A few days later Nestel approached Frank Robertson, the Patrick Partner who was virtually running Castlereagh's share portfolio at this time. Nestel and Robertson, who knew each other well, discussed a possible pooling of operations. Then, on September 17, Roy Hudson secured John Gorton's consent to the Nabarlek Ordinance, limiting overseas holdings in QM and KI.

Nestel told the Rae Committee that immediately after this had been proposed "very active discussions (with Castlereagh) took place along the lines that this very valuable deposit was to be preserved by an almost unprecedented step of legislation for it to remain in Australian hands. The view then was openly discussed that Mineral Securities and Castlereagh Securities were two Australian-controlled companies."

The September 17 statement by Gorton certainly must have kindled the enthusiasm of both Minsec and Castlereagh for a joint bid for a stake in Kathleen, as the bulk of their buying was done in the next two months. It is not clear why. By limiting the stake overseas companies could take in the Nabarlek companies, the government was also limiting drastically the number of potential bidders. It could now be taken over only by Australian companies, and few were rich or interested enough to try. There has never been any evidence that anyone, apart from Minsec and Castlereagh, was seriously considering acquisition of a controlling interest in Kathleen. The introduction of the Nabarlek Ordinance should have meant there was less competition for the stock. Further, the Nabarlek Ordinance meant that if Minsec and Castlereach ever wanted to sell their holdings there would be a correspondingly limited number of potential buyers. This, in fact, became a major problem after Minsec collapsed and the liquidator wanted to sell to overseas companies.

Nestel and Robertson agreed to buy Kathleen shares in concert through Patricks. Sometimes all the shares bought on one day were split between the two companies and sometimes Castlereagh would buy on one day and Minsec on the next. Sometimes they stayed out of the market altogether. Minsec also bought shares in Queensland Mines. This was a curious decision because it broke a fundamental rule of investment by buying into the same situation twice. Dowling and Robertson saw the situation more clearly and concentrated their buying on Kathleen. Control could only be gained through Kathleen, which held just over 50 percent of Queensland Mines. Minsec, according to McMahon, wanted Queensland Mines as both a long-term investment and as a trading stock. As a trading stock it might have been a defensible purchase but for long-term investment it would have been better to concentrate on Kathleen. Apart from being poor investment strategy, the buying of Queensland Mines represented an important diversion of funds for Minsec. Between August 31 and December 31 Minsec poured $16.6 million into Queensland Mines shares.

Minsec financed its purchases by borrowing. Ultimately the

concept was evolved that the combined Castlereagh-Minsec holdings of Kathleen and Thiess would be put into a company called Power Resources of Australia Ltd, PRA would be floated publicly, the funds raised by the public float would revert to Castlereagh and Minsec as the sponsors and Minsec would be able to repay at least part of its borrowings. But PRA never happened. The partners-to-be broke up in acrimony, with different versions of what had transpired.

The possibility of the joint venture had arisen as a logical result of the joint buying. The discussions went far enough for a draft prospectus for PRA to be prepared by Robertson. This draft was presented at a meeting held on December 3 attended by Robertson, Dowling, Jim Millner (chairman of Castlereagh), McMahon, Nestel and Ross Tompson (assistant secretary of Mineral Securities). The draft prospectus envisaged PRA having a holding of 1.8 million Kathleen, representing about 20 percent of that company's issued capital, and 1.4 million Thiess, representing about 12 percent of that company. Minsec had originally proposed that they and Castlereagh should seek outright control of Kathleen by buying 51 percent. Dowling dismissed this as a "pipe dream".

Dowling said: "I rejected the idea entirely because they were proposing a purchase which could cost $90 to $100 million, which seemed to me ridiculous as it was far beyond the resources or ambitions of Castlereagh to maintain anything like a 50 percent share and seemed to be much beyond the financial resources of Minsec." Minsec and Castlereagh subsequently agreed to set up a joint company called Mincast Holdings Pty Ltd which would be the vendor of the shares to PRA.

The meeting on December 3 between members of both the Castlereagh and Minsec boards agreed in principle to four propositions in respect of the PRA float. These were that:

- Mincast would hold a minimum of 70 percent of PRA and PRA would aim at holding about 20 percent of Kathleen Investments before floating.
- PRA would have to have at least two major holdings, one in uranium and one in coal, and have a further major holding in prospect before going to issue. The market would have to be steady enough to ensure both major holdings would stay stable during the foat.
- As an incentive to subscribers, PRA would have to buy its Kathleen and Thiess holdings from Minsec and Castlereagh at prices below market. However, Minsec and Castlereagh should not lose money on the deal. In other words the market would

have to rise far enough in these stocks for the recently-acquired holdings to be sold at a profit, but below market prices.

- As soon as PRA was considered possible to float, Hudson should be approached to join the board.

The proposed statement by directors in the draft prospectus described PRA as "a corporate base for large-scale investment in the production and marketing of Australian mineral fuels and other mineral resources". The proposed statement said Patrick Corporation Ltd was investment banker to PRA and had undertaken to provide the company with short-term loans as required. It also said: "As a matter of long-term development policy, the company will maintain an intensive search for additional projects concerned with the production and marketing of minerals". The draft prospectus was by no means a firm document, but it is worth looking for a second at what was envisaged. Nominal capital was proposed to be 300 million 50 cent shares, of which the following would be allotted:

- Mincast, 100 million paid to 10 cents, plus 16 million fully-paid issued for cash. Paid value, $18 million.
- Patrick Corporation, 5 million 10 cent paid and 3 million fully issued for cash. Total value, $2 million.
- 14 million fully paid issued for cash to a financial institution. Total paid value, $7 million.

The draft prospectus was for an issue of 30 million fully paid 50 cent shares of which 20 million were to be reserved for prior right of application by the shareholders of Castlereagh, Mineral Securities, Aberfoyle and Cudgen. The other 10 million shares would presumably have been offered to the general public in the prospectus. These 30 million shares would have raised a total of $15 million.

Mincast, owned 50-50 by Castlereagh and Minsec, would have held 69 percent of PRA under the prospectus. This seems low in view of the partners' stated intention of holding at least 70 percent, but obviously all the figures in this rough draft would have been liable to alteration. The important point is that however the numbers were altered, it is difficult to see how this prospectus could have significantly eased Minsec's liquidity problems.

At December 31, Minsec held $13.4 million worth of Kathleen and Thiess at cost. Let us presume that Minsec was able to sell these shares to PRA at cost and that this would have represented a satisfactory discount on the then market price of the shares. Minsec would have got back $13.4 million cash but would have had to subscribe $9 million to the float (half of Mincast's $18 million). It would have received a net cash injection of $4.4 million and instead of holding Kathleen and Thiess shares it would now hold 50 percent of a company which held 116 million PRA shares. The $4.4 million would have been welcome, but a drop in the bucket compared to its total short-term borrowings by that time, and it would have lost direct control of its shareholdings which would mean it could not sell them to gain liquidity.

The Minsec group would have faced the problem of arranging long-term loans against its Mincast holding, which would probably not have been earning dividends. This scenario presumes a strong recovery in the market in December. In fact, Kathleen shares were looking decidedly doggy by this stage, and it was for this reason that the float never proceeded.

Nestel later alleged that Patrick Partners had failed to honor a promise to provide $30 million in bridging finance. He claimed that the promise had been made during early talks between himself and Frank Robertson. Nestel said Robertson had phoned him in late September and indicated quite specifically that $30 to $50 million would be made available.

In the event all that was ever provided from the Patrick stable was a $5 million advance from the associated company, Bill Acceptance Corporation Ltd. Dowling denied flatly that Patricks or any of their associated companies had ever promised such support.

Robertson said that after first discussing the bridging finance with Nestel he had spoken to the managing director of Bill Acceptance Corporation and next day called Nestel back and told him that "the liquidity pool available for acceptable first-class paper appeared to be in the vicinity of $30 to $50 million and that, subject to Mineral Securities paper being able to meet the standards required of the market and a firm proposition being put for the amount he required, if acceptable to the bill company, it seemed to me that it was possible that he would raise up to $35 million." The Rath investigation of Minsec in NSW examined the claim by Nestel and concluded there was no evidence that any promise was made to provide bridging finance and there was also no evidence of a promise to provide finance of the order of $30 million. Certainly there was never any agreement in writing, and it would hardly appear logical for Patricks or its associates to have undertaken financing for a greater amount than would have been necessary for Minsec's half of the PRA project.

Nestel was bitter at what he regarded as a broken promise. "I have often made deals involving vast amounts relying solely on a man's word," he said to the Rae Committee. "This is the first time that a man whom I trusted has reneged on his promise."

One feature of this dispute is that Minsec appears to have lost touch with the realities of the money markets of Australia. Originally they had proposed a $100 million plan to acquire 51 percent of Kathleen. This was a huge amount of money, but so was $30 million. After Minsec went into liquidation a consortium of Australia's mightiest companies was only prepared to pool $35 million to aid the liquidator.

Another aspect of Minsec at this time is that, despite its prowess as Australia's most aggressive investor, the group was in a strangely impotent position over Nabarlek. It could not send a geologist of its own to Nabarlek, although it tried for months. It did not have the money to buy the shares and was running out of lenders. It was also relying on the wrong people. McMahon and Nestel were relying on Dowling to obtain access to Nabarlek for them, but Dowling was on bad terms with Hudson who effectively controlled access to the site. Nestel was negotiating for finance with Robertson, but Robertson himself was not going to provide finance. While Minsec continued to buy shares, the rest of the deal was drifting loosely.

Minsec also had little to do with the Thiess side of the deal. Thiess was a Brisbane-based civil engineering company which had diversified into coal mining in Queensland, both on its own and as a partner in Thiess-Peabody-Mitsui Pty Ltd at Moura. Thiess had 11.4 million 50 cent shares on issue and now ranked as a heavyweight in the coal industry. At the end of 1970 its shares stood at $4.65, having hit a peak of $5.60 a couple of months before under buying pressure from the PRA partners. Thiess shares had eased toward the end of the year but the losses being sustained by Minsec in this area were relatively small.

Dowling was handling negotiations in this area, having approached the company's chairman, Sir Leslie Thiess, in December and sounded him about the PRA project. Sir Leslie's reaction to the project was not unfavorable. McMahon said it was hoped the Thiess family would sell its stake to PRA for a mixture of shares and cash. The family were understood to have agreed to sell at $6 per Thiess share, which would seem a likely figure on the market prices at the time.

The partners' concept also included a third company in PRA. This was never more than a vague concept, but it would have had to be an oil company to balance PRA. One story I heard was that if PRA had been successful, it might have turned its sights on Woodside Oil.

Meanwhile Minsec had been proceeding on the third front of its expansion campaign. This was the takeover of the shares it did not already own in the Aberfoyle group of tin and wolfram mining companies and in the beach-sand miner Cudgen RZ, which in turn held a controlling interest in Consolidated Rutile. Originally Minsec had planned a bid for Consolidated Rutile too, but could not get unanimous approval from the Cons Rutile board.

These bids were part of the expansion program envisaged by Minsec's board on September 9, but it was not until November 30 that Minsec announced its intention in a statement to the Sydney Stock Exchange. By this time, Minsec had done the bulk of its buying in both the Robe River and PRA acquisitions. The Aberfoyle and Cudgen bids did not involve any cash outlays because Minsec shares would be swapped for shares in the two mining companies.

In a memorandum entitled "Financial Evaluations", drawn up in September or October, Nestel had asked "Who will buy group companies during currency of bids?" and "Who will buy MSAL up to December 1 to ensure that MSAL/group relation pricewise is reasonably maintained?" This second question is always a problem in any bid with a large paper element. Australian companies are forbidden under law from buying their own shares or from lending money to subsidiaries to do the same.

If the bidder's own stock is not supported while the takeover offer is in progress there is always a danger that its price will weaken — either from a natural market reaction to the prospect of the dilution of capital or from an active bear raid on the stock by opponents of the bid. Sometimes the company making the bid has a staff superannuation fund which suddenly conceives on all-consuming desire for its own company's stock. More often the bidder seeks friends and organises a little support.

Another source of support is a friendly mutual fund, of which Minsec had two. The First and Second Australian Growth and Income Funds had the same directors and investment committee. The management company was Equity Funds of Australia Ltd, which was a subsidiary of Mineral Securities, and in which Turnbull Krishnan & Co Pty Ltd also had a stake. The principals of the second company were investment advisers Shann Turnbull and Krishnan Tatparanandam.[9] Turnbull, Krishnan, Nestel and McMahon were all on the EFA board and the only other director was the chairman Eric Spooner, another director of Minsec.

Turnbull later said[10] that he and Krishnan had very little control over the investment policy of the funds. "Minsec's attitude was 'You've been allowed on board and we've pulled up the gangplank. Just sit there and don't worry'," he said.

Such an attitude was understandable. Every wheeler-dealer in Australia wanted to be taken on board by Minsec.

The funds were set up by Minsec early in 1970. Minerals Securities held one million fully-paid shares in the First Fund and 2.5 million shares in the Second. These shares were subscribed for at a price of $1, plus service fees, which meant Minsec had put in more than $3.5 million. Within a few months subscriptions to the funds exceeded $14 million, public enthusiasm for them being due to the Minsec association. The prospectuses for these funds included a statement on investment policy which gave the impression that they would not be investing in areas closely associated with the Minsec stable. The statement read as follows:

"The fund will be of a 'general' investment type, limiting itself to neither industrial equities nor mining stocks so that investors may participate in superior investment opportunities arising from any sector of Australia's growth and development. The objective will be to obtain consistent long-term, above-average performance of both capital appreciation and income...

"Mineral Securities Australia Ltd has now substantially reduced its long-term investments in companies in which it does not have management control and has become a mining house with 13 listed subsidiaries. It is intended that the fund will direct its activities to long-term portfolio investment *not associated with management control and extended to include securities of Australian public companies outside the mining sector.*"[11]

In fact, the funds invested heavily in stocks associated with Minsec and at times were helpful to the Minsec group. The First Fund had barely been floated when it began buying Cudgen and Aberfoyle shares. By the end of April, about $1.4 million of the $9.6 million raised in its float had been spent on shares in Mineral Securities. During October 1970, when Mineral Securities was preparing its bids for Aberfoyle and Cudgen, the First Fund spent $500,000 buying Mineral Securities shares on market and the Second Fund spent $750,000. By this time the First and Second Funds had respectively committed 60 and 56 percent of their long-term investment to companies in the

Minsec group or to companies which Minsec was planning to control. The two funds had also traded heavily in Minsec group companies in their short-term portfolio.

The funds were a welcome source of support for Minsec because they were buying with equity capital subscribed by the public, not with money that Minsec had to borrow in the short-term market. These borrowings had now reached very large proportions. A few professionals close to the share and money markets gleaned that Minsec was borrowing to buy KI, QM and Thiess. But no one realised until the eve of the collapse how large these borrowings had been.

Minsec had subscribed $14.4 million to Robe River in the first half of the year. The bulk of this was financed by the placement of 587,500 ordinary shares at a premium of $13.70, raising $8.6 million. The rest of the Robe subscription was funded from accumulated profits. All this occurred before balance date of June 30, 1970.

It is difficult to gauge the group's true position at December 31, 1970. In the last half of calendar 1970 the group borrowed massive amounts to fund its PRA purchases. The best available guide to the group position is the combined balance sheets of Mineral Securities, Minsec Investments Pty Ltd and Norausam Pty Ltd (another share-trading member of the group). The table alongside, showing the position of these three companies at June 30 and December 31, 1970, was prepared by Nestel as part of the interim report to the Mineral Securities board.

In this table the $6.8 million overstatement of profit for the December half-year (from Mineral Securities' sale to Minsec of Robe River shares) had been included in the shareholders' funds figure of $38.8 million and the investments figure of $97.7 million. Whether one deducts this overstatement of profit or not, the table shows the large and dangerous borrowing program undertaken by the Minsec group over the last six months of 1970. In half a year the group had trebled its short-term debt. After deducting the $6.8 million the gearing of shareholders' funds to outside liabilities had changed from 3:2 to a ratio of less than 3:8. And there is no long-term debt at all. Of the current liabilities, $20.9 million was owing at call, nearly all unsecured, and a further $31 million was owing on short term.

Minsec executives must have been fully aware of this malignant growth because their accountants, A.H. Dickins & Co, provided a daily or near-daily liquidity report showing Mineral Securities' position with respect to inter-group

MINERAL SECURITIES AUSTRALIA LTD
MINSEC INVESTMENTS PTY LTD
NORAUSAM PTY LTD
Combined balance sheets

30.6.70 ($ million)			**31.12.70 ($ million)**	
		Issued Capital:		
	4.5	Ordinary	4.5	
	5.7	Preference	5.7	
	25.5	Reserves	28.6	
35.7		Total Share Capital and Reserves		
		Represented by:-		
	23.7	Current Liabilities	86.4	
		less:		
	3.8	Current Assets	23.9	
19.9		Net Short Term Indebtedness		62.5
		Other Assets:		
	52.2	Investments	97.7	
	3.6	Convertible Loan to Robe River Ltd	3.6	
55.8		Total Other Assets		101.3
35.7		Loan and Investments less Short Term Debt		38.8

company loans, brokers' accounts and loans raised on the short term market. Extracts from these liquidity reports showed how Minsec's short-term deficiency kept growing:

Date	**Owing by Minsec ($ million)**	**Owing to Minsec ($ million)**	**Deficiency ($ million)**
1.7.70	20.7	2.9	(17.8)
24.7.70	16.6	5.1	(11.6)
21.8.70	17.3	9.2	(8.2)
28.9.70	39.6	15.0	(24.6)
28.10.70	55.6	17.4	(38.2)
25.11.70	71.4	20.4	(51.0)
29.12.70	79.7	22.6	(57.1)
27.1.71	73.8	12.3	(61.4)

The upsurge in borrowings, and the resultant higher gearing meant that, as the borrowings had been spent on shares, Minsec had become increasingly vulnerable to any weakening in the share market. Such a weakening was now under way.

In the six months to December, Mineral Securities had spent $16.6 million buying Queensland Mines, nearly $11 million on Kathleen and $2.5 million on Thiess — a total of $30.1 million. At the end of the year the market value of these holdings had slipped to $24.6 million, representing an unrealised loss of $5.5 million. This was only the loss in Mineral Securities Australia Ltd. The funds and other companies in the group buying these shares would also have been losing money.

Mineral Securities Australia Ltd's statement of investments at December 31 showed that the total of investments in subsidiaries, controlled companies and long-term investments had a market value of $68 million, slightly above cost of $64.2 million. However, this valuation depended heavily upon Robe River Ltd. More than 10 million Robe shares were included in this total at a market price of $2.35. That market price had been largely set by the buying of the Minsec group itself. Ten weeks later the companies' liquidator was unable to get $1.15 a share for Robe.

There was no festive spirit among the Minsec directors as they faced the New Year. Whispers were beginning to spread about Nabarlek, weakening the price of Queensland Mines and Kathleen Investments. This in turn was having a bearish effect on Minsec's own shares, which had closed 1970 at $10, compared with a peak of $23 during the year.

Worse, Nestel had received during December a report from Chris Castleman, the managing director of the group's merchant banker, Hill Samuel Australia Ltd. Castleman, a lean, dark-haired young Englishman with a precise City accent, may well have been the most capable of all the overseas merchant bankers who arrived in Australia in the late 1960s.

His report clearly delineated the weaknesses of the existing Minsec structure and, while cloaked with some tactful phrases, gave the company an unpleasant glimpse of its possible future. Discussing restructuring Minsec's debts from the short to the long term, he said: "We believe that this combination within Mineral Securities of share trading and other activities may mitigate against the acceptability of MSAL as a borrower. Rather than regarding share trading as a potential source of cash flow with which to support debt payments, medium to long-term lenders will tend to view this activity as being one which may at any time, and in certain market conditions, lead to operational losses which in turn could prove prejudicial to debt-servicing and the overall strength of the borrowing vehicle."

Castleman urged that the group reduce its debt and replace at least part of it with equity. But this would take time and the report raised the possibility that Minsec would have to meet unfunded commitments of $32 million until the end of its current financial year. Castleman noted that on the basis of Minsec's current exposure it was probably the biggest user in Australia of the inter-company and unofficial money markets, with the exception of financial institutions. While this market had been easy in the past, it was now facing a tight money period between March and June 1971.

Looking at the longer term Castleman proposed a drastic reduction in the size of the share-trading portfolio and, by implication, in share-trading activity. It suggested that the group had reached a scale where too much could be at risk in its speculative activities. The report said: "We believe that Minsec stands to lose substantially more through incurring market losses than it would gain by earning similar profits."

In the short term Castleman's formula for survival was chilling. He pressed for reducing borrowings by at least $20 million. Ways of doing this included selling its holdings in the mutual funds, completion of the PRA deal and further reductions in share holdings. The report estimated that these actions would leave the group with an exposure of about $8 million to the unofficial money market which Castleman considered "the absolute maximum which should be envisaged".

Even more chilling was Hill Samuel's attitude toward forming a consortium of merchant banks to provide further finance for Minsec. The report noted that Minsec had already arranged $15 million in bill finance through to September 1971 which was "substantially more than could be made available through these markets to the great majority of substantial commercial and industrial companies" and went on to say: "If we were to set out to form a syndicate to provide additional facilities for Minsec we think there would be a distinct danger that, as it will be necessary to reveal full information regarding Minsec's current position and commitments, its very great temporary dependence on the market might become widely known and discussed with the effect that its own operation in this area could be prejudiced."

Castleman's report was an excellent summary of Minsec's weaknesses, with one exception. It did not mention the time element. Minsec was due to fold on February 3 and it was now December 11. Yet there was nothing in Castleman's report which appears to stress the urgency of the measures proposed. The reason was that Castleman felt if he pressed it too strongly the report would be unacceptable to directors. Castleman viewed the matter so seriously that he wrote to John Alexander of Hill Samuel in London, who was on the Minsec board, asking to be appointed Alexander's alternate so that he could personally urge that the measures be taken.

Hill Samuel did not have a large exposure to Minsec itself, holding only a few shares. However it had put a number of its clients into the stock. Luckily, just after Minsec crashed Wall Street began rising and Hill Samuel had put a number of the same clients into US stocks.

Nestel received the report from Castleman on December 11. Ken McMahon pencilled in the margin "Obviously not happy with us. Will HS London know? Will potential 'funders' know?" On the front page he wrote: "Rather a frightening document — but what we need to make us think. However we should study situation, make up our minds correctly and quickly and move fast and definitely." The report was distributed to directors, but it was now Christmas. In a memorandum dated December 29 and addressed to the board Nestel said he found the Castleman report stimulating, "but it proposes philosophies significantly at variance with those that have characterised MSAL to date." He suggested it be used as a basis for discussion when Castleman returned from holidays. He also suggested that "some of Castleman's recommendations can be implemented gradually".

One move in the direction Castleman had indicated was made at the board meeting on December 30. Nestel suggested and the directors agreed that a preference issue for $16 million be made, to be underwritten by Ralph W. King & Yuill in Sydney, Wilson & Co in Brisbane and Leonard G. May & Son in Melbourne.

This equity raising would have been a significant step toward resolving Minsec's financial difficulties, but another action taken next day was to lay the trail to disaster. On New Year's Eve the Aberfoyle and Cudgen takeover notices were sent out in the mail. Amid the thicket of routine legal declarations in these documents was one small note, included under the requirements of the Companies Act. This read: "There has not been any material change in the financial position of Mineral Securities since the date of the last balance sheet, namely June 30, 1970, laid before it in general meeting, except that the company has substantially increased the value of its share portfolio by $34.6 million financed by outside borrowings to the extent of $31.2 million."

A lay investor would probably not even have attempted to read something as generally incomprehensible as a takeover document, let alone pick up the significance of this item. But to analysts throughout Australia, it set the alarm bells ringing. The news that Minsec had borrowed $31 million to buy $35 million worth of shares in a weakening market spelled trouble. Brokers began easing clients out of the stock. Worse, Minsec's lenders became aware of the note and started worrying. None of them had realised the size of its total debt.

Nestel knew that the note would be picked up quickly by analysts and the news would spread quickly in financial circles, holidays or no holidays. Minsec prepared to put on a bright face, as it had three bull points to announce. First, there was a $3.5 million profit for the December half-year, second, there was the underwritten preference issue, indicating an intention to scale down its debts, and third, there would be a share split and bonus issue. The bonus had been foreshadowed in November. It would not raise any cash for Minsec but would be a gesture of confidence to shareholders.

Minsec increasingly needed to show the brightest face possible, because its problems were tightening like a noose. Some $23 million of its short-term debt was secured on assets. The terms of security varied but most commonly the collateral offered by Minsec was scrip with a market value amounting to 150 percent of the loan. However, as the market weakened in several of the stocks which had been offered as a security, lenders were demanding that Minsec put up more collateral. Typically a company may have lent Mineral Securities $1 million on security of $1.5 million worth of Queensland Mines. As Queensland Mines shares began falling in price, the lender would demand that more shares be pledged as security to maintain the 150 percent cover.

This drain on scrip was in turn limiting Minsec's opportunity to raise fresh secured loans. In mid-January two merchant banks, Westralian International Ltd of Perth and Ord-BT Co Ltd of Sydney, were considering proposals to raise medium and long-term finance to restructure Minsec's debt. Then on January 25 Minsec announced its half-year profit, together with the preference issue. In a final burst of cheer the statement notified shareholders that interim dividend had been lifted to 10 percent.

Within a few days, the empire of Mineral Securities crumbled. On January 27 Ord-BT talked to McMahon and turned down a proposal to lend Minsec $30 million. By that time the prospects of success with Westralian were remote. The short-term position had become grave, resulting in a detailed review over the long Australia Day weekend from January 30 to February 1.

The tightening cash squeeze, with lenders calling in funds and demanding more collateral, had induced a sense of desperate urgency in Minsec's top men. Suddenly everything was for sale. An early morning meeting of the Minsec board on Friday, January 29, had been told that Angus Moir, chairman of the NSW Permanent Building Society, had said he was not interested in buying the group's 50 percent stake in Invincible Life & General Insurance Co Ltd. The board also decided in principle to redeem their $3.5 million investment in shares in the two mutual funds.

An hour later McMahon and Nestel saw Frank Robertson and offered their Thiess and Castlereagh shares to him. They said they would also be interested in offers for their Kathleen and Queensland Mines shares. But by that time Castlereagh had enough Thiess and Kathleen itself.

A "study team" of Minsec workers spent that Saturday and Sunday in McMahon's office assessing Minsec's position. The gravity of their conclusions was reflected in a report by Nestel put to the board on Monday. His report recommended stringent controls on outgoings, deferment of payments and liquidation of assets that "were not of a really critical nature".

Monday was a public holiday for most of Australia but for Minsec's board and advisers it was a long, apprehensive day. They met at 10 am and did not break up until 7 pm. The board approved Nestel's recommendations and added further suggestions. The decision to redeem the fund shares was confirmed.

This was grim enough work but for some directors there was yet another unpleasant surprise in store. It was not until the end of the meeting that McMahon, Spooner and Richard Swift learned exactly how the half-year profit had been made. They resolved to seek legal advice on whether the Robe shuffle had resulted in a legitimate profit.

Nestel, Swift, Spooner and Tompson saw Mr Bainton, QC, the next morning. After briefing him and receiving advice, all except Nestel were convinced the Robe River deal was an inter-company transaction and the profit would have to be altered.

Nestel wanted to seek further legal advice, but the other directors wanted a revised profit statement to be made to the Stock Exchange as soon as possible. After working out that this would mean announcing a $3.3 million loss instead of the $3.5 million profit (which had been poorly received by the market anyhow) a meeting was held in the McMahon and Partners' office that night.

By this time a sense of doom was pervading the discussions as several of those present began to realise that the group was facing the end. Spooner described this meeting as "a sort of desultory series of gatherings with people coming and breaking up into groups". Five of the board were there, including McMahon and Nestel. At various times Bogg, Castleman, brokers from King & Yuill, D.A.T. Dickins and Mark Acheson from Ord-BT sat in. Occasionally people went out for drinks or dinner. It was so disjointed that afterward Spooner had the impression it had been spread over two or three rooms, whereas it was in fact all in one office.

Of those who attended, Castleman and Bogg objected to the proposed announcement, but Bogg did not even know Castleman had been there until I showed him the Rath Report seven years later. Castleman, who seems to have been fated to play Cassandra in this tragedy, said the company was finished as soon as it went out and reversed the profit. Castleman also urged Spooner to keep King & Yuill "on the hook" for the $16 million underwriting agreement.

On February 3 the boards of the two funds met and Minsec redeemed its shares, raising $2.87 million. This was lower than the amount it had subscribed because the redemption prices were geared to the funds' asset backings and these had suffered because they were so heavily invested in the Minsec group. Minsec received prices of 85 cents a share for the First Fund holdings and 91 cents for the Second Fund. In 24 hours the funds' holdings of Minsec scrip would be valuable only as wallpaper, but the directors appear to have accepted without question the contention that the move was to aid Minsec's liquidity problems.

McMahon later justified the redemptions, in evidence to the Rath inquiry, by saying: "I was aware at the time, or I thought at the time that the best thing that could happen to the funds was that Mineral Securities survive." The Rath Report concluded that Nestel, McMahon and Spooner did not use reasonable diligence in the discharge of their duties as directors of the funds in relation to the redemptions. The Rae Report said: "The company which had been responsible for seriously and lastingly damaging the finances of the funds was the one shareholder in the funds to be given exemption from bearing a share of the consequences."

The redemptions had a particularly severe impact on the Second Fund, from which Minsec drew $2 million. Originally Minsec had subscribed some $2.5 million of the total $5 million to this fund. Another $2.3 million was subscribed by a group called Securities Management Co Ltd, which meant that the public subscription to this fund must have been very small, amounting to no more than a quarter of a million dollars. Securities Management was incorporated in the Bahamas for tax reasons, but was controlled by a large Geneva-based investment group. The manager of Securities Management, Jean Beaurang, had been prompted to invest in the Second Fund because of his personal contacts with McMahon. He was furious at the redemptions.

By February 3 one portent of the gravity of the situation was that Nestel, always the epitome of the strong, silent trader, was showing signs of nerves. Frank McHugh, of Hattersley and Maxwell, who saw him late that morning, said: "He was not the Mr Nestel I had known over the last couple of years. He was shaking and hesitant in his speech and a bit halting and I could tell he was in trouble."

At 2.50 pm, just before the trading banks closed, Minsec stopped accepting cheques. This was on Acheson's advice. At 4.30 pm McMahon sent a letter to the Sydney Stock Exchange requesting suspension of Minsec shares. Next morning McMahon went to the office of the exchange in O'Connell Street and saw the exchange's chairman Jim Cooper and general manager David Butcher. He gave them a notice prepared at 11 pm the previous night announcing the $3.3 million loss and detailing the holdings of Kathleen, Thiess, Queensland Mines and Robe River. Although he and Nestel retained hope for some time that the company could return to solvency, this was, in fact, the end.

King & Yuill withdrew from the underwriting agreement, claiming it had been breached by a material change in Minsec's affairs. This shut Minsec off from access to $16 million that might have saved the group. But if there had been a substantial shortfall in the issue, King & Yuill could have had trouble meeting its share of the commitment. A significant proportion of its assets was tied up in loans to Minsec, a fact which was to put it in grave financial stress very shortly.

On February 11 three of Minsec's major creditors — Mobil Oil, Westralian International and Commercial Continental — petitioned the NSW Supreme Court for Minsec to be wound up. The court appointed Jim Jamison as provisional liquidator.

A lingering belief persisted that Minsec could have survived. If the profit had not been reversed, if the preference issue had gone ahead and PRA had floated, Minsec would have survived in the very short term but would have still been facing immense difficulties. Whispers were spreading about the dubious nature of the profit, and shareholders could have questioned McMahon on the point at an extraordinary meeting set down for March 1 to approve the preference issue, share split and bonus.

Even if the profit was legal (and on a later court judgment it was held to be) it remained an artificial profit. It might have been a profit in the eyes of a lawyer or even a judge but it would have been regarded as artificial by analysts. It might be defended in a court (as it was) but it could not have been defended behind closed doors with the financial institutions from whom Minsec was trying to borrow money. Minsec shares would have weakened and borrowers would have pulled out. Borrowings would have run down further as the tight June quarter approached. Simultaneously reports of the downgrading of Nabarlek were filtering through the market, depressing QM and KI and creating a further serious source of weakness for Minsec.

PRA could only have been floated at heavy losses to both Minsec and Castlereagh, if indeed it could have been floated at all. By not reversing its profit statement, Minsec might have forced King & Yuill to honor the underwriting agreement, but, given King and Yuill's state at the time, this might not have aided Minsec much.

Also, as Minsec battened down the hatches it would have had to curtail buying Robe River on-market. The consequent fall in price in these shares, say to $1.80, would have eroded Minsec's assets and the lenders' security. Minsec would still have had to find scrip to fund its 150 percent cover.

Minsec's withdrawal from the funds would have raised doubts about the viability of both Minsec and the funds. It was obviously an act of desperation. Even if Minsec could have secured the $16 million from the preference issue, it would have plugged the drain in money-market funds, but not for long.

A concise epitaph was written by Ric Dowling. In a statement to the Rae Committee, he wrote:

"Shortly, the fact is that those who controlled Minsec wrote their own death warrant in classic terms:

● They borrowed short and they invested long.

● They did this in equities in a falling market.

● They forgot that gearing loan funds to equity increases profits in a rising market but aggravates losses in a falling one.

● They forgot that you don't buy anything until you can pay for it.

● At the time of their collapse their total borrowings were $64 million. Shareholders' funds at June 30, 1970, were $24 million. This ratio of three to one is wholly unreasonable in a company of this kind. One-half to one would be reasonable."

Dowling was admittedly a hostile witness, but very similar conclusions were reached in the Rath Report, which said: "The cause of the crisis was primarily the improvident management of MSAL, particularly the implementation of a program of large investments without adequate financial arrangements. It may be that the successful flotation of PRA might have delayed or perhaps even forestalled the crisis; but such a successful flotation would have required market conditions that never occurred. MSAL speculated, perhaps more grandly than it had done before, and its speculation failed because its expectation regarding the Nabarlek discovery did not materialise."

---

[1] Interview with author, March 1970.

[2] Colloquially, Mineral Securities Australia Ltd was widely known as Minsec. There was also a subsidiary called Minsec Investments Ltd which was involved in some important share transactions. When referring to the latter company in this chapter I have given its name in full. Otherwise the term Minsec refers to Mineral Securities Australia Ltd.

[3] The National Mutual Life, the Colonial Mutual Life and the MLC.

[4] A pseudonym. His real name is George Goodman.

[5] This is typical of bull profits. A consistent bull will go through long, lean periods then make large profits in a very short time, but he must not overstay.

[6] The NSW Government appointed a team of inspectors to investigate the Minsec collapse. The inspectors were A.F. Rath, QC, P.C.E. Cox and H.A. Collum. I refer to it as the Rath Report.

[7] Nestel's emphasis.

[8] If Minsec converted its loan holding into shares it would have had a little more than 50 per cent of Robe River Ltd at the time.

[9] Insofar as any individual can take credit for starting the funds, it would be Krishnan Tatparanandam. He had frequently suggested the idea to Nestel.

[10] Conversation with author, June 1978.

[11] Author's emphasis.

CHAPTER SEVENTEEN

Panic in the Boardrooms

The collapse of Minsec set off one of the greatest financial panics in Australian history. Not only was the collapse of a major company sending shock waves through the stock market, the collapse of a major borrower was wreaking havoc in the money markets. For a couple of nervous weeks, the crash of Minsec seemed likely to bring down a number of other companies and financial institutions in a chain reaction. The panic was so widespread that it provoked the intervention of the Prime Minister.

The year 1969 had not only been a great one for mining stocks; it had been a vintage year for merchant banking. Revelling in a prospering economy and enticed by a prospect of boundless mineral sources of wealth, financial institutions had been breeding at a prodigious rate. Consortia of high prestige were formed by local banks and financiers in partnership with leading overseas banks until it seemed that every major financier in the world had a stake in an Australian merchant bank. The term "merchant bank" became fashionable, carrying connotations of sophistication and high financial expertise. It was a term capable of wide definition. Australia had institutions such as Darling & Co and Development Finance Corporation which had for years fulfilled most of the functions of merchant banking, such as lending, underwriting, corporate financing, money-market dealing and portfolio management. Now they faced a massive invasion of foreign merchant banks, and the establishment of local consortia with strong foreign links.

The concept rapidly caught on among local financiers. Soon it was a backward broking firm that did not boast its own money-market operation and merchant banking expertise. Fly-by-night operators with nothing more than a company shell and a shingle were calling themselves merchant bankers.

The formation of so many small exploration companies

helped to fuel the fire. Immediately after floating each company had some cash, ranging from a few hundred dollars to a few million, for which it had no immediate use. An obvious solution was to place the funds on call or at term in the money market, where the multitude of new, competitive merchant bankers were bidding eagerly for business.

Before continuing to explain what happened in the Minsec affair, it is necessary to give a thumbnail sketch of the Australian money markets. At any given time in Australia there are a number of companies, institutions and other organisations with surplus cash. Also at any given time there are a number of companies and institutions who need cash. The money market is a mechanism by which the surplus companies lend their money to illiquid companies for set terms and rates of interest.[1] There is no physical marketplace as such, like the floor of the stock exchanges. There are just a lot of brokers, businessmen and merchant bankers calling each other on the telephone.

A couple of simple examples will show how it works. Let us say Ford Australia is short of cash. The company is perfectly sound because it has masses of assets, including its factories and all those cars out in the paddock, but it does not have any cash at the moment and it has a bill to pay. It may have to meet the weekly payroll or pay the windscreen supplier. Let us say it needs $1 million. Ford could get the money from its bank, but it may be on its overdraft limit, or the bank may be charging a high interest rate. So the company secretary rings around a few companies to find one with a spare million. Hamersley may just have received payment for an iron-ore shipment and is looking for somewhere to place the money until it pays its next dividend. So when Ford's company secretary rings Hamersley's they are in a perfect position to do a deal. Hamersley might lend the money at call, which means that the loan can be recalled in 24 hours, or it might lend the money for a term. The usual terms are 30, 60, 90 and 180 days.

In practice, it would be highly unlikely that the two companies would intersect in this way. Each of them would ring a money-market dealer instead. These dealers sometimes work as individual outfits but more often as an adjunct of a financial institution such as a merchant bank or broking firm. The operators in these offices spend all day on the phone trying to match borrowers and lenders. By letting these dealers place the money for a small commission, Hamersley has a much better chance of getting its money out and also of getting the correct prevailing rate of interest. Rates of interest on call money are usually lower than those for term-money and term-money normally costs more as the term lengthens.

A deal such as the Ford-Hamersley one above is called an inter-company loan. After it has been arranged between the two secretaries, or more probably through a dealer, the lender sends the borrower a cheque and the borrower sends the lender a slip of paper acknowledging the terms of the debt. A million dollars, or far more, can change hands in this way with a maximum of speed and a minimum of formality. This efficiency is the beauty of the unofficial market. A company can raise large amounts on the strength of its name alone far more quickly than it could borrow the same sums through the banking system. This provides an incentive to use the money market even if the rates there are a little higher than the overdraft rates.

In the exercise described above the loan was unsecured. If, when a debt falls due, the borrower is unable to repay, the lender can put him into liquidation. But in a liquidation the unsecured creditor ranks behind some other claimants. The claims against a company in liquidation rank as follows:

1. Liquidator's fees and legal expenses.
2. PAYE tax.
3. Employees' wages, holiday pay and long-service leave.
4. Commissioner of Taxation (company tax, sales tax etc.)
5. Secured creditors.
6. Unsecured creditors.
7. Preference Shareholders.
8. Ordinary shareholders.

A normal lender to a company cannot be an employee or the Department of Taxation, but he does have the option of being a secured or unsecured lender. Secured loans normally bear lower interest than unsecured, which means the lender has a choice between maximising his safety or maximising his return on the loan.

A loan may be secured on any assets of the company. It may be secured on a specific asset, such as a block of land or a parcel of shares, or it may be a floating charge over all the assets of the company. A company wishing to borrow can also issue a commercial bill. In the above example, Ford might have decided it needed $1 million for six months and to do this might have issued a bill for 180 days at, say, 9 percent. Typically it would issue the bill to a dealer, who would pay Ford the $1 million and rediscount (resell) the bill on the market. Such a bill might be rediscounted up to a dozen times before falling due. Whoever held it at that time would then present it to Ford for payment of principal and interest. If the market desired greater security, Ford could have the bill endorsed by a bank. This virtually means that the bank guarantees repayment of the loan should Ford default. Bills such as these are the daily fodder of the money markets of Australia, along with short-term notes and debentures and government securities such as Commonwealth bonds and Treasury notes.

What has been described here is called the unofficial money market. There is also an official market comprising nine firms authorised by the Reserve Bank. These official dealers operate mostly in bonds, T-notes and semi-governmental authority debt, but can also hold small amounts of commercial bills and private sector debt. The difference between these dealers and the rest of the market is that the official (sometimes called authorised) short-term dealers have lender-of-last-resort facilities with the Reserve Bank. Very broadly, this is tantamount to a Reserve Bank guarantee of their debts, but in return the nine dealers must operate within very strict guidelines and under constant Reserve Bank scrutiny.

This is by no means an exhaustive account of Australia's money-market structure but it will suffice for a layman to understand what happened in the Minsec affair. Minsec had been Australia's heaviest borrower in these unofficial markets. At the time of its collapse in early February, Minsec's borrowings from these unofficial markets totalled some $70 million, of which $50 million had been borrowed between August and December 1970 to finance the KI-QM purchases. This was a massive sum by the standards of the day. Many a small, newly-fledged mining company, dazzled by Minsec's glamor and new to the inter-company market, lent the group large sums unsecured.

For months before its collapse Minsec had been luring deposits by offering interest rates as high as half a point over the prevailing market. If call money was 6¼ percent elsewhere, Minsec would offer 6¾ percent. This is a classic danger sign of a hungry company but it was ignored by sophisticated and unsophisticated lenders alike.

The undisclosed expansion of borrowings by Minsec was a secret, but not a very secure one. If any of the major institutions or dealers who had advanced money to Minsec ever began comparing notes, hints would immediately begin circulating about the great change in Minsec's gearing. The same things could also occur if any tightness in the money market created a drain on call-money. A lender normally leaves call-money with a borrower until he needs it back. Many of the lenders to Minsec would need their money back between April and June to pay their annual income tax levies. The June quarter was traditionally a tight money quarter for this reason. In 1971 it was expected to be tighter than usual. Tax payments would be higher because of larger company profits. From the end of 1970 onward this approaching tightness was of major concern to Minsec and its advisers.

The first pebble was dropped into this unstable pond on New Year's Eve when the takeover offers for Aberfoyle and Cudgen went out in the mail. The compulsory Part B statement attached to the takeover documents contained the note revealing Minsec's borrowings.

Unease spread. Lenders began calling money back from Minsec or demanding greater security. The interim profit announcement did little to stem the ebbing tide of confidence, which was reflected by the downward drift in Minsec shares. By the end of January they stood at $7.70, only one-third of their peak value for 1970.

One person who became acutely concerned was Mark Johnson. Johnson was one of the bright young men of Darling & Co, and widely tipped to have a big future in merchant banking.[2] Astute, determined, and possessing a strong sense of humor, Johnson was a protege of the redoubtable John Darling, one of Australia's top financiers. Johnson was largely responsible for Darling & Co's underwriting of the float of Robe River Ltd. He sat on the Robe River board, which included McMahon and Nestel, and was becoming increasingly concerned by Minsec's apparently desperate need for money. He feared that Minsec may have raised some money by pledging assets of Robe River. If this were so, Robe River Ltd would be imperilled because all its funds were needed for future commitments to the big iron-ore project.

On February 2, the Tuesday before Minsec crashed, Johnson decided to take control of Robe. To do this he needed the votes of three directors, of whom two had to be physically present to convene a legal board meeting. Johnson obtained a proxy from the chairman, Bruce Macklin, who was about to go overseas. Then Johnson went down to the office of another director, David Burns, of Burns Philp & Co Ltd, and persuaded him to go around to the Robe River office and convene a board meeting. As Johnson and Burns were walking down the street, Burns asked how much Minsec was supposed to have borrowed. "I hear $60 million," said Johnson. Burns, whose company was one of the oldest, largest and soundest in Australia, gasped. "Do you know," he said, "we've only got an overdraft limit of $2 million with our bank — and they only extended it to that last year?"

Johnson and Burns held their meeting and acquired the records of the company. They found that a number of Robe River securities had been given as security by Minsec. Johnson and Burns set about trying to get the money back as quickly as possible and managed to salvage about $2 million before Minsec went into liquidation. This reduced the total owing to Robe River to about $9 million. In legal theory this money might have been demanded back by the liquidator. The law, broadly, states that where a creditor knows a company is insolvent and receives preference in repayment of a loan, the liquidator can subsequently force him to put the money back into the pool.[3] But when a company is sinking its creditors invariably take the attitude of "sauve qui peut". Johnson was primarily concerned to get the money back into Robe River first and argue about the legal niceties later.

Minsec had continued its money-market operations right up to the death. These operations were conducted by Rod Stringer, with the assistance of Brian Payne, under the general direction of Nestel. For a group which became Australia's biggest corporate borrower on the unofficial market this was a tiny money-market department. Stringer did not get much assistance from anyone else in the organisation and once asked Nestel to employ a senior money-market operator. Nestel told him to continue the work he was doing, explaining that if he hired anyone more senior it would appear to clients such as Patricks and King & Yuill that Minsec was trying to start its own money-market operations.

Stringer had to balance the daily cash-flow, which meant to get enough money in each day to balance the amount flowing out. He achieved this feat quite successfully. Right up to the point where Minsec crashed, he was able to balance the books on a day-to-day basis. However, the group's larger financing problems and strategies were well outside his responsibility. He did not know what funds were required for particular projects and was not concerned with payment of interest on Minsec's mounting debt. By the end of Minsec's career, the ever-increasing short-term borrowings were being used to finance brokers' accounts and loans.

While Stringer was responsible to Nestel, his authority and limitations never appear to have been accurately defined.

Stringer was left to raise the money as he saw fit. "There is little doubt that he was given responsibility without adequate supervision," the Rath Report says. "His principal task was to raise funds for the day-to-day requirements of Mineral Securities and this task he performed with considerable success, but with little regard to the interests of those who owned either the funds or securities deposited in his custody. He seems to have acted on the simple principle that if some particular dealing was in the interests of Mineral Securities then that was sufficient justification for it, without any regard to the particular interests of the other companies concerned."

The most questionable arrangements were those made with the funds of Robe River Ltd. Under Robe's trust deed the maximum amount which Minsec could have borrowed from that company in 1970 was $1.5 to $1.7 million. However, Minsec had also been given authority to "place" Robe's funds for a brokerage rate of one quarter of 1 percent. In addition, there was an approved list showing the people to whom Robe was prepared to lend money, and how much.

As Johnson began checking after February 2, he discovered that some borrowers had not been on the approved list, that others had borrowed beyond their limits and that loans which should have been secured were not.

One example was Ord-BT, which had borrowed $4.5 million although its limit was $3 million. Bill Acceptance owed $2 million although its name did not appear on any of the lists (the approved list was altered from time to time) seen by the Rath inquiry.

In some cases, dealing had been "back-to-back". This meant that Robe would lend money to a company in exchange for an equivalent loan to Minsec. Ord-BT claimed that it had acted only as a "clearing house" in such back-to-back transactions and that it would demand recourse from Robe River if there was any shortfall in the repayment it received from Minsec. In another case Minsec had lodged $3 million of Robe River's cash with King & Yuill Investments Ltd with strict instructions to on-lend the money to Minsec.

When Johnson tried to recoup the Robe River loan to King & Yuill Investments he was told that Robe could not get the $3 million until it had come back from Minsec. King & Yuill Investments was at this time run by John Northcott, who was general manager, and his brother Tom, who was assistant general manager. On Wednesday February 3, Johnson went to Tom Northcott's office and gave 24 hours notice that he was calling back the $3 million and told Northcott the loan had breached Robe's trust deed. It was a dramatic confrontation.

"That was the first time I had met Mr Johnson," said Northcott. "He burst into my office. He was shaking from the fingertips down to the knuckles."

Johnson said he wanted his money back and Northcott said he would pick up the phone and ring Minsec. Johnson did not get the $3 million. Several months later a deal was arranged between King & Yuill, Robe and liquidator Jamison under which Robe would be repaid when King & Yuill was repaid.

The shortcomings of these back-to-back loans are obvious. A back-to-back is a loan from one company to another with a financial intermediary acting as agent. Almost the only logical reason for doing a deal this way is to circumvent some legal difficulty such as a trust-deed requirement. If the lender has no recourse against the intermediary (a legal point which could become quite complicated) his loan may well be at risk. Whatever the legality of this type of on-lending, the morality of the device is dubious.

The Rath Report leaves the impression that McMahon and Nestel preferred not to know how their deals were being funded. Stringer said Nestel had remarked to him in January 1971: "I understand you probably juggled things around a bit but that is not of my concern." Of McMahon, Stringer said: "Mr McMahon was frequently overseas and unavailable and on some occasions I tried to speak to him about the problems of raising funds; at one stage I waited for months to see him because he was always away on business of different types..."

Even at the death-knock Minsec had enough prestige to raise large sums on the inter-company market. On February 3 — before Minsec told the Sydney Stock Exchange its profit had changed into a loss — the group opened its doors for money-market operations as usual. Stringer, ignorant of the impending announcement, was quite pleased with the day's raisings because he had attracted two new large lenders for the first time.

Both Western Mining Corporation and Mobil Oil had lodged money with Minsec. Stringer, with the optimism of innocence, told Nestel that the fact both these large companies had lent money to Minsec for the first time indicated acceptance of Minsec as "a company in the better part of the market".

At about a quarter to three that day Nestel and Bogg came into Stringer's office and told him to cease money-market operations pending a statement to the Stock Exchange. They looked at his book and told him they could not accept the Western Mining and Mobil deposits. Stringer told them that the cheques had been banked. Nestel tried, unsuccessfully, to stop the cheques.

Western Mining could be regarded lucky. The company had deposited $300,000 with Minsec and the cheque had been banked. At 3.30 that afternoon Western Mining offered a further $700,000 to Minsec, but by this time Stringer was under orders to stop and he declined the money.

The Mobil transaction was tinged with black humor. All Mobil's Australian money-market transactions had traditionally been controlled by the parent company in New York. For years the Australian management had been chafing under this control. They had pleaded with the US head office to be allowed autonomy to place money in the market themselves. Finally New York conceded. Independent at last, Mobil Australia made its first money-market transaction by placing $1 million unsecured with Mineral Securities on February 3. Minsec banked the cheque and two hours later closed its books. After the long struggle to gain its cherished autonomy, Mobil Australia had put $1 million at risk only two hours after its first transaction.

It was the one good laugh the money market had all week. Mobil's Australian management was mortified and the oil company's head office was incredulous. The head of international security for Mobil flew to Australia to investigate the mystery of the missing million. His suspicions were by no means allayed when Nestel, Spooner and Swift — each shell-shocked by Minsec's collapse and with problems of his own — offhandedly gave him different stories. "He simply couldn't believe what happened," Bogg said with a grin, years later. "He wanted to stay the weekend with one of us. I told him he could come out to my place and mow the lawn, and he agreed. He never turned up, though."

In Minsec's case, as in many liquidations, the major problem was to distinguish between the secured and unsecured creditors. Minsec had been pledging scrip as security for loans. In the event of Minsec failing lenders holding such security could, if their charges had been properly registered, regain their money by reselling the shares on market.

Let us say a lender had lent Minsec $500,000 and had legal proof of the debt and a properly registered charge over a parcel of Pexa shares. If Minsec could not repay the debt when due that lender had a right to sell his block of Pexa on market, even if the liquidator did not want him to. Presuming that the lender was owed $510,000, including interest, he could sell the shares to recover his debt. If he received only $400,000 for the shares he would rank as an unsecured creditor of Minsec for the balance of $110,000. If he received $550,000 he had to turn the $40,000 surplus over to the liquidator for distribution among the other creditors.

As it became increasingly obvious in late January and early February that Minsec was going to fail, a panic began among creditors. At first they tried to get their money back, and then they tried to get greater security. In this period, several had their loans converted from unsecured to secured so that they would rank higher in any liquidation.

The arithmetic of a liquidation works like this. Let us say a liquidator is appointed to a company which has secured creditors who are owed $500,000 and unsecured creditors who are owed $1 million. If, after meeting the claims of employees, his own costs and the tax people, the liquidator has $500,000 left, the secured creditors will be paid and the unsecured will get nothing.[4] If he has $750,000, the secured creditors will be paid in full and the $250,000 will be divided between the unsecured, who will get 25 cents in the dollar. If he has $2 million left both classes of creditor will be paid in full (depending on the terms of their loans, they may also receive interest) and the remaining $500,000 will be split between the shareholders. Among the shareholders, preference stock must be paid in full before anything accrues to ordinary shareholders.

As Minsec tottered, creditors scrambled to obtain better security, or any security, to place themselves higher up the ladder of priority. In its last few days of life Minsec converted a number of loans from an unsecured to a secured basis. In some cases there were simultaneous "repayments" and "new loans", with the fresh loan being secured.

After Minsec collapsed, one constant nightmare for everyone was the prospect of a dispute being taken to the point where the security of the creditors claiming security was challenged. This could have led to a mass of lengthy and complicated legal actions as everybody contested the validity of each other's security. This would have taken time and been fearfully expensive, swallowing a large portion of the funds available for distribution. A conspiracy of silence developed on the grounds that it was better if the whole rats' nest were left undisturbed. There was still a possibility that an unsecured creditor might test the validity of the secured loans claimed, but, thanks largely to Jamison's strength of character, this threat never developed.

The real panic started in the money markets just before

Minsec crashed. Creditors desperately tried to withdraw money from Minsec and failed. Then they in turn became targets. A rumor that a company had been "caught" in Minsec would spark a run by its creditors as they too frantically withdrew call money or demanded to be paid on accounts. As such rumors spread more companies became suspect. Wholesale panic set in as virtually all but the very strongest companies became suspect overnight. Perfectly safe and respectable second and third-line companies found all their borrowings being withdrawn in the general panic. These borrowings were almost unreplaceable because lenders had gone into a blue funk, including a couple of trading banks. Through no fault of their own, many small companies were facing insolvency.

In the midst of this panic, which began on February 2, major creditors had begun meeting to discuss what could be done. The timetable of the creditors that week gives the impression that the top men in Minsec and the major creditors did nothing else but meet each other for a whole week.

The most important of the first meetings were organised by a young American, Mark Acheson of Ord-BT. Ord-BT had strong motives for wanting Minsec to survive, or, failing that, the most orderly liquidation possible. Ord-BT was one of Sydney's leading merchant banks and one of Minsec's largest creditors, being owed nearly $8 million including interest. Minsec had approached Ord-BT in January for a loan of $30 million as part of the long-term refinancing suggested in the Castleman report. Ord-BT rejected this request on January 27 but continued to take part in financing discussions with Minsec.

Acheson was trying to put a lending consortium together until the afternoon of February 3, when he rang McMahon and told him he had had no luck in his attempts. At Acheson's suggestion, McMahon then told Nestel not to accept any more money or bank any more cheques. After speaking to Acheson, McMahon told another director, Max Moorfield, that Minsec was "going to the wall".

Another major creditor of Minsec was the Perth-based merchant bank, Westralian International Ltd. Acheson rang Westralian's managing director, Anthony Van Toll, on February 3 to tell him that Minsec was going to announce that the interim profit statement would be changed to a loss. Ord-BT had been empowered to act on behalf of Minsec and Van Toll was invited to come to Sydney to participate in discussions. Westralian was owed $2.5 million, of which $2 million was unsecured. According to Van Toll's affidavit in support of his petition to wind up Minsec, Acheson had said: "The days of Mineral Securities as a share trader are over. The possibility of repaying 'upstream loans' is out and the dividend income over the medium term will be totally insufficient to meet external borrowing payments. Creditors can only get out by a realisation of assets. Those depositors who have money at call may call but Mineral Securities cannot cover."

This was 9 am Perth time. That afternoon Van Toll called $600,000 of the $2.5 million Westralian had on deposit with Minsec. Acheson called Van Toll again and said: "You should not have called that $600,000. You have precipitated the situation we were trying to avoid." Acheson handed the phone over to Nestel who told Van Toll that Minsec was in a very difficult situation and would appreciate any help that could be given.

Van Toll flew to Sydney that night and next morning attended a meeting of a number of large Minsec creditors in Ord-BT's office. This meeting included representatives of King & Yuill, Elders Finance & Investment Co Ltd, Patrick Partners, Bill Acceptance, Trans City Discount Ltd, the ANZ Banking Group, Chase-NBA Group Ltd, Commercial Continental Ltd and Waltons Ltd. Several of these and other creditors had been busy in the preceding days trying to obtain greater security, which Minsec had granted wherever it could. Their security was dubious, however. Jamison later said[5] that in all but two cases it had not been properly registered. One of the two cases was a small exploration company called Esperance Minerals, which had lent $500,000 to Minsec. If Jamison was right, Esperance had thereby upstaged half of the allegedly sophisticated merchant bankers in Australia.

At the February 4 meeting the Ord-BT representative stressed the shattering consequences for both the equities and fixed interest markets if there was a "disorderly liquidation" of securities by individual creditors. Acheson proposed a scheme under which all the major secured creditors would put more money in to enable Minsec to meet its call position. When the situation had stabilised, Minsec would sell some of its assets and the lenders would be repaid.

When this proposal was put to the meeting, one merchant banker agreed. Some others agreed also, although the ANZ Bank was not enthusiastic. The ANZ was owed $5 million. Then Bob Thomas, the managing director of Elders Finance, said he disagreed. Thomas, an experienced banker and financier who ran the merchant banking arm of Australia's largest pastoral house, did not think the scheme would work and declared that Elders would not be in it. After listening to Thomas' arguments, the first merchant banker to declare support for the scheme changed his mind and said he too was opposed. The meeting ended inconclusively, with creditors haunted by the spectre of a stock-market collapse.

That weekend saw a series of discussions between various groups. On Saturday, February 6, the head of the giant UK Rio Tinto Zinc group, Sir Val Duncan, flew into Australia to take part in the talks. He held discussions with representatives of Peko, MIM and the Wales. The constant talks and secret meetings such as these gave the impression of portentous activity, but, in fact, there was little effective action.

That same Saturday Van Toll phoned George Todner, the chairman of Trans City Securities Ltd, and said he was sorry to see Trans City was one of the creditors.

Todner said: "We are now secured."

Van Toll asked: "When did you get your security?"

"Last Tuesday," replied Todner.

"I think that might be the sort of security you could paper your wall with," said Van Toll.

"I don't know what right they had to give it to us," said Todner, "But we have it and we are sticking to it."

Trans City was not the only creditor to get security late in the piece.

Another petitioner to wind up Minsec was Commercial Continental. An affidavit in support of this petition detailed the new charges which appeared on Minsec's security register in the last few days of its life. The creditors who obtained security in this time are shown in the table on p.284.

The secured creditors between them held controlling interests in Cudgen, Aberfoyle, Robe River and large blocks of stock in Thiess, Queensland Mines and Kathleen. The prices of these shares were already retreating under the knowledge that large blocks would be up for sale. A panic sale by one or more of the large secured creditors could push prices even lower. The temptation to be first out was offset only by the consideration that there might be a higher price if the stock could be offered in one block as a controlling interest. Top Sydney accountant Ken Humphreys (later Sir Kenneth) was called in as a prospective liquidator. He was one of Australia's leading "company doctors", and also held board positions in many large companies.

Sunday February 7 was one of the most nervous days of the

Creditor	Date		Secutity
MINERAL SECURITIES AUSTRALIA LTD			
Asian Development Foundation	Jan. 21	45,000	Kathleen
State Government Insurance Office	Jan. 21	100,000	Robe River
Bill Acceptance Corp.	Jan. 26	25,261	Queensland Mines
North Aust. Uranium Corp.	Jan. 26	1,500,000	Robe River
Patrick Partners	Jan. 26	1,000,000	Aust. Antimony
		80,00	Brins Australia
		2,000,000	Castlereagh
		19,600	Queensland Mines
		800,000	Westmoreland
		32,600	Queensland Mines contribs.
Selected Securities Ltd	Jan. 26	130,000	Robe River
Ord-BT	Feb. 3	107,000	Aberfoyle
Elders Finance	Feb. 3	150,000	Robe River
King & Yuill Investments	Feb. 3	173,000	Aberfoyle
		86,000	Cons. Rutile
		80,000	Cudgen

MINSEC INVESTMENTS LTD

Creditor	Date	Security	
Ord-BT	Jan. 28	200,000	Aberfoyle
Ord-BT	Feb. 3	2,000,000	Cudgen
King & Yuill Investments	Feb. 3	768,000	Cudgen

panic. On a summer Sunday afternoon, the heart of downtown Sydney is normally a desert of concrete canyons, the only life in sight being a few idling youths and derelicts. If these flaneurs had been better versed in high finance, they would have recognised some of Australia's top businessmen converging on Wynyard Square.

Humphreys had called a meeting in an upstairs function room of the Menzies Hotel. Those present were predominantly top mining men, with a sprinkling of financiers. Those present were Fred Deer, general manager of the MLC; John Dunlop (later Sir John), chairman of CSR; Jim Vernon (later Sir James), CSR's general manager; Jim Foots (later Sir James), chairman of MIM; Alan Coates, investment manager of the AMP; Noel Kirby, joint general manager of EZ Industries; Brian Massy-Green (later Sir Brian), chairman and managing director of CGFA; Jim McNeill (later Sir James), managing director of BHP; Sir Robert Norman, chief general manager and a director of the Bank of NSW; John Proud (later Sir John), chairman of Peko; Charles Rennie, general manager of the ANZ; Mark Turner, (later Sir Mark) deputy chairman of Rio Tinto Zinc; and Rod Carnegie (later Sir Roderick), managing director of CRA. Humphreys was assisted by Bob Stevenson, senior partner of Allen, Allen and Hemsley.

News of the meeting had leaked out beforehand, with the result that Minsec and Humphreys suffered a minor plague of telephone calls from other businessmen wanting to attend. In some cases their main motive was the prestige that would attach to sitting at the table with the cream of Australia's business community. Some of those who were excluded lurked in the lobby downstairs, watching those who came and went.

There were a lot of comings and goings because the meeting invited various people in from time to time for questioning, discussions or suggestions. McMahon and Nestel were kept waiting for a long time before being called in, together with a few of their senior employees. The meeting's discussions quickly centred on the money market. They interviewed the Northcotts, who were very nervous, and Acheson. Everyone was deeply worried that the next day might see a major panic begin on the money markets of Australia. These markets had grown explosively over the last few years. Now they were facing their biggest crisis. However, the meeting could not agree on any plan of concrete action. The talks had begun just before 3 pm and dragged on until early morning. They did not break for dinner. "I don't think anyone wanted to eat at that stage," one of them said later. Nothing was achieved. "It was a night of absolutely futile discussion," said one participant. "The most immediate concern was that King & Yuill Investments was in dire risk of failing. The trouble with this and other meetings at the time was that everyone wanted to be a big fellow by being in the discussions, but nobody wanted to put cash up." The meeting broke up after 1 am, having done nothing except instil a very real fear in all present that the panic would grow greater. And the money markets would reopen in less than eight hours.

While this high-powered conference had been in progress, a much smaller one was proceeding a few blocks away at the top of Martin Place. Keith Phillips and Bill Shugg, the senior partners of the broking firm of King & Yuill, were in conference with senior executives of the Bank of New South Wales. The men from King & Yuill knew that their group would be in the front line, taking the brunt of any money flight the next day.

King & Yuill had been one of the broking firms most closely associated with Minsec during the boom. They had not only bought and sold shares for Minsec but their associated merchant bank had been a virtual banker to the group, providing large lines of credit. It was a noteworthy feature of Minsec's relationship with brokers that Minsec took its business where it could get the best credit. Brokers who were earning large commissions out of Minsec's leviathan trading were not loth to provide finance, either from their firms or their associated merchant banks.

The largest exposure among brokers and merchant bankers when the crash came was by King & Yuill Investments Ltd ($8.26 million); Ord-BT Ltd ($9.3 million); and Patrick Partners ($2.5 million) and its associate Bill Acceptance Corporation ($5 million). Other brokers heavily involved included Ord Minnett T.J. Thompson & Partners and Johnson & Wedell, each owed more than $300,000. When Kurt Wedell was finally paid out he came around to Jamison's office with a bottle of champagne.

The panic run actually took more money out of the markets than the Minsec crash itself. Jamison told the Rae Committee that about $67 million had been called on one finance house in three days and that one stockbroker had been paying out at the rate of $12 million a day. Considering that these were merely the two most lurid examples, it is obvious that the total run on the market in those nervous days was far greater than the $70 million which had been frozen in Minsec.

The board's official list of creditors was not published until March 1. Well before then word was spreading about which companies had been caught. Several companies who had not been creditors had to issue statements denying they were. Others, under pressure, revealed their involvement. The worst sufferer was the King & Yuill group.

The partners in the broking firm had an effective controlling interest in Australian Investment and Development Ltd, which was a listed public company. Other major shareholders in AID were the Bank of NSW Superannuation Fund, the Mutual Life and Citizens Assurance Co Ltd and the Producers and Citizens Life Insurance Co. AID had a 51 percent interest in King & Yuill Investments Ltd and in one of Australia's nine authorised dealers, Short Term Acceptances Ltd. The remaining 49 percent in these two companies was held by the partners of King & Yuill.

On Monday February 8, creditors pounced on the group. The run could not have occurred at a worse time for King & Yuill Investments, which was then abnormally exposed through other transactions. At the time of Minsec's failure, the gearing of King & Yuill Investments was abnormally high because the company had made a special loan of $26 million for four weeks to the broking firm so that the brokers could buy rebatable Commonwealth bonds, receive interest payments on them and gain special tax rebates — a practice known as bond washing. King & Yuill itself had borrowed the money to buy the bonds and sell them to the broking firm. This had increased the deposits (and the corresponding liabilities) of the investment company to some $46 million. Its paid capital was only $300,000.

On the day after the public announcement of Minsec's difficulties, depositors made heavy withdrawals from King & Yuill Investments. It was now the Northcotts' turn to have white knuckles. The investment company could not replace this drain with fresh money because — in the vicious circle of a money run — the securities it held were not acceptable as collateral for further borrowing. King & Yuill Investments finished the day with an immediate cash shortage of $1.2 million. This was made up by the broking firm, which deposited that much money with the investment company on the security of Minsec bills, which would have been commercially unacceptable anywhere else in the market. King & Yuill itself had raised the $1.2 million from another broking firm.

The pressure on King & Yuill and its associate companies was immense. This was the most crucial point of the panic. If the broking firm or any of its associates had closed their doors the panic would have spread and accelerated. At this vital moment the saviour was the Bank of New South Wales, which stepped in and backed the investment company. The bank took over, as a first charge loan, a secured credit amounting to more than $6 million which King & Yuill Investments had with Minsec. This was a heavy commitment by the bank and a bold step in highly nervous times. Doubtless the shareholding of the bank's superannuation fund in AID was a factor in the decision, but the Wales also saw the necessity for stopping the panic. The Wales deserves great credit because if the panic had been allowed to spread there is no telling where it would have stopped. John Northcott later told the Rae Committee that if the bank had not stepped in "we would have suffered a further run and the domino effect may have started."[6]

King & Yuill's business that day was conducted in an atmosphere of tension. Some creditors who had called their money that morning kept telephoning the merchant bank during the day to ask when they would be paid. King & Yuill Investments told them to be patient and assured them everything would be all right. By 5 pm the creditors were sitting in King & Yuill's office. They waited there until 6.30 pm, when the bank cheques from the Wales arrived.

Many other financial institutions acted in the worst way possible and were largely responsible for the panic. Large institutions called substantial funds from the marketplace generally and redeposited the money in the banking system or with authorised dealers.

Apart from the Wales, John Gorton can take some credit for helping restore confidence. The meeting he called at Kirribilli House did much to calm fears, thanks to the well-kept secret that the participants had no plan that would have been effective in the circumstances. The meeting on February 9 was kept a secret at first, and details of it became known only gradually. Gorton, a strong economic nationalist, had been taking a keen interest in the Minsec affair. His meeting appears to have been called at short notice.

The meeting illustrated the erratic aspects of Gorton's character. With typical energy he had summoned a group of Australia's top businessmen to be present that evening; then he kept them waiting an hour while he finished dinner.

At dinner with Gorton were Sir Val Duncan, Rod Carnegie and Sir Cecil Looker, chairman of the Melbourne Stock Exchange. Also present, according to my sources, was Ainsley Gotto, Gorton's petite, pretty and fearsomely efficient secretary.

Having dined well, Gorton went into conference with the businessmen, at 9.15 pm. They had been waiting since 8.15 pm. Apart from those already mentioned, the men present were Sir Alan Westerman of the Australian Industry Development Corporation; Humphreys; Rennie; Keith Steel, general manager of the AMP Society; Geoffrey Bowen, chief general manager of the Commercial Banking Co of Sydney; Sir Norman Young, director of Elder Smith Goldsbrough Mort Ltd and the Bank of Adelaide; Sir Robert Norman; Philip Miskin, managing director of Bill Acceptance; Acheson; Asa Davis from Ord-BT, Mark Turner; John Phillips (later Sir John), Governor of the Reserve Bank; and Ted O'Halloran of the leading Sydney law firm Freehill, Hollingdale and Page.

Gorton had three main areas of concern. He wanted to convey the government's concern that a possible domino effect did not develop through the Australian financial world. He said the government was anxious that overseas confidence in Australian investment would be damaged as little as possible (a paradoxical standpoint for an economic nationalist), and he wanted a consortium formed to buy Minsec's assets.

The consortium idea appears to have originated within the RTZ-CRA stable. The plan as announced to creditors after the meeting was that a consortium of trading banks and other organisations, led by the Wales, would make an offer to the liquidator of a line of credit of up to $35 million. The consortium was as powerful as could be formed within Australia. Its members were the six major trading banks (Commonwealth, Wales, ANZ, CBA, CBC and National), CGFA, CRA, MIM, the AMP, the MLC, Australian United Corporation Ltd, Development Finance Corporation Ltd, Ampol, CSR and the AIDC.

This announcement engendered some return of confidence the next day. Financiers were impressed by the strength of the companies in the consortium as well as by the actual offer. The fact that the Prime Minister had convened the talks also stiffened morale as financiers implied that the Federal Government was prepared to take some strong but unspecified action to stop the rot.

As often happens in time of panic, what was believed to be happening was more important than what actually was happening. It was not known that the plan had been criticised at the Kirribilli House meeting, and was in fact impracticable. When Gorton put the consortium plan to the meeting it had been as though he were announcing a fait accompli. Unfortunately his enthusiasm was stronger than his grasp of the financial mechanics involved. The proposal would have required secured creditors to sacrifice their security and many were not prepared to do so.

Two or three of those present, including Steel of the AMP,

asked why the group should not respect the position of those creditors who had been prudent enough to seek security. These creditors could be accused by their shareholders of not acting in the best interests of their own organisations if they now gave up their preferred position. Sir Norman Young was particularly critical of the plan. It is notable that Elders and the Bank of Adelaide did not lend their names to the consortium. The CRA group offered to put up $10 million of the line of credit, but could still not get unanimous support. Miskin, of Bill Acceptance, pressed by Gorton for a viewpoint, said his company would be reluctant to surrender security. Turner and Sir Val Duncan tried to persuade the critics to change their minds, but to no avail.

Gorton, becoming heated, said he would make no bones about publishing in or out of the House, the names of those who by refusing to yield security had frustrated the creation of the consortium. Despite this threat, the creditors would not agree because the gap between them and the consortium was too wide. As one observer wrote in a board memorandum afterward: "It was clear, and understandably so, that no member of the consortium was prepared to accept a scheme which might transfer any risk of ultimate loss from the creditors to the consortium."

At a hastily convened meeting of creditors after the Kirribilli House discussions, Sir Robert Norman announced the consortium's offer of a $35 million line of credit to the liquidator. The news leaked out to the financial press and its publication helped restore confidence.

What the public did not know at the time was that the liquidator, Jamison, had decided he could not use the offer because of an attaching condition. This condition stipulated that the liquidator would have to satisfy himself, and accept responsibility for deciding, that the security held by the creditors of Minsec whom he paid out was good for the amount of $35 million. In the market conditions of the time, nobody could have guaranteed this with any confidence.

When Jamison informed the secured creditors privately that he was unable to draw on the $35 million for this reason, he asked them not to make this public. They kept the secret well and it did not become public knowledge until Jamison revealed it in May — well after the panic had subsided.

The consortium offer is open to criticism. Sixteen of Australia's largest institutions would have made an average commitment of a little over $2 million cash each. Considering the dire threat feared to the whole financial system, it was a small enough offer. Their risks of loss would have been only a small proportion of this amount at worst. But the main point is that the offer was in fact made and that money markets were reassured. The main need was to halt the panic and that was done. As nerves steadied, some of the more panicky trading banks began venturing back into the marketplace. This was helpful, because it enabled Jamison to persuade more banks to support Minsec creditors.

The Reserve Bank added to the feeling of reassurance by delivering a message to the trading banks. In its Delphic way, the Reserve Bank did not mention Minsec by name, merely referring to "current events in the financial markets". The Governor said he wished to be kept informed of any important developments in the markets, told the banks to exercise their commercial judgment in dealing with customers and said the Reserve Bank "stands ready to support as necessary the actions of the banks". As usual, this statement from the bank was open to more than one interpretation, but it was read correctly by the banks as meaning they should extend overdraft and bill facilities to hard-pressed Minsec creditors.

The Reserve Bank had made reassuring noises but again the impression was more reassuring than the facts. Although the bank said "we are particularly concerned to keep in touch with the situation" they never contacted Jamison at any stage. Like any reassurance in this nervous week, it was welcome.

Such reassurance was particularly timely after two disturbing events on February 8 and 9.

On February 8 three creditors, Westralian, Commercial Continental and Mobil, had petitioned to wind up Minsec. The hearing was set down for Friday, February 11.

On February 9, VAM Ltd announced it was seeking voluntary liquidation, thereby disappointing any speculators who were waiting for a tree mine to be developed on the Burdekin. Fortunately VAM's collapse added little to the existing level of panic. VAM was a large mining house which at one stage had a market capitalisation of nearly $40 million, but it was never in the same league as Minsec. Its main prospect, the Gunpowder copper mine in Queensland, was known to be in financial straits. The shares had been marked down heavily on the exchange in the days before the announcement of its liquidation.

One might even say that the Minsec collapse was a stroke of good fortune for VAM. The concentration on Minsec's spectacular crash and the near-collapse of the money markets diverted public attention almost totally from VAM, which was never exposed to the investigations and publicity that attended many other failures of the era.

On February 11 a packed hearing in the Equity Court heard Mr Justice Street appoint James Hardie Jamison as provisional liquidator of Minsec. Shortly afterward he was appointed liquidator. The blaze of publicity given to Jamison in Minsec, followed by his role in the Mainline and Patrick Partners crashes, have made Jamison one of the best-known liquidators in Australia. It is hard to remember that at the time of the Minsec appointment he was virtually unknown outside his profession. Jamison was a partner in the accounting firm of Cooper Brothers, later Coopers & Lybrand, and had acted as liquidator in 1966 for Dreyfus & Co. He had also acted as trustee in the bankruptcy of former Sydney sharebroker William Hewson. During this trusteeship he clashed frequently with a solicitor representing one of the other parties. That solicitor respected a tough opponent and was now acting for one of the petitioners who were seeking to wind up Minsec.

The solicitor rang Jamison and said: "You're tough enough when you're working against me, do you reckon you can be the same working for me?" He asked Jamison if he would take on the Minsec liquidation. There are two lists of accountants qualified to handle liquidations, an A and a B list. Mr Justice Street had told the petitioners that if they could not agree on a liquidator he would appoint one himself. On a job of this size he would certainly pick one from the A list and Jamison was only on the B list. Jamison said he would take the job and the solicitor said he would persuade the other parties to accept him.

Jamison was then aged 58 — a tall man with a bald patch surrounded by a tonsure of white hair. His two main characteristics could be described as pragmatism and autocracy. He is more concerned with getting things done than with the letter of the law, which in the Minsec situation was just as well. He is a tough-minded man capable of browbeating an opponent when necessary. In a liquidation, he is in charge and he behaves like it. Reactions to this attitude differ. I have heard him criticised by former top men in Mainline and Minsec but he has been highly praised by former Patrick Partners.

Jamison has a strong sense of humor and a down-to-earth manner which is refreshing, not to say surprising, in a top accountant. He would need all these attributes to handle the

Minsec job. It needed a strong-minded man to control and coordinate the creditors, who included some of Australia's most egotistic financiers. It needed a practical approach to avoid the potential morass of legalities which would result from large-scale testing of creditors' security. And the job certainly needed a sense of humor.

Jamison remembers going to the court on February 11 and sitting down on a bench alongside a little man who was an individual creditor. "It's all right for these fellows with their millions," said the creditor. "but I'm owed $300 and that's a lot of money to me." When the brief hearing was over the creditor asked "Do you know this fellow Jamison? Point him out to me." Jamison pointed to himself. The creditor said: "I'll go home and tell the wife not to worry, because I know the liquidator." "That fellow used to phone me every month or so," Jamison reflected later.[7] "When he received his final payout he brought me a bottle of wine."

Jamison decided the first thing to do was to go to the ANZ Bank, where the scrip was held and ensure that none of it was sold. He walked out to the lift through a throng of financial pressmen asking each other, "Who's Jamison?"

He had invited friends to dinner that night. Mrs Jamison rang his office to say that the phone had been running hot with calls from the press. "Shall I put off our guests?" she asked. "No," said Jamison, adding inaccurately, "I'm not going to let this change our life." He drove home at about 7 pm and pulled up to meet a TV cameraman coming down his drive.

"When's your Dad getting home?" asked the cameraman.

Jamison, flattered, said: "I don't know. Mum doesn't expect him for a while."

"Ah well, I've missed the news anyway," said the camerman and stumped off.

On the front porch Mrs Jamison had divided the callers into separate parties, guests on the left and press on the right.

"The press started on beer but now they're drinking Scotch," she whispered to Jamison. "Have we got enough?"

Jamison managed to cope with the press questions before the Scotch ran out and then started a belated dinner. His guests had been delighted by the whole occasion. They had been right in the centre of what was happening, and the problem of finding a topic of dinner conversation was solved.

Although Jamison could not use the $35 million consortium offer in its original form, he though he might be able to make it usable by turning Gorton's interest in Minsec to good account.

This meant he needed a pipeline to Gorton. As soon as he had taken over the role of liquidator, Jamison had received a host of telephone calls from well-wishers offering to help. One had been from Sir Robert Crichton-Brown, chairman of the Institute of Directors and a prominent member of the Liberal Party.

What Jamison was looking for was a form of government guarantee for the $35 million. With government underwriting, probably through one of its instrumentalities such as the AIDC, the offer would not only be usable, but would represent more tangible reassurance. Jamison, who always prefers to deal with the men at the top, rang Crichton-Brown and asked if he could put Jamison in touch with the Prime Minister. "He said he could get me as far as Ainsley," recalled Jamison. "I said: 'That's even better'." Crichton-Brown rang back to say: "I've got an appointment with the Prime Minister for us on Monday next."

"That Friday was the night of the long knives," said Jamison. "Gorton was out and Bill McMahon was in." Jamison rang Crichton-Brown once more. Crichton-Brown spoke to the new Prime Minister, who had formerly been Treasurer. Crichton-Brown rang back with the message that McMahon believed it was a matter for Treasury. He would not have anything to do with it unless Treasury agreed.

This would have taken time, which was exactly what Jamison didn't have, and probably would not have achieved anything. Neither the secured creditors nor the market would hold together for that long, so Jamison was forced to go ahead with the liquidation.

"We always maintain in liquidation work that you do your best work when you have your back to the wall," said Jamison afterward. "I think perhaps it was just as well there was a change in the Prime Ministership because it forced us to do something else."

Jamison formed syndicates of secured creditors. Everyone holding Queensland Mines scrip, for instance, was in the Queensland Mines syndicate, and so on. Some creditors were members of more than one syndicate. The arrangement was that Jamison would negotiate the sale of the shares in block, the proceeds would reimburse the syndicate concerned and any surplus would go toward the other creditors in order of preference. The major problem was blocking the secured creditors from selling off their scrip individually, but Jamison managed to hold them together.

"I think it is clear that when, as a general principle of liquidation, you feel and your legal adviser tells you the security is no good, you know that the lawyers on the other side know that too," he later told the Rae Committee. "I think this was the reason in this case why nobody actually took action to sell the scrip over which they claimed security."

It took some force of character to keep the secured creditors in line, but Jamison had one card up his sleeve. If any creditor had tried to sell Jamison would have challenged him in court instantly. He had already had "a quiet cup of tea" with the Chief Judge in Equity, Mr Justice Street, and spoken to him of the dangers of a disorderly liquidation.

A greater peril was that some unsecured creditor would challenge the security claimed by the "secured" creditors. Jamison kept this threat at bay by reassuring unsecured creditors that they would be repaid in full. He also formed an informal consultative committee of 10 of the largest unsecured creditors who between them held more than one-third of the unsecured debt. One unsecured creditor which refused to cooperate was Trans City Securities, which was owed a total of $1 million and could arguably have been described as secured.

To be fair, Trans City also tried to be constructive, putting up a scheme under which creditors could have been paid out by a series of bills totalling $60 million. The bills would have been made available to Jamison, drawn by each creditor, endorsed by a bank and discounted by Jamison. The scheme would have put Jamison at some risk but he described it as "a genuine effort".

Jamison quickly realised that some of the creditors were in trouble. Those with the worst problems were generally those who had come in toward the end and had taken Kathleen Investments or Queensland Mines scrip as security. Hattersley and Maxwell, who were owed $278,000 by the Minsec group, had a severe liquidity problem. "They were reputedly in a pretty dicky financial position," he told the Rae Committee.[8] Several creditors were embarrassed by having money frozen in the Minsec liquidation when it was needed for other purposes. Pokolbin Winemakers had lent Minsec $50,000 for six months, putting the money on deposit until it was needed. Now they needed the money to build their winery. Western Compass, the speculative miner floated by Cam Cooper, had put $100,000 on deposit with Minsec until it was needed to pay for work on a farm-out. Now Cooper had to put the money up or lose his interest in the prospect. In both these cases Jamison helped persuade the banks to carry their clients until the money from Minsec could be freed.

He also persuaded creditors claiming security to accept payment in full on the basis that they might have to repay 10 cents in the dollar if there was insufficient in the rest of the distribution to cover the claims of unsecured creditors. The legality of the manoeuvre was dubious, but the creditors agreed.

The reverberations of the crash were felt in the City of London. During the first week of February the collapse of Minsec came as an even bigger shock to London than to Australians. British brokers had elevated Minsec almost to the status of a blue-chip stock. They had been concerned by the revelation of the borrowings in the Aberfoyle and Cudgen takeover documents but had been reassured again by the $3.5 million profit announcement.

The subsequent crash, coming on top of previous calamities in Australian mining stocks, caused a disgusted retreat from Antipodean shares. Prices of Australian stocks were slashed, although heavyweights recovered a little toward the end of the week. The London Stock Exchange Council made a blanket request to members to reveal their Australian commitments. The check came at a bad time because two small London firms, F.G. Osborne-Smith and Co and Brown, Bliss and Co, had recently failed due to dealings in Australian shares. Rumors sped around the City that two or three more firms could fail because of the Minsec collapse.

As one unkind Australian broker commented: "We can't beat them at cricket, but we can really bash them on the bourse." Thankfully for Australia's reputation in the City, the odium of Minsec was outranked by that emanating from the crash of Rolls-Royce on the same day.

The controlling interest in Robe River was Jamison's biggest headache. Jamison thought, logically, that BHP might be interested because the iron giant had the bordering deposit at Deepdale. He wrote to BHP, but the company rejected the suggestion out of hand. Interests associated with Thyssen of Germany were interested for a long time. Another possibility was Lang Hancock, who made the memorable comment: "We are the only people who can save Robe." He added: "But we are not going to do anything until the vultures clear from the Mineral Securities carcass." Jamison solved the problem after a fashion by selling the Robe stake to WA entrepreneur Alan Bond on terms. Bond had difficulty meeting the payments and the Robe holding was not satisfactorily bedded down until it was bought by Burns Philp six years later.

Another embarrassment to Jamison, in a financial sense, was the Rath inquiry. The NSW Attorney-General appointed the three inspectors to investigate Minsec on February 10, 1971. It is not widely understood that the cost of such investigations is usually borne by the company being investigated, which is invariably in poor financial shape to afford the expense. The $250,000 bill for the inquiry was presented to Jamison at a time when he was planning a $3 million distribution to creditors. As a result he had to trim the promised distribution.

When it came down to cold cash, very few major Australian groups were prepared to buy parts of the Minsec empire. As liquidator, Jamison's duty was to sell to the highest bidder, whether Australian or foreign. He was dismayed by the reluctance of established Australian mining groups. MIM and Peko had been prominent in early talks, but neither took the opportunity to buy any mining properties. Nor did CRA buy anything despite its prominence in the consortium. From interviews while he was in Australia, Sir Val Duncan appears to have been mainly interested in the Nabarlek shareholdings, but the ordinance blocked him from bidding for the whole of them. The Collins House group appears to have taken little interest.

The controlling interest in the Aberfoyle group was sold to Cominco of Canada, and that in Cudgen to Union Corporation of South Africa. The blocks of stock in Kathleen were sold to the AMP Society, Newmont Mining and Noranda. The AMP was the only Australian institution to take a big chunk of the stock that became available. It is no consolation to know that their decision was wrong because the overstatement of Nabarlek's reserves was not known at the time.

Considering the subsequent slump in the mining industry, Australian institutions were probably smart to stay out, but they proved that enthusiasm for Australian ownership of Australian resources can evaporate quickly when money has to be spent.

Despite all the early talk of rescue operations and support, Jamison was left to solve the problems almost entirely by himself. Considering the nature of liquidations and private enterprise, this is as it should be. But it was no comfort to Jamison at the time. The Rae Report concluded: "A number of witnesses, in referring to the psychological value of the consortium's offer of standby credit, said that it could be taken as a sign that 'something was being done' to control the situation. In reality, anything which was being or to be done was left to the liquidator and his assistants. The liquidator was left to make decisions on priorities and relative responsibilities and to exercise his ingenuity to achieve his objectives, in a situation that prominent businessmen considered to be of national importance."

One clue to the attitudes of the business establishment might be gleaned from a conversation Jamison had much later with one of Australia's most senior stockbrokers. The broker accused Jamison of "letting the side down" because he had not sold Minsec's assets at bargain prices to local institutions.

Whatever the results of the sale, it was fairly obvious from the start that there was little hope for shareholders. The ordinary shareholders were not beset by panic or anguish. It was brutally clear they would get nothing and that they were helpless to control what was happening. When a company fails, the shareholders have no say in the disbursement of its assets. All they can do is sit and wait, filling in time by papering the walls with their scrip.

In Minsec's case the shareholders did not even receive the interim dividend which Minsec had proposed as a way of bolstering confidence. The dividend was withdrawn on February 19, a week before books were due to close to determine entitlements.

Preference shareholders were luckier, but only if they were fleet-footed. The company actually sent out the cheques for its preference dividends, but Minsec's account was frozen a few days later. Any cheques which had not been cashed by then were returned and their holders ranked as unsecured creditors.

The preference shareholders with the most reason to complain were the English clients of Hill Samuel who had taken a placement of preference shares at the huge premium of $14.70 per share the previous June. Within eight months these shares were a wipeout. Hill Samuel were embarrassed, but as mentioned earlier had put many of the same clients into Wall Street at the right time, so were able to point to some offsetting gains.

The two mutual funds survived the crash. Their assets had shrunk convulsively but as their liabilities were insignificant, they were still alive. It was just as well they did not collapse because the results could have been catastrophic. Each of the funds was an unlimited company, which means the personal liability of the shareholders is literally unlimited. Basically there are three kinds of company — no liability, limited liability and unlimited liability. If a no liability company fails, the shareholders have no liability for the deficiency. If a limited liability company fails the shareholders are liable only for the uncalled capital, if any, on each share.

Unlimited liability companies, however, have potentially horrible consequences. Let us say an unlimited liability company has 1 million shares and fails with a deficiency of $5 million. Each shareholder is liable to meet the deficiency, which means they have to put in $5 per share. This is bad enough, but individual shareholders can be saddled with unequal portions of the debt. Let us say that half the shares are held by a $2 limited liability company. That company can default for $2 and the remainder of its debt must be borne by the rest of the shareholders. And as successive shareholders fail, the debt is loaded on to the survivors. It is theoretically possible for a shareholder to lose all his possessions in such a company.

Despite these potentially dreadful consequences, unlimited liability is widely used in Australia as a structure for mutual funds. The reason is that this structure permits the redemption of existing shares in the funds without the delays involved in obtaining court confirmation of a special resolution for reduction of share capital which is required when limited liability companies propose similar action. There must be thousands of investors in Australia who hold shares or units in a mutual fund without realising the full extent of their potential risk. A responsible manager of such a fund would, of course, make sure that it incurs only small debts in relation to its subscribed capital, or no debt at all.

What happens when such a fund fails was illustrated in the crash of the empire of Dr Michael Garretty in Melbourne. When this happened in 1971 individual shareholders in the Dividend Fund Incorporated were saddled with that company's debts. Some shareholders said afterward they had bought the shares from salesmen who had sold them door to door and that they had not even been told it was an unlimited liability company.

On May 29, 1973 the Liberal MHR for the Victorian seat of Chisholm, Tony Staley, attacked Dr Garretty's handling of this fund. Speaking in Federal Parliament, Staley said: "Using a siphon shaped like a cheque book Dr Garretty and his professional fellow managers had arranged for Dividend Fund to lend nearly $640,000 to four other Garretty companies." Staley said that after the crash three of these companies had only repaid amounts ranging from four to 20 cents in the dollar. Shortly before the Dividend Fund collapsed several shareholdings, amounting to about 100,000 shares were transferred to a company called R. Theo Hopewell Investments Pty Ltd.

Staley said: "Dividend Fund was created as a company with unlimited liability; a characteristic ... which meant that when Dividend Fund went broke ... the shareholders were faced with the problem of meeting the deficiency.

"Such shareholders generally were not closely linked to the company or its directors, and were in most cases merely innocent, often elderly, investors.

"Having lost their life savings ... some of these innocent people are now threatened with court action if they do not contribute amounts like $500, $2000 or $4500 towards the debts of the company.

"Despite the obvious responsibilities which the directors of the company and its managers would have to such investors, they abused the trust, they engaged in dubious investment activities, they siphoned money to other private companies, they set up a round robin of company financing and in the final analysis they simply gambled with the money in the riskiest of investments during the mining boom of a few years ago."

Shareholders in the Minsec funds were more fortunate. Their assets may have shrunk but at least they were not threatened with extra bills as well.

The collapse of Minsec illustrates a common feature of large company failures. When a large company fails, the cause is usually a business misjudgment by its top management. Such misjudgments are an unfortunate part of the system. They are certainly not criminal, nor are they preventable. When criminal prosecution eventuates after such a collapse it rarely has anything to do with the error which caused the collapse. Typically any criminal prosecution will revolve around some last-minute stratagem undertaken in an attempt to stave off disaster. So it was with Minsec. Directors were widely criticised for their action in redeeming from the funds and were unsuccessfully prosecuted for having made a false interim-profit statement. But these were eleventh-hour stratagems. The seeds of destruction lay in the expansion program of September 1970, and particularly in the buying of the Nabarlek stock.

Minsec was suspended from the stock exchange lists on February 4, never to return. One of the last buyers had been Nestel, who picked up a parcel of shares in January. Asked why, he said: "I had confidence in the company and its future. I still believed in the philosophy that had carried the company along so successfully."

[1] Individuals can place money in this market but in practice only a few have the funds available. The minimum amount dealt in is $50,000, and more often $100,000.

[2] He did, too. Johnson is now manager of international investment division of Hill Samuel in London.

[3] Companies Act, Section 293.

[4] A secured creditor also has the right to appoint a receiver to take charge of the assets over which he has security. The liquidator has no rights to these assets until and if the receiver relinquished them. Where a company has given a floating charge over all its assets, a receiver can take control of everything, leaving nothing for the liquidator. This is a highly contentious and unsatisfactory point of commercial law.

[5] Interview with author, April 1978.

[6] Rae Committee transcript of evidence, Part II, Volume 2, p. 1545.

[7] Interview with author April 1978.

[8] Rae Committee transcripts of evidence, Part II, Volum 3, p. 2972.

CHAPTER EIGHTEEN

The Destruction of Patricks

It seemed for a while as though Patrick Partners would survive their own errors, the collapse of the mining market and the general chaos of the broking industry after the boom. But it was not to be and Patricks closed its doors on July 27, 1975. By this time the boom was long dead, but as Patricks were the biggest of the boom-time broking firms and the seeds of their destruction were sown in the boom, the fall is relevant to this book. The story also sheds further light on the Minsec crash.

The founder of Patricks was one of the legendary figures of Sydney's financial history, Patrick R. Levy. A cheerful, warm-hearted character, Levy became a member of the Sydney Stock Exchange in 1924. He was shrewd, he began in hard times, he survived many more and at his peak he was a millionaire. Levy spent 50 years in the tooth-and-claw world of stockbroking and at the finish of that time had no enemies — an achievement that speaks for itself.

His first partner was Dick Allen, cousin of the former Test cricketer and English Captain, Gubby Allen. In the 1970s Gubby had a house at St Johns Wood in London. Whenever Pat or Dick were in London and a Test was playing they would go around to Gubby's, park their car in the drive, have a gin or two with the great cricketer, then walk out through his back gate, which by special arrangement opened directly on to the Lords ground.

After a World War Two in which both were taken prisoner of war, Pat and Dick resumed broking in the late forties. Their biggest coup was to corner the business of Australian Fixed Trusts. AFT was controlled by the Development Finance group. AFT was then, and still is, the biggest unit-trust group in Australia. Executing its buying and selling orders was a lucrative business, sometimes generating commissions as high as £1200 a day. This was a lot of money in the late forties and early fifties. Then Allen and Levy merged with two other brokers, Arthur Perkins ('Perk') Reynolds and Max Richard Laidley ('Ric') Dowling, who had the firm of Eccles, Reynolds and Dowling.

These four, Levy, Allen, Reynolds and Dowling, were the cornerstones of the new firm. The name of Patrick & Co was chosen as a combination of the first names of Levy, Dowling and Allen. Reynolds was a great stock salesman. He would sit down at the telephone each morning, ring the institutions and go through the whole share list until he had sold each of them a stock. He might have to go clear through from ARC Industries to Yellow Cabs before he got an order, but he always got an order. Reynolds was of the same vintage as Allen and Levy, but Dowling was younger by a generation.

The three older men were not only partners but close friends, playing poker together each week. As they grew older they sold a percentage of their respective interests to Dowling, who became the major equity partner in the firm. As partners came and went his equity varied, but at all times relevant to the boom Dowling would have held between 32 and 36 percent of Patricks.[1]

As Levy and Allen gradually retired from the scene and Reynolds died, Dowling became the dominant partner. While he was younger than the rest of the founding four, he was older than most of the other partners and had a long and highly successful record in broking. All these factors combined to increase his dominance. His authority was undisputed within the firm. Tall, with blue eyes and thinning hair, Dowling radiated an aura of Olympian assurance. Behind his back, he was nicknamed God. He could be disarmingly charming at times, but his presence at any meeting was still formidable. Junior partners with less experience and stakes of around three percent in the business were in a poor position to argue with him, although ultimately some of them did.

There were two classes of partners in Patricks, which might be called the owners and non-owners. The owners were the equity partners. They owned the assets of the firm. The extent to which they were responsible for its liabilities became a burning question later. The owners also divided among them, pro rata, 40 percent of the profits each year. The remaining 60 percent was split pro rata among both the owners and non-owners.

Dowling had two lieutenants, Roy Course and Frank Robertson. Again the percentage of equity varied from time to time but Robertson would have held 17 percent and Course slightly more during the boom. Both supported Dowling on all important matters decided within the partnership. Robertson was Dowling's right-hand man in Sydney and we will return to him later. Course was head of the Melbourne office. He had joined Patrick & Co after a long and highly respected career as a financial journalist. He had been deputy finance editor to the legendary Jack Donovan on the Melbourne Sun and then became finance editor of the Age. After many years in that post he was approached by Dowling to join Patrick & Co. Like all the other partners he had great respect for Dowling. As the rest of the partners concentrated on their own specialist areas in the times of heavy activity, Dowling was virtually the only one who ever spoke to Course in Melbourne, or visited him there. As Dowling had recruited Course and had also appointed him to boards of companies associated with Patricks such as Norseman Gold, Course's loyalty to Dowling was natural.

Patricks employed a number of respected financial journalists. One notable partner, albeit junior, was Michael Baume, who started Speculator's Diary in The Bulletin and also worked for the Financial Review. Course's offsider in Melbourne was Robert Gottliebsen, former chief of the Financial Review's Melbourne office. Later he returned to the Review to write the back-page Chanticleer column, and in 1977 he won the first Graham Perkin memorial award as Newspaperman of the Year. Another fine financial journalist to work in Patricks Sydney office was Malcolm 'Wombat' Wilson, who subsequently joined the National Times and is now finance editor of the Sydney Morning Herald.

One spin-off from the appointment of so many distinguished journalists was that Patrick & Co was on good terms with much of the financial press. The journalists who joined the firm retained their social links with old colleagues in the newspapers. These former journalists also knew how to communicate with the financial press, who are always grateful for information. Some of the newspapermen became clients of their former colleagues, which meant they had occasional small allocations in Patricks' lucrative floats.[2]

This generated a legacy of goodwill in the press toward Patricks during the boom, a time when Patricks were generating a lot of exciting news. But a newspaperman's first duty is to the facts, and as the facts began to emerge about the weaknesses within Patricks, they were covered as thoroughly as anyone else's.

The partnership had strong associations with the Australian mining industry. Course was close to Ken Craig and Reg Hare, who ran the Aberfoyle and Metals Exploration groups. This also gave Patricks an association with Hare and Craig's young protege, Ken McMahon. In Sydney the partnership had strong links with the Longreach group of exploration companies. Like Minsec, Patricks were in the right area at the right time.

As the mining boom of the late sixties bubbled upward, so did Patricks' turnover. They enjoyed mushroom growth, becoming the largest broking house in Australia at the height of the boom. As a house with strong mining associations they were often used by mining companies for placements and underwritings of issues. For the same reason, promoters and prospectors with interesting leases sought Patricks' aid in floating new mining companies. Patricks rejected many, but accepted enough to make them the largest mining underwriter of the boom.

As mining floats were eagerly sought by clients, and small speculators scrambled to buy stocks in the roaring mining market, Patricks' client list expanded hugely.

The volume of business was larger than they could handle. The ambition of every speculator was to become a favored client of Patricks. It was a lucky client, however, who could even get through their constantly jammed switchboard.

It is difficult to put an accurate total on the amount of business they handled, but their turnover during the boom on a daily basis would have accounted for about 25 to 30 percent of trading on the Sydney Stock Exchange. Some days it might have run as low as 15 percent, but on other days it reached 40 percent. On their biggest trading days Patricks were generating more than $50,000 brokerage. Working on the basis that brokerage would have averaged around 1½ percent on each buying and selling order, after taking big lines into account, this would mean the total value of scrip handled by Patricks would have averaged around $3 million on a big day.

Total turnover on the Sydney exchange in 1969-70 was $1900 million. If Patricks averaged 25 percent of this, the firm must have handled $500 million worth of business that financial year. At first sight this seems an impossibly high figure, but it averages at $2 million per trading day.

The biggest single deal ever done by Patricks, however, was not in the high period of the boom and not even in a mining stock. It was the placement in November 1968 of 1.3 million shares in Washington H. Soul Pattinson & Co Ltd at $9 premium each, making a deal worth a total of $13 million.

Patricks' burgeoning growth was regarded with awe. They branched into the money markets, both through the firm and through an indirect 60 percent holding in Bill Acceptance Corporation Ltd. They traded heavily in London and their business was eagerly solicited by leading London houses. They had branches in Hong Kong and Brussels. Their partners sat on a number of company boards. They had close links with top Sydney businessmen. When Castlereagh Securities Ltd was formed its board included the chairman of Pioneer Concrete Services Ltd, Sir Tristan Antico,[3] the managing director of Thomas Nationwide Transport Ltd, Sir Peter Abeles,[3] and the chairman of Washington Soul, Jim Millner.

One weakness, which first appeared in the guise of a formidable asset, was the creation of Mining Traders Ltd. Like many an incipient financial disaster area, this started out as a tax avoidance scheme. In 1967, Patrick & Co had the opportunity to take a placement of some 600,000 shares in the beach-sand miner Western Titanium NL. The effect of this offer was to crystallise the thoughts of Dowling and other partners about the long-standing problem of the partners' investment and share-trading activities. At the time they were earning good salaries and were consequently liable to high tax rates.

The obvious solution was to form a company. As shareholders in a trading company, the partners could minimise tax, apart from dividends, and shore up capital gains. The short-term share trading and long-term investments would be under a neat, separate corporate roof and Patricks would be able to take advantage of market situations as they arose. The payoff would come when the shares were sold or the company was liquidated. Dowling suggested the Western Titanium placement be used as a nucleus for the company, so Mining Traders was formed.

There was a second reason behind the formation of Mining Traders, one which testifies to the farsightedness of Dowling. He feared that one day brokers might be barred from acting as directors or underwriters. He had already had an argument with John Marks, the founder of Development Finance Corporation Ltd, on whether brokers should underwrite issues — an argument so substantial that DFC took the business of Australian Fixed Trusts away from Patrick & Co. When Mining Traders was formed, one long-range purpose was to ensure that the firm could still command the same range of activities in future. If a ban were imposed on brokers underwriting issues, half the firm's partners could resign from Patricks and become executive directors of Mining Traders, which could handle the underwriting.

The birth of Mining Traders was not without trauma. It

issued a prospectus for six million 50 cent shares paid to 20 cents each in early 1967. The issue was offered to clients on the basis that the stock would not be listed for six months. More than $1 million of the $1.2 million raised by the prospectus would be used to buy the Western Titanium parcel. This meant that after formation expenses Mining Traders was left with only $84,000 in working capital. In the prospectus Mining Traders stated: "From time to time there appear in the mining scene opportunities to obtain substantial parcels of shares which, although not amounting to a significant percentage of a company's listed capital, nonetheless constitute a considerably larger parcel than is normally held in one block. It is the company's principal aim to purchase sizeable interests in mining ventures where the quality of the investment is attractive and the price is favorable."

This attracted fire from Bob Blackburn, an executive of the NSW Branch of the Australian Shareholders' Association. Blackburn pointed out that Mining Traders had been able to pick up the Western Titanium block, amounting to 11 percent of that company's issued capital, only because the vendor was Aberfoyle Holdings Ltd, with which Patricks had strong connections. Mining Traders had been offered the block below market price, but the potential profit on this deal might not be typical of the company's future experience.

The prospectus issue was filled with some difficulty and Mining Traders later listed. It was basic to the whole scheme that the partners of Patricks should at all times hold a controlling interest in Mining Traders. To this end, the partners took a block of 45 percent of the stock in Mining Traders, to which they contributed pro rata. This represented effective control.

This led to a very important agreement between the partners. The agreement was that whenever a partner retired, the other partners would take up his stake in Mining Traders. As partners retired in the late sixties, when Mining Traders was making large profits and its shares climbed over $2 on the market, the remaining partners were faced with a series of heavy bills to maintain their equity in the company. Younger partners could finance their commitment only by borrowing large sums from the senior partners, who then controlled the Mining Traders stock.

The agreement was not legally enforceable, but tacit. Theoretically, any of the remaining partners could have refused to take up his share of the action. In practice, this would have been almost impossible. The junior partners had no desire to break the accepted practice. Also there was a strong boom running in which both Patricks and Mining Traders were prospering. Up until mid-1970 at the very earliest it would have taken an abnormally percipient and courageous small partner to opt out of the agreement.

At prices around $2, the retiring partners were receiving about four times the net tangible asset-backing of the stock. The main reason for Mining Traders shares selling at such a premium was the expertise and track record of Patrick & Co, which had to some extent been provided by the very partners who were leaving, notably Dick Allen.

This situation was the key to Patricks' eventual collapse. The broking firm mostly operated at a profit, or at tolerable losses. However, the assets of the partners were increasingly bound up in their investment in Mining Traders. This equity holding had been devised in a boom market for the purposes of minimising income tax. When the market turned sour it became a source of ultimately fatal weakness. In his official report on the partnership, George Masterman, QC, said: "The undue concentration of Patrick Corporation Ltd[4] shares, a speculative stock, as part of the assets and liquid capital of the broking business was clearly a major cause of the ultimate failure of Patrick Partners." Most of the partners, if not all, would agree with this.

If the market had kept rising, of course, Mining Traders would have remained profitable. Alternatively, if the men running Mining Traders had better foresight, the investment of the partners would have been protected. Neither of these things happened. The first step in the eventual disaster was taken in early 1970 when Patrick Partners floated another company called Castlereagh Securities Ltd.

The rationale of Castlereagh was that it would be a mining investment and finance company which would be the equivalent of a unit trust. Its formation appears to have sprung from a genuine desire to provide an investment avenue for small investors. Castlereagh was also designed to provide a form of self-protection for Patricks. At this stage of the boom small investors were besieging brokers' offices, and particularly Patricks. Many wanted to invest only a few hundred dollars at a time when processing any order of less than $500 was uneconomic. But if these small investors were unable to get a piece of the market action through the broking system, there would be increasing pressure on the exclusivity of dealing claimed by the broking fraternity.

Castlereagh started out in life as an attempt to resolve these difficulties. It was floated with so many shares, 76 million 25 cent fully paids, that it was virtually open-ended. Any small investor who wanted a piece of boom-time action could put his few hundred dollars into this company, which would apply its expertise to investing in mining. The concept was backed by a strong sales drive as partners of the broking firms spent their nights and weekends speaking to RSL clubs, superannuation funds and anywhere small investors gathered.

The result was that Castlereagh raised $19 million on an empty prospectus.[5] Castlereagh had no assets: indeed it had a net tangible asset backing slightly below par once flotation expenses were taken out. The sole incentive to subscribe was the magic of Patrick & Co, which had just changed its name to Patrick Partners. The catch was that Castlereagh itself as yet possessed none of the expertise of Patrick Partners. It had little more than cash, a board and an office.

At first Castlereagh adhered to the "unit trust" concept. The prospectus included the statement: "Castlereagh will provide for investors a diversified interest in exploration, mining, refining and marketing companies operating within the Australian mineral industry." However, any commitment to diversification of investment was abandoned in September 1970, when Castlereagh began buying heavily in two stocks — Kathleen Investments and Thiess. As previously explained this was coordinated buying between Minsec and Castlereagh. The principal initiators of the plan were Tom Nestel of Minsec and Frank Robertson of Patricks.

Robertson's role is interesting. He did not gain the same notoriety as the other principals in the Power Resources of Australia affair. He is mentioned often in the Rae Report, in the Rath Report on Minsec and in the Masterman Report on Patrick Partners, but somehow he is never in the full glare of the spotlight in the way that Roy Hudson, Ric Dowling, Tom Nestel and Ken McMahon were. Nevertheless, Robertson was very much at the centre of events.

Robertson came to Patricks from the Commonwealth Bank. He was a brilliant financial operator with the capacity to envisage really large deals. He was a go-getter. He never sat around the office waiting for a deal to come in: he went out and hunted for them. He generated a lot of activity for Patricks and was a valuable asset to the firm in many ways. Dowling had a high opinion of Robertson. "This man will make you all rich," Dowling once told the other partners. Robertson's office was next to Dowling's and the two were in close liaison.

Dowling's admiration for Robertson was not shared unanimously. Some of the smaller partners and employees disliked him. They were disturbed by an aspect of his share dealings and one partner was disturbed enough to raise the matter with Dowling. Reluctantly, Dowling admonished Robertson. It was nothing illicit, but an internal matter within the partnership.

This incident was followed by a larger one with important ramifications — the float of Hall's Peak Australia Ltd. The float of Hall's Peak would rank among the most controversial of the boom. It is also the most misunderstood. It has been criticised for the large amounts paid to the vendors and the comparatively small amount left in working capital after the company floated. However, all this was clearly stated in the prospectus. What was not shown in the prospectus was the full identity of the vendors.

The company's only asset was a group of 421 mineral leases south-east of Armidale in the New England region of NSW. Some small mines had already operated in the area producing silver, lead, zinc, copper and a little gold. Its potential had first been realised by prospectors in 1896 and a lode containing some rich silver-lead-zinc grades was discovered in 1913. In recent years a small amount of high grade hand-picked sulphide ore had been extracted. Hall's Peak Australia was floated to buy these areas and raise funds for further exploration.

The prospectus was for the issue of 22.24 million fully paid 10 cent shares, raising $2.2 million. Of these shares 4.5 million went to Westmoreland Minerals Ltd — an Alexander Barton company — and 3 million went to Amad NL, part of the Minsec family.

In a separate issue, Patricks underwrote an allotment of 7,765,000 shares to shareholders in Westmoreland, raising a further $776,500 for Hall's Peak. The Westmoreland shareholders' entitlement issue occurred a little after the public float, which was also underwritten by Patrick & Co.

The two issues raised a total of nearly $3 million for Hall's Peak. Of this, a total of $1,875,000 was paid to the vendors of the leases. The leases had been owned by a company named Hall's Peak Mining Co Pty Ltd. The vendors who received the $1,875,000 were as follows: John Mahony, $375,000; Iris Cratchley, $421,875; Sir Peter Abeles, $375,000; Norman John Mannix, $140,625; Ross Cribb, $46,875; and Wynstay Investments Pty Ltd, $515,625.

This is a very interesting string of names. Sir Peter was one of the most prominent businessmen in Sydney, being managing director of Thomas Nationwide Transport. Cribb was one of his top executives. Iris Cratchley's address was given as c/- F.W. Millar in Suite 3107 of the Tower Building, Australia Square. This was Fred Millar, the present chairman of TNT. Mannix was the Minister for Justice in the former Labor Government.

Of equal interest were other names connected with the prospectus. The chairman of the company was to be John Osborne Bovill, of Bellevue Hill. Bovill appeared on a number of boards with Alexander and Thomas Barton. The registered office was to be Level 43 of the Tower Building in Australia Square, where Alexander Barton had his offices. The secretary was Francis Pieterse, who later became involved in the Antimony Nickel fracas. The solicitors were McCaw Johnson & Co, who were solicitors to the Barton group of companies. This fact caused frequent embarrassment to the Liberal Attorney-General of the day, Sir Kenneth McCaw. His department was responsible for policing the corporate scene in Sydney and Barton was a frequent target of criticism. McCaw had left the partnership many years before but it still bore his name. Auditors of Hall's Peak Australia were Hungerford Spooner and Kirkhope, who audited Barton's public companies.

Brins Australia Ltd, another public company in the Barton group, had agreed to provide management services for Hall's Peak. Hall's Peak was also connected to the Mineral Securities group. Amad subscribed to the float and held large areas bordering Hall's Peak. Consulting geologists to the underwriters of the float were Kenneth McMahon and Partners.

Hall's Peak was heavily criticised at the time for the very large proportion of cash raised in the float which was paid to the vendors. After deducting preliminary expenses, the company had only $900,000 working capital. This meant the 30 million shares in the two issues would have a cash backing of only three cents each after the float. Any higher price the market put on the shares would be "blue sky". It should be added that the vendors agreed to lend the amounts of the vendor-payments back to the company. However, these amounts were paid again to the vendors in October 1969, before the prospectus was issued.

All the above facts have provided ammunition for various critics of the Hall's Peak float, but they were all shown clearly in the prospectus and nobody was forced to buy shares in the company. Such was the hunger for new floats and the hysteria of punters that Hall's Peak went to fantastic premiums after listing. The 10 cent paid shares, which were plainly revealed to have a cash backing of only three cents (until the Westmoreland entitlement issue was completed the backing was only three-quarters of a cent) closed at 50 cents on the first day of trading.

Later Hall's Peak shares rose as high as 75 cents. At that price they were valued at 25 times their cash backing. To put it another way, a company with a group of leases and $900,000 was being valued by the market at $22.5 million. That was an awful lot of blue sky.

So much could be worked out by anyone who cared to read the prospectus. The same reader, however, would not have known the full identity of the vendors. Thus the association of Robertson with the float was not fully revealed. Robertson was a partner in Patricks at the time of the float, which they underwrote. A company in which Robertson's wife was a director was a vendor to Hall's Peak.

The list of vendors included the name of a private company, Wynstay Investments, which received $515,625. Wynstay's address was given as the 31st level of the Tower building, where Millar had his office. As at July 11, 1969, Wynstay was a company in which the beneficial shareholders were members of the family of Millar. This could have been shown by a search of corporate records. What would not have been shown by such a search was that the bulk of Wynstay's investment was held in trust for a company called Pinduro Pty Ltd, and that the beneficial owners of that company included Robertson's wife and accountant.

Wynstay received 27.5 percent of the total vendor consideration of $1.9 million. Of that, only 2.5 percent was directly attributable to Wynstay. The rest, $468,750, went to Pinduro.

Not only were the public not told this. It was also kept from the knowledge of other partners in Patricks.

Under the additional statutory information in the Hall's Peak prospectus, paragraph 9(c) read: "There are no transactions in respect of property acquired by the company or its subsidiary completed within two years preceding the date of this prospectus in which any director, proposed director or promoter has had any interest, direct or indirect." As Patricks were promoters of the float, this statement was incorrect if Robertson had a direct or indirect vendor interest.

I have also been reliably informed that there may have been another undisclosed vendor interest in the Hall's Peak float. According to one source, George Somgavyri Summers had an effective 10 percent overall interest in the float. Summers was the founder of Vale Corporation, which during the boom spun off Abaleen Minerals and Munga Creek as public floats. Vale Corporation collapsed a few years later and at the time of writing the whereabouts of Summers is unknown.

Hall's Peak was floated in November 1969 and listed in December. It was not until several months afterward that two partners in Patricks, Michael Baume and John Keir, had their suspicions aroused and began investigating. When they discovered the Robertson family interest in the float they were both enraged and alarmed.

Baume and Keir informed the other partners of their discovery and the equity partners held a meeting to discuss the matter. Robertson defended himself on the grounds that the vendor interest was his wife's and not his. He was neither a shareholder, director nor officer of Pinduro. Baume and Keir maintained that the interest still should have been disclosed in the prospectus. At the very least it ought to have been disclosed to Robertson's fellow partners. The only reference in the prospectus had been in paragraph 14(b) of the statutory information, in which Pinduro was named as a company agreeing to deposit money with Hall's Peak. Baume and Keir demanded Robertson's resignation from the firm.

None of the participants has ever revealed what happened in the ensuing argument — indeed Robertson's link with Hall's Peak has never been known publicly at all until now — but the argument must have been traumatic for everyone involved. Breaking up a partnership is akin to breaking up a family. Eventually a compromise was reached. It would be put to a full meeting of partners, both equity and non-equity.

The date of the meeting is uncertain but must have been in mid-October 1970. It was held in Patricks boardroom on the top floor of the P & O building. It appears to have been planned carefully. All the partners except Robertson and Dowling assembled, then Dowling appeared to open the meeting a quarter of an hour later. Robertson did not attend.

The rest of the partners gathered around the board table,[6] a long truncated oval of blond wood. It was only then that the non-equity partners learned what the compromise was. Robertson was to resign from the partnership, but he was to become managing director of Castlereagh Securities.

Dowling had still not arrived. His argument was put by Baume and Keir. Briefly, it was that Robertson was a brilliant dealer, that Castlereagh needed such a man and that no chief executive had yet been found for Castlereagh. At least one of the non-equity partners objected saying that he wanted Robertson removed altogether, not placed as chief executive of a company closely linked with the firm and for which the partners had just raised $19 million. Baume and Keir replied that they had agreed to the compromise because it would have been embarrassing to Dowling to remove Robertson entirely at this point.

The non-equity partners were not given further explanation of the source of this embarrassment, but it must have been the proposed formulation of PRA. At this point Robertson and Nestel were heavily engaged in the buying of Kathleen Investments, indeed the buying was at its peak in mid-October. Robertson had been acting as de facto chief executive for Castlereagh in this deal. Nestel and Robertson were very close ("Like that", said one former Patrick partner, holding two fingers together.) Dowling may well have feared that sacking Robertson altogether might bring the Minsec link unstuck and jeopardise the PRA float.

As said before, none of this had been known previously to the non-equity partners around the long, pale table ("I never did like that table", one said later.) While they were digesting the news that Robertson was to become chief executive of Castlereagh, Dowling entered and took his place at the head of the table. Dowling put briefly to the partners the proposition that Robertson would resign and go to Castlereagh.

One young non-equity partner spoke against the appointment. He said the Castlereagh prospectus had been sold to the public largely on the Patrick name and that if Robertson was unfit to be in the partnership he should not be in a company so closely linked.

Dowling may or may not have been listening. He sat with his head slightly bent, looking over the top of his spectacles at a point about half-way along the table. He had a mannerism, when posed like this, of toying with his notepad. He would take the top sheet and turn over the bottom edge of it in a meticulously straight line. He would smooth it down with his thumb, then tear it off in a neat, straight strip. Starting at one end, he would then roll up the strip tightly into a small cylinder like a cigarette butt. Then he would throw it away, sometimes on the table and sometimes on the floor. He did this all the time the partner was speaking.

When the speaker finished, Dowling neither looked up nor replied. After a very short pause, he said: "It is agreed, then?"

It was barely a question. Nobody else spoke.

On October 23, a few days after the meeting, Patricks

announced the retirement of Robertson from the partnership to take up his post at Castlereagh Securities. Financially, the move was beneficial to Robertson, as he collected a payout of just over $400,000. The remaining partners in Patricks paid $72,600 for his share in the partnership and $335,000 for his equity in Patrick Corporation, whose shares were over $2 at the time.

The purchase of Robertson's equity in the broking firm was calculated on the basis of the profit which the partners expected to earn in the financial year 1970-1. In October when Robertson was bought out, the firm was doing well. In the latter half of the year, however, the firm did badly and the expected high profits did not materialise. Consequently Robertson had to pay a large proportion of the $72,600 back again. He did so after some protest.

An intriguing aspect of the affair is the role of the law firm Allen, Allen and Hemsley. Allens — one of the largest firms in Sydney — were solicitors to the underwriters of Hall's Peak, the underwriters being Patricks. Allens also retained Fred Millar as a consultant. His name was on the letterheads of the firm at the time, along with the names of the partners. It was Millar's family company which had acted as trustee for Pinduro, which meant that Millar must have been aware of the link between Pinduro, Hall's Peak and the Robertson family. Another Allen partner who must have known was John McGlinchey, who wrote to Robertson in a letter dated November 14, 1969, saying that as at July 11, 1969, members of Robertson's family had been the beneficial owners of the shares in Pinduro. Allens confirmed in that letter, however, that on July 11 Robertson was not, according to Allens' records, a beneficial owner entitled to the profits received from the sale of the shares in Hall's Peak Mining. So a partner and a consultant of Allens knew the background to Pinduro at the time of the float and at the same time Allens acted as solicitors to Patricks in the underwriting.

These may have been some of the thoughts on Dowling's mind when, asked about Hall's Peak by the Rae Committee, he said: "It caused me more worry than any float we have ever made."

So Robertson went to Castlereagh and the PRA project continued. The link between Nestel and Robertson was important for this purpose but the companies were becoming bound closer to each other anyway by their mutual involvement in Kathleen Investments. As the largest share trader in mining stocks, Minsec had naturally been channelling some of its business through Patricks. To accommodate Minsec, Patricks had created a special carry account at the Commercial Banking Co of Sydney. This account normally stood at several million dollars.

Carry accounts involve the deposit of securities with a broker in return for an advance, usually in the ratio of 150 to 100. Minsec's carry account with Patricks was on the terms of 160:100, which meant that Patricks would extend $100 worth of credit to Minsec for every $160 worth of scrip, at market value, lodged in the account. It is almost certain that no other broking firm in Australia could have provided Minsec with facilities on the same scale. Without it, Minsec's titanic share dealings would have been on a lesser scale.

After Minsec collapsed and the details of PRA gradually came to light, the position of Castlereagh came under pressure. Its holdings in Kathleen and Thiess had been equivalent to Minsec's. Patrick Partners was holding most of the Minsec Kathleen and Queensland Mines shares as security against advances. The firm managed to place these shares with Noranda and the AMP Society. But Castlereagh was still left with its own holdings in a market which had already fallen greatly and would fall still further in KI and QM as doubts spread about the value of Nabarlek.

Castlereagh hung on to its own holding in Kathleen Investments. This was the biggest single mistake ever made in the Patrick group. Castlereagh may even have been still buying some Kathleen in the first half of 1971, because by June 30, 1971 the company held 829,389 shares in KI. Minsec held only 750,000 KI when it crashed in February and Castlereagh's holding ought to have been roughly the same at that time. Castlereagh appears to have been "averaging down" in the bear market of early 1971. This technique reduces the average cost of a holding, but increases its total cost. If you hold 1000 shares at $1 and they fall to 50 cents, you can buy another 1000 and average down your cost to 75 cents. But if the shares go to zero you will have lost $1500 instead of $1000.

The slide in Kathleen's price was crucifying Castlereagh shareholders. Castlereagh's investment portfolio had cost $18.8 million. By June 30 it was worth only $10.5 million and by the time its 1971 annual report was published in mid-October it had suffered a further sickening drop to $5.4 million. (The real Nabarlek figures had been published on August 12). A company which had raised $20 million in a public float had wiped out $13.4 million of it by the time its first annual report had been published.

Nearly all this loss was attributable to the Kathleen fall, which means that Castlereagh had spent the bulk of its investible funds buying one stock. As the value of its largest investment fell, Castlereagh's own shares began falling on the market. By the end of 1971 they fell as low as three cents, recovering to around six and seven cents in early 1972. This exposed a dangerous new flank. The float of Castlereagh had induced such euphoria at Patricks that in an optimistic but rash gesture Patrick Corporation Ltd had subscribed to 20 million shares in the float. These were a special issue of contributing shares paid to five cents each, which meant that Patrick Corporation had subscribed $1 million cash. There was still a 20 cent unpaid liability on these shares, totalling $4 million. Patrick Corporation had undertaken not to sell its contributing shares until they were fully paid. They had also undertaken to pay the balance of 20 cents per share by June 30, 1975. In other words, Patrick Corporation had an outstanding liability of $4 million to Castlereagh and the day was slowly approaching when it would fall due. In the deteriorating market, and considering the sick state of Castlereagh, this would be throwing good money after bad.

Castlereagh's low share price and its contingent asset of $4 million made it possible prey for a corporate raider. Should someone else take over Castlereagh, this $4 million would be literally given away to hostile parties. In September 1972 exactly such a raider appeared when Ian Murray, acting through top Sydney lawyer Aleco Vrisakis of Dawson Waldron, bid 11 cents a share for half of Castlereagh. Murray, a cheerfully brash entrepreneur who had formerly headed Slater Walker in Australia, would certainly not let Patrick Corporation off its $4 million hook. Castlereagh and Patricks rallied enough support to beat off the bid. Then to forestall any similar threats in future they announced the merger of Castlereagh and Patrick Corporation.

This manoeuvre was strongly criticised by the Australian Shareholders Association and the financial press. An independent valuation prepared for the directors by Hungerford Spooner valued Patrick Corporation shares at a net tangible asset backing of $1.08. The ASA reckoned they were worth only 73.6 cents n.t.a. at maximum and challenged the accounting firm unsuccessfully in the courts.

The merger effectively removed the $4 million liability[7] but it created new underlying problems. The merged group, called Patrick Corporation, underwent further problems which lie outside this history. It suffered from steadily worsening liquidity. The board tried to alleviate this position by selling its 60 percent stake in Bill Acceptance Corporation Ltd, but this task was the exclusive responsibility of Phil Miskin, who was both chairman of the new Patrick Corporation and chairman and chief executive of Bill Acceptance.

According to the Masterman report, Miskin promised Patrick Corporation's board he could get $6 a share for Bill Acceptance. No sale was ever achieved and a number of directors believed Miskin was reluctant to sell the controlling interest in a company where he was chief executive. Originally he had hotly opposed a proposal that Patrick Partners should negotiate the sale.

The failure to sell Bill Acceptance was of no benefit to that company. Ultimately, well after Patrick Corporation and Patrick Partners had crashed, Bill Acceptance was sold to Australian Guarantee Corporation Ltd for one cent per share.

Patrick Corporation was an unhappy ship after the 1973 merger. Some of the directors had, after the merger, opposed the election of Dowling as chairman of the new group. Miskin was appointed instead. From the evidence given to Masterman, some of the directors were unhappy with Miskin's failure to sell Bill Acceptance. The years 1973 to 1975 were difficult ones for the mining industry and the money markets as the Australian economy moved into its deepest recession since the thirties. Patrick Corporation's board and management had a wealth of experience and expertise but were disunited. On the evidence Masterman concluded that Patrick Corporation became legally insolvent around January 1975, some seven months before a liquidator was appointed.

The company had continued to make investment errors. It had moved into property at the top of the market and found itself locked into properties that were difficult to resell. The group developed increasing problems and went into liquidation in August 1975. The provisional liquidators, W.J. Hamilton and Ian Ferrier, estimated that in the period up to September 30, 1974, Patrick Corporation and its subsidiaries had incurred losses totalling $28.8 million.

The deteriorating financial position of Patrick Corporation and the consequent slide in the value of its shares were fatal to Patrick Partners. The main assets of the partners were shares in Patrick Corporation which ultimately became worthless. The debts on the shares were beyond the capacity of the partners to repay. It might be called a financial version of the domino theory.

The overstatement of the Nabarlek reserves had set up the dominoes as Nestel and Robertson, encouraged by their respective organisations, bought Kathleen Investments. The fall in the value of Nabarlek caused a consequent fall in the value of Kathleen Investments and toppled Minsec and weakened Castlereagh. The weaknesses of Castlereagh (which admittedly had other causes also) were a major factor contributing to the weakness of Patrick Corporation, whose insolvency caused the collapse of Patrick Partners.

After Patrick Partners' collapse, Jim Jamison became trustee of the firm. At that time there were seven equity partners and five non-equity partners and a major point in dispute was whether the non-equity partners, who had never had a stake in the firm's assets, were now liable for its liabilities.

The assets of all partners fell well short of the firm's liabilities. The affairs of Ric Dowling were arranged so that he held only one-hundredth of his Darling Point home and his personal fortune turned out to be not a fortune at all but a deficit of $81,777. By late 1978 it appeared that the dominant secured creditor, the Commercial Banking Co of Sydney, would be paid in full but that unsecured creditors would get little.

Dowling, along with another former partner Rupert Burge, joined Boris Ganke's Alexanders group. Michael Baume, one of the five non-equity partners, had formally resigned on February 24, 1975 — before the crash — to devote his full time to a political career. Under an agreement with Jamison he repaid $106,000 debt claimed by the firm in return for him giving up an indemnity granted by other partners. Since 1975 Baume has been Liberal MHR for Macarthur.

The only partners to remain in sharebroking were Bill Edwards, John Keir and Lionel Milligan, all with the Sydney firm of Clinton Jones & Co. All had to fight long battles with the NSW CAC to retain their licences as dealers in securities. As part of their income is committed to the trusteeship, it would have done creditors no good if they had been forced out of the business.

The worst sufferer was probably Roy Course, who at the time of the crash had left a net $160,556 — virtually everything he had made — in assets. All this went to meet the partnership's debts. He operated as Melbourne representative for Clinton Jones for a time, then retired because of ill-health.

In considering the reasons for the rise and fall of Patricks it is necessary to consider the strengths and weaknesses of the senior partner. Dowling, a big man and a lawyer by training, was an awesome figure to outsiders. As the head of Australia's most successful broking firm, he bore an aura of wealth and power. He was an optimist who held visions about the future of the Australian mineral industry. His bullish outlook persisted well after the boom had turned into a bust and this had a lot to do with Patricks eventual failure. But on the way up his optimism had been a great asset to Patricks. It enabled the firm to exploit a boom market as no other firm in Australia could.

His forte was raising venture capital for people with new ideas or discoveries. If an entrepreneur could get to see Dowling, which was seldom easy, and persuade him that he had a worthwhile enterprise or mining prospect, Dowling could solve his money problems. Once Dowling was convinced, a company could be formed and the necessary funds raised through Patricks. In their heyday Patricks would have to rank as the greatest sellers of stock the country has seen. At the height of the boom, of course, their prospectuses sold themselves. But they were able to sell some borderline propositions during the boom because they had put clients into good companies earlier. Their most enduring achievement would be the launch of Pioneer Concrete, for which the firm raised £300,000 in the early fifties.

In hindsight, partners now say Dowling's worst weakness was that he was a poor judge of character. He had some good men around him, but they were not always the ones he listened to.

"Heaven help me that I should say this as a criticism of the man," says one former partner, "but Ric was gullible, because he believed in people. He trusted people, but he had no judgment of people."

[1] Patrick & Co changed its name to Patrick Partners on July 1, 1970. Throughout this chapter they are referred to as Patricks both before and after the change.

[2] For the record, I received small allocations in about 10 Patricks' floats and until this book was published I was still a friend of several former partners.

[3] Then Mr Antico and Mr Abeles.

[4] Mining Traders changed it name to Patrick Corporation in 1970.

[5] $20 million including a $1 million subscription to contributing shares by Patrick Corporation Ltd.

[6] Being a partnership they did not have a board, but this seems the best way to describe the furniture.

[7] Patrick Corporation paid the call but Castlereagh bought an asset in exchange. As Patrick Corporation became a wholly-owned subsidiary of Castlereagh (which, confusingly, changed its name to Patrick Corporation) the net position remained the same.

CHAPTER NINETEEN

After the Ball

Sanders of the River, the memorable fictional character created in a series of books by Edgar Wallace, says in one story: "It seems to me that I may divide my chiefs into three parts, saying this part is made of rogues, this part of fools, and this, and the greater part, of people who are rogues in a foolish way."

District Commissioner Sanders used to lay down the law in Colonial West Africa, but he would have felt equally at home among the inhabitants of the share market jungle in 1969-70. One popular image of the stock market depicts it as a den of robbers, but, in truth, very few of the participants in the mining boom were outright thieves. Many were optimists and nearly all were gamblers of a kind. A few were visionaries who ignored fundamentals. Others ran companies which were reasonably honest but looked after themselves well in the process. There were some appalling misjudgments made by highly respected men. There were directors who were quite unfit to be custodians of public money. The closer one studies them, the more human their grandeurs and follies become.

In this potpourri, it was not always possible to separate the thieves from the optimists. However, far more money would have been lost through optimism than theft. Someone who genuinely believes he has a big mine can pave the way to a far bigger disaster than a thief who is merely trying to make a quick killing. Some of the worst deals occurred in the smallest companies.

Nor was there any maestro orchestrating the whole story. Individual manipulators and would-be manipulators fell into traps of their own making or ran foul of other groups working at cross-purposes. It is vain to seek for any grand design behind the boom and crash. As the English historian A.J.P. Taylor once said of the historian's craft: "We are writing to shape into a version a tangle of events that was not designed as a pattern."

As the boom rolled past a lot of victims were left strewn in its wake. They included speculators, stockbrokers and promoters. One casualty was a theory, sometimes found in stock exchange literature, which holds that the stock exchange is a mechanism for the efficient allocation of resources. The theory holds that by establishing freely determined prices for shares and other securities, which includes the investment community's views of the future prospects of different industries, the stock market helps ensure that new funds flowing into the capital market are directed into the most profitable outlets. This is because the success of companies in raising funds in new issues is related to the market standing of their present securities or, if they are previously unlisted companies, of similar companies. Therefore the share market performs a useful and efficient allocative role.[1]

This view of the stock market's allocative role may well be true in normal times (leaving aside the question of what constitutes normality on the stock exchange) but it was certainly not true of the 1969-70 mining boom. Vast amounts of money were raised during the boom by companies which never found a viable orebody. Not one major deposit was discovered on boom money, although some were developed on it. The share market was certainly a highly efficient medium for raising money, but it was a most inefficient medium of allocating it.

The accompanying tables list all the amounts raised by mining companies from July 1968 to June 1971, the three years during which the boom was at its highest. On these figures — based on some assumptions which will be explained later — up to $900 million was raised by mining companies during 1968-71. This is $200 million more than the net proceeds from Commonwealth loan raisings over the same period.

Before commenting further it is necessary to explain the basis for the tables. Companies which were listed industrial, such as Capital Mining and Properties Ltd, have been excluded with the exception of Comalco. The first column of figures in the tables shows the initial capital of 241 companies which floated in the 36 months. Their initial capital totalled $476 million, including $62 million in vendor shares. Deducting the vendor figure, we find that 241 companies raised $414 million in cash.

One large float just missed being in the list. Bougainville Mining Ltd raised $29.3 million on a prospectus issued in May 1971 but the shares were not listed until after June 30. The total amount raised in cash, including Bougainville but excluding vendor shares, would be $443 million. The three largest floats shown in the tables are Comalco ($36 million), Robe River Ltd ($29.2 million) and Woodsreef Mines Ltd ($40 million, inc. $23 million vendor shares). Comalco raised money to help develop Weipa, the greatest bauxite deposit in the world. Robe River Ltd raised its money to help develop the Robe River iron ore deposit in WA, and the company has had a long struggle to become profitable. It had only managed to pay eight cents per share in dividends by the end of 1978. Woodsreef raised its money to develop the Barraba asbestos mine in northern NSW. At this point it would have to be counted a failure.

The best long-term prospects of any company floated proved to be those of Utah Mining Australia Ltd, which provided the Australian public with an indirect minority interest in the huge Queensland coal developments of the San Francisco-based Utah Construction and Mining. Ironically, Utah was one of the stickier floats of the era. Public acceptance of the prospectus was slow and the underwriters — Ord-BT of Sydney and Australian United Corporation Ltd of Melbourne — had some difficulty placing the stock. The shares sold at a discount soon after listing.

The success of the rest of the floats can be assessed by a glance at the dividends column, which does not include payments by companies before June 1968 or by companies after they shifted to the industrial lists. The last column tells briefly the fate of each company. In the vast bulk of companies, those that changed to industrial or were taken over were either destitute or inept. Certainly they did not find mining worthwhile. Some shifted quickly. Central Mining & Finance Ltd, floated in October 1970, was formed as a miner and investor, but could hardly have paid for a shovelful of earth to be turned over. Its management rapidly decided the boom was over and switched the company into property, changing its name to Central Management and Finance. Several of the late listings went the same way.

The final column shows the highest share price to June 30, 1978. Since then, some companies have risen further as newly fledged diamond prospectors. Most of these companies have been extensively reconstructed since flotation.

Of the $414 million cash subscribed to floats of new companies in 1968-71, a total of $338 million (excluding vendor shares) was raised by companies that have never since paid a dividend as a miner. This figure would be $29 million higher if Robe River had not managed to pay its three cents.

Some of the companies had remarkably short lifespans. We

have already seen how Samin Ltd lasted less than a year before getting a takeover bid from Poseidon. Others did not last much longer. Overland Mining NL raised $650,000 and listed at Christmas 1969, but was delisted within three years. Kimberley Mineral Holdings Ltd, listed in October 1970 with an initial capital of $750,000, was delisted two years later. Percy Marks Mining Co Ltd, the spin-off of a Sydney jewellery store, listed in December 1970 with $286,000 in paid capital and in April 1972 became a backlisting vehicle for Thomas and Coffey Constructions.

This raises the point that the majority of companies floated during the boom were undercapitalised. After eliminating the three largest floats in the tables — Comalco, Woodsreef and Robe River — the other 238 companies raised $238 million excluding vendor shares. This means their average paid capital was just under $1.4 million each. As noted earlier in this book, Poseidon had nearly $2 million cash in mid-1970 but was still quite unable to finance Windarra by itself. If the average small company had struck another Windarra it would have been in an even worse position than Poseidon.

This comment would not apply to some types of mines. Some minerals, such as tin, tungsten, rutile and gold, can be mined on small scale with small amounts of capital and still be profitable.[2] Tasminex in 1978 was engaged in exactly such a venture, mining tungsten in northern Tasmania. Nevertheless there is a prima facie case that most of the mining companies were too small to be viable as miners. Of course, the promoters had formed companies with capital deliberately kept small so their personal stake in a company would be greater and its price could be more easily manipulated. Some of them were obviously aimed only at mining the big paper quarry in the share market.

(It might be noted that diamond mining calls for large amounts of capital. In 1978, the cost of setting up a crushing and separation plant to process bulk samples was $3 to $4-million. This would be in addition to operating costs.)

Apart from the floats, there was a great deal of money raised by companies which existed before July 1968. The second and third columns of the tables show the amounts raised during the financial years ending June 1969, 1970 and 1971 by these mining companies. The second column is limited to cash issues only, as far as they can be defined. In this column are cash issues (no bonus issues), placements, call moneys raised and share options exercised. Some of the placements and allotments made by small companies during the boom may have been more in the nature of a vendor consideration than a cash issue, but it is impossible to ascertain the true nature of them all. All placements which appear to be non-cash have been omitted. The total is still formidable. The total cash raised by companies formed before the boom and by companies after floating amounted to $433 million.

Added to the previous figure of $414 million raised in floats excluding vendor shares, this gives a total of $847 million. There is an untraceable amount of double counting in this figure because some of the newly-floated companies subscribed to placements or issues by other companies. Conversely, some money raised by existing companies would have been subscribed to floats. But the double counting might not be substantial. The bulk of the money raised in issues was by heavyweight miners and the speculatives did not contribute much to those.

The figures in the third column show amounts issued in vendors shares by companies after listing and the paid value of shares issued in takeovers. The scrip issued in these forms was a highly liquid commodity and could be resold on market. The definition here of vendor shares has been very broad. Shares were issued during the boom for a range of purposes, sometimes in lieu of payment of debts. Perhaps the oddest example was Westralian Nickel's issue of 3532 contributing shares as the purchase price of a home unit. The par value of that parcel was only $706 but if the vendor was smart enough to sell them promptly he would have received a great deal more.

Also there were a few takeovers involving share issues. Pine Vale Mines took over Resources of Australia Ltd with a scrip issue. Scamander Mining Corporation NL took over Sub-Oceanic Minerals NL in the same manner. The scrip issued to shareholders of the merged companies was immediately resalable. The raider companies were acquiring assets with scrip, which was a liquid commodity. The total amount of such vendor shares and takeover shares comes to $36 million.

If this figure seems low, it is because wherever possible 'illiquid' share issues have been eliminated. By this I am referring to defensive share placements between companies, the biggest of the era being a swap between Great Boulder Gold Mines and Western Mining. There were also swaps between Lefroy-North Deborah, Seeko-Sipos, Quest-Murchison and a number of other small miners.

The $62 million worth of vendor shares omitted from the total of the first column should also be considered. These vendors were restricted from selling their shares on-market — in theory — for varying periods, usually a minimum of three months. They were not usually listed until the time limit had expired and a separate application had to be made to the exchange before they could be listed. But there were rumors that some vendor scrip was pushed on to the market before time. More than one broker discovered to his surprise that he had bought vendor shares for a client.

The biggest item in this $62 million was the $23 million vendor consideration for Woodsreef, which was not resold. An unknown proportion of the remaining $39 million was resold.

Adding all the figures together on the basis of these assumptions, we would have to put the total raised by mining companies during the three years at between $800 and $900 million. A median figure of $850 million would probably be the most accurate estimate.

One must make a concession to the efficient market theory by admitting that the heavyweight miners used the boom market with the utmost efficiency by exploiting their large share premiums to raise the maximum amounts possible. The CRA group were easily the best exponents of this technique.

CRA itself raised $58.3 million in two rights issues. Hamersley raised $31.5 million and New Broken Hill Consolidated Ltd raised £10 million. The Comalco and Bougainville floats raised a total of $65 million. This makes a total of over $170 million raised by this group, but a significant proportion of this would have been subscribed by overseas shareholders. Other large raisings of the era were $12 million by Peko, $11 million by CGFA and $9.3 million by Western Mining. On the other hand, four companies which have since disappeared from the lists — VAM, Mineral Securities, Surveys and Mining and Harbourside — raised $54 million between them.

The shrewdest effort was by Woodside. In mid-1968, Woodies raised $1.9 million by calling the remaining unpaid 10 cents on some 40 cent shares. This left them with 29 million fully paid shares. At the start of 1969 the company had a one-for-one issue of contributing shares, raising a further $2.9 million. These shares were then called to 20 cents in 1970 and 30 cents in 1971, raising a further $5.8 million. Woodside's chairman Geoff Donaldson was very experienced in financing exploration.

Indeed, it is a notable point of the list that the oil companies survived better than the miners. Several boom floats such as Reef, Basin and Vamgas managed to win a small equity in a producing field.

There is a whole family of oil stocks who not only survived the boom but pre-dated it. Alliance Oil Development has been around since 1960. It has never paid a dividend but it does have a small interest in the Cooper field. Beach has survived since 1961 and Exoil since 1962, neither having ever paid a dividend. Santos Ltd and Woodside date back to the oil boom of the early fifties. In both cases the original shareholders' equity has been eroded by the Burmah takeover, but the companies have survived with stakes in valuable, or potentially valuable, fields. Most amazing of all is Oil Search Ltd. Founded in 1929 this veteran of the boards has never struck a paying field but is still alive. This is a truly remarkable achievement. It has lived for 50 years, whereas some boomtime disasters couldn't last more than two.

Another point worth noting in the tables is the appalling performance of the investment companies. The only rationale for these companies is that their management possesses superior skill or knowledge which enables them to invest shareholders' funds in mining stocks and situations more wisely than the small investor could by himself. On all the evidence from the boom, the small investor has a better chance on his own.

There were three glamor mining investment houses during the boom — Minsec, Mining Traders and Mining Finance. None now survive. We have already seen what happened to Minsec and Mining Traders (Patrick Corporation).

Mining Finance Corporation Ltd was a spin-off by Slater Walker in Australia and enjoyed some big profits in the early days, making $1 million profit on one trade in the unlisted Nickel Mines of Australia Ltd. Later it became infected by the troubles that beset the rest of the Slater Walker empire, switched into property (as St James Properties) and was eventually taken over by the parent (named Oakbridge Ltd by then) to become a wholly-owned subsidiary.

A small investor could have justifiably put his money into these companies because they often showed considerable flair. They had some spectacular successes before they hit their sticky patches. The rest of the investment companies performed poorly.

Castlereagh Securities, having listed with $20 million cash and no liabilities in August 1970 was in deep trouble in less than six months. Corporate Mining, spun off by Mining Finance, was sold to Alexander Barton who resold it to Bela Csidei. By 1974 its shares were worthless. Smaller investment companies such as Munduna and Mincorp entered and left the lists without ever paying a dividend.

The most dreadful performances were those of three companies formed by Australia's best-known investment advisers of the time. Major Brian Douglas, James Cowan and Dr Michael Garretty were the main forces behind Selected Mining Holdings, Australia-Wide Mining Corporation and Trendex Mineral Corporation respectively. Selected Mining was a disaster for its promoters, being taken over by a raider within days of being listed. Australia-Wide gave its shareholders a small flurry with the shares hitting a peak of 80 cents before going downhill. The company passed through various hands, at one time being part of the Ian Murray empire. It was delisted in 1975. The Victorian Government appointed an inspector into its affairs.

Trendex is too complicated a story to relate in any detail here. This public company was involved in extensive share dealings with Garretty's family companies, including Stock Options of Australia Pty Ltd. Trendex listed in a weak market in January 1971 with an ominous underwriting shortfall. The prospectus had been for the issue of two million shares, of which one million were underwritten by the London broking firm of Citron, Williamson, Croft & Co. A private Garretty company, Trendex & Co Pty Ltd, had sub-underwritten three-quarters of a million of these, but rescinded the sub-underwriting agreement when Citron, Williamson, Croft & Co asked it to take up the shares. As a result the London broker was left holding half a million shares.

After listing Trendex Mineral Corporation became a tempting target for a takeover raider because the shares were backed by $1 cash (it had floated on an empty prospectus) but were only 60 cents each on market.

Such a raid soon began as Glomex Mines and Gordon Barton's Australian Mutual Growth Fund teamed up and started buying Trendex shares. As a defensive manoeuvre, Trendex Mineral Corporation took over Selected Mining. But the raiders persisted and Garretty sought refuge in yet another takeover, this time of Rimibo Resources Ltd. Rimibo was a small public company in which Selected Mining held a substantial interest. One of Dr Garretty's techniques in this fight was to stack a Rimibo meeting with more than 100 university students, equipped with Selected Mining proxies and voting on Selected Mining (Garretty) instructions. The students were paid $1250 for their services.

Melbourne stockbroker Richard Parsons, who had underwritten the Rimibo float and was a director of the company, managed to stall the raid long enough to thwart Garretty's plans. There was then a general collapse of the Garretty family companies. Trendex was a sick company for a long time afterward. It is still listed under new owners and named Lemarne Corporation.

The Rae Committee, having investigated the above affairs and several episodes associated with them, said "It would not be difficult to heighten the note of farce in a full account of Dr Garretty's activities: the shufflings of money so freely and without scruple from one listed company under his control to another or into buying shares in a company where his control was under challenge; the stacking of company general meetings with university students, and to no avail; the repeated failure of the inventor of so many devices to achieve or construct anything permanent as a result of his busy stratagems."

But the episode bequeathed one memorable passage of testimony before the Rae Committee, when Garretty was justifying the actions of Trendex. The following exchange occurred:[3]

Dr Garretty: You questioned me yesterday as to the consistency of what the company was doing with the objectives as stated in the prospectus. I think I have now referred to reasons why there may have been some apparent departure from following them to the letter, and I am reminded that when somebody asked a loaded question of Christ 2000 years ago relating to working on the Sabbath, I seem to remember from 30 years ago that Christ's answer was something like this: 'What man among you, learning on the Sabbath that his sheep has fallen into a well, does not go and attempt to pull it out?' So the actions of Trendex Mineral Corporation in this period that we were discussing yesterday, were affected by very strong forces from outside.

Senator Wheeldon: You would say they were Christ-like?

Dr Garretty: The word may have been used at the time, but probably not in that direct relationship.

One point can be made about the Trendex episode which is applicable to a number of mining investment companies in the dying stages of the boom — there's no mine.

However overblown the prices of companies such as Endurance, Poseidon and Tasminex may have become at least it was accompanied by the excitement of exploration. With the investment companies there were only pieces of paper. There was no outdoors adventure to quicken the pulses. In the process, the investment companies proved that speculators could lose as much money in paper as they could in the ground.

Nevertheless, the share markets may serve a useful economic role in a mining boom. By mopping up disposable funds in the hands of the general public, the stock market provides a safety valve for incipient inflation, as the price of shares is not a component of the Consumer Price Index. Some of the money might be permanently removed where a promoter, having acquired shares for $100,000 and sold them for $1 million, transfers the $900,000 profit to a Swiss bank. This would represent a small but real reduction in money supply at a time of inflationary pressure.

DIGRESSION V: A Few Words From Our Sponsors

"Those who controlled Minsec wrote their own death warrant in classic terms."

— Ric Dowling.

"It's easy. You can breathe, can't you — you can make money."

— Millie Phillips.

"I am not going to be a party to seeing some widow lose a lot of money because I make some statement that I can't fully back..."

— Roy Hudson.

"Profit making is neither illegal nor immoral. The means of achieving it can be."

— John Maguire.

"An investor who proposes to ignore near-term market fluctuations needs greater resources for safety and must not operate on so large a scale, if at all, with borrowed money..."

— Lord Keynes.

"There is still plenty of scope for the alert prospector."

— Sir Harold Raggatt.

"Short selling is a practice that should only be gone into by people who are either very sophisticated or who are gamblers."

— Gordon Barton.

"Because of the widespread implications of Minsec's collapse the situation was one of near-crisis in Australian securities marketing and a threat to Australia's investment image abroad."

— Australian Industry Development Corporation.

"I saw bank booms, land booms, silver booms, Northern Territory booms, and they all had one thing in common — they always burst."

— Banjo Paterson

"It (Hall's Peak) caused me more worry than any float that we have ever made."

— Ric Dowling.

"It was felt by some of us — I in particular felt quite strongly — that it would not be in the best interests of Trendex Mineral Corporation to do what was fairly common, and grab a pick and shovel and some lease pegs and run around the country and secure ground, dig a few holes and so forth."

— Dr Garretty.

"I have come to the conclusion that in spite of all previous theories, the mines of Colorado belong to the following classes:

1. Those that always did pay and always will.
2. Those that didn't once and do now.
3. Those that did once but don't now.
4. Those that don't now but will.
5. Those that never did and never will.

Of the first class, I regret to say neither history nor experience gives us an example. Of the four remaining classes the last is the only one upon which I can speak with certainty."

— US mineral engineer F.C. Kruger.

---

[1] One clear statement of this view is given by Professor R.L. Mathews in the first chapter "Studies in the Australian Capital Market," ed. by R.R. Hirst and R.H. Wallace (University of Adelaide, 1964).

[2] Associated Minerals Consolidated Ltd, the world's biggest beach-sand miner, went through the boom with capital unchanged at slightly $2.3 million. However, the company had a strong profit history and total shareholders' funds amounted to $7.6 million. Its infrastructure costs had been mostly incurred in cheaper times and its majority shareholder was the Consolidated Gold Fields group, providing ultimate protection.

[3] Rae Committee transcript of evidence, Part II, Volume 3, pp. 2440-1.

Amounts Raised by Mining Companies from 1969/70 — 1971/72

AMOUNTS RAISED ($'000)

Company	Initial Capital (Raised between July 68 & June71	Placements, Issues, Calls & Option Payments	Vendor Shares inc. takeovers	Highest Price per share (unadjusted)	Divs Paid	What Happened
A1 Consolidated Gold NL	-	240	741.1	$6.00	-	Delisted 1975
Abaleen Minerals NL	605.7	401.2	263.3	$2.80	-	Delisted 1975
Aberfoyle Ltd	-	3903.9	196.7	$3.80	24¢	Still listed
Academus Minerals NL	1500	-	-	.50	-	Delisted 1976
A-Cap Development Ltd	440	22.4	-	.80	-	Listed industrial
Acmex Holdings NL	-	1714.9	-	$5.20	-	Listed industrial
Acmin Explorations NL	2500	-	10	.22	-	T/O by Acmex
Acre Minerals Ltd	704.3	-	4	.16	-	Delisted 1976
Alkane Exploration (Terrigal) NL	991.6	703	-	$1.28	-	Still listed
Alliance Oil Development NL	-	3191.3	-	$1.20	-	Still listed
Alliance Petroleum Australia NL	-	2359.0	-	.96	-	T/O by AOD
Allied Minerals NL	1200	-	-	.90	-	Still listed
Allstate Exploration NL	650	2067	10	$2.80	-	Still listed
Alpha Minerals NL	1450	-	-	.70	-	T/O by Tjuringa
Altarama Minerals NL	1097.5	-	56.6	.16	-	Now Altarama Ltd. Listed industrial
Amad NL	-	5444	-	$5.10	-	Still listed
Amber Gold Ltd	1221	9.1	120	.80	-	Still listed
Ampol Exploration Ltd	-	6561.8	-	$4.70	44¢	Still listed
Anglo Range NL	520	-	-	.65	-	Suspended 1975
Antimony Nickel NL	999	-	150	$2.00	-	Suspended 1971
Apollo International Minerals NL	2300	300	-	.80	-	Suspended 1976

Arcadia Minerals NL	761	98.1	-	.80	-	Suspended 1976
Archean Mining Investments Ltd	119.6	-	-	.18	-	Listed industrial
Ashburton Oil NO	2000	236.1	-	.35	-	Liquidated 1974
Atherton Antimony NL	300	79.9	2	$1.13	-	Still listed
Audimco Ltd	1000	-	-	$2.45	-	Still listed
Aurora Minerals NL	906.4	175	-	$1.40	-	Delisted 1975
Australasian Mineral Search & Investments Ltd	3100	-	-	.40	-	Suspended 1976
Australasian Mining Corp Ltd	1200	-	-	$1.35	-	Delisted 1974
Australian Antimony Corp NL	1700	1455	-	$3.50	-	Delisted 1975
Australian Consolidated Minerals NL	3401	191.1	684.4	$1.60	-	Still listed
Australian Continental Resources Ltd	6005	-	-	.50	-	T/O by Aston Developments
Australian Copper Mines NL	4250	-	-	.23	-	Delisted 1976
Australian Development NL	-	1890	-	$5.10	29.5¢	Still listed
Australian Oil & Gas Corp Ltd	-	1765.5	-	$2.95	7¢	Still listed
Australia-Wide Mining Corp Ltd	2000	-	-	.80	-	Delisted 1975
Australis Mining NL	567.5	-	-	.25	-	Still listed
Austral Mining Co NL	-	193.8	-	$1.90	-	Now Quay Corp. Listed industrial
Austral Pacific Mining Corp Ltd	900	-	-	.15	-	Now Dainford Ltd. Listed industrial
Bamboo Creek Gold Mines NL	2834	-	5	$1.00	-	Still listed
Barrier Exploration NL	-	922.4	-	$3.20	-	Still listed
Base Minerals Ltd	650	100	-	$1.45	-	Delisted 1975
Basin Oil NL	2000	940.5	150	.58	-	Still listed
Beach Petroleum NL	-	750	-	$4.80	-	Still listed

Company	Initial Capital (Raised between July 68 & June71	Placements, Issues, Calls & Option Payments	Vendor Shares inc. takeovers	Highest Price per share (unadjusted)	Divs Paid	What Happened
Beaver Exploration Australia NL	3700	6	-	.25	-	Listed industrial as Beaver Australia
Binti Exploration Ltd	1250	100	-	.90	-	Now Cord Holdings. Listed industrial
Boomalli Ltd	320.4	-	-	.35	-	Delisted 1974
Bounty Oil Ltd	5000	3517.7	-	.29	-	Delisted 1974
Bridge Oil NL	3300	278.9	-	$1.10	-	Still listed
Broken Hill Metals NL	1501	50	73.6	.70	-	Still listed
Broken Hill South Ltd	-	1922.8	-	$7.30	62¢	Now BH South. Still listed
Buka Minerals NL	3735.9	-	-	.50	-	Still listed
Carr Boyd Minerals Ltd	1169	2680.3	100	$9.00	-	Still listed
Castlereagh Securities Ltd	19000	-	-	.39	-	T/O by Patrick Corporation
Centamin NL	950	-	4	.34	-	Still listed
Centaur Mining & Exploration NL	630	-	-	.57	-	Still listed
Central Deborah Gold Mining Co NL	-	90	-	$1.70	-	Delisted 1975
Central Mining & Finance Ltd	1000	-	-	.28	-	Now Central Management & Finance Ltd. Listed industrial
Central Norseman Minerals NL	359.3	-	-	$13.00	-	Now Kemple Mines. Still listed
Central Pacific Minerals NL	-	2543.5	-	$9.00	-	Still listed
Chapman Mining NL	550	-	-	.16	-	Still listed
Charterhall Finance Holdings Ltd	1116.3	1370.0	56.3	$1.90	-	Now Charterhall Ltd. Still listed

Citadel Mining & Investment Ltd	3001	-	-	.10	-	T/O by Corporate Mining
Clarence River Basin Oil Exploration Co NL	-	420	-	.90	-	Liquidated 1975
Cliffminex NL	267.9	-	-	.26	-	Still listed
Clifford Minerals Ltd	501	-	-	.50	-	Still listed
Coastal Rutile Ltd	700	-	-	$1.35	-	Receivership 1973
Comalco Ltd	35750	-	-	$6.10	52.75¢	Still listed
Combined Resources Ltd	2000	-	-	.65	-	Now N & K Properties. Listed industrial
Command Minerals Ltd	871	-	-	.30	-	Suspended 1977
Comstock Minerals Ltd	3200	-	-	.40	-	Delisted 1974
Consolidated Gold Fields of Australia Ltd	-	11000	3744.8	$11.40	139.5¢	Still listed
Consolidated Mining Industries Ltd	1342.0	10.9	29.8	.98	-	Liquidated 1974
Consolidated Rutile Ltd	-	1009.0	-	$4.60	117¢	Still listed
Conwest Exploration (Australia) Ltd	5250	-	-	$2.65	-	Delisted 1977
Conzinc Riotinto of Australia Ltd	-	58295.5	9.5	$25.00	93¢	Still listed
Coopers Creek Mining & Exploration NL	940	-	-	$1.30	-	Still listed
Corella Minerals Ltd	851	-	-	.34	-	Now Craigmoor Wines. Listed industrial
Corporate Mining Investments of Australia Ltd	2500	510.0	-	$1.60	-	Delisted 1974
Cosmos Exploration & Minerals NL	1586.7	-	20	.34	-	Still listed
Crusader Oil NL	3717.3	7.6	-	.43	-	Still listed
Cudgen R.Z. Ltd	-	2900	-	$4.10	58¢	Still listed

Company	Initial Capital (Raised between July 68 & June71	Placements, Issues, Calls & Option Payments	Vendor Shares inc. takeovers	Highest Price per share (unadjusted)	Divs Paid	What Happened
Cultus Pacific NL	-	450	-	$2.10	-	Still listed
Day Dawn Minerals NL	2201	350	-	.40	-	Delisted 1975
Delta Minerals NL	1675.2	-	861.5	.30	-	Delisted 1975
Devex Ltd	500	302.5	-	$1.65	-	Still listed
Dominion Mining NL	1000	-	-	.90	-	Still listed
Durack Mines Ltd	8500	-	-	.25	-	Merged into Planet Resources
Eagle Exploration Ltd	780	-	-	.23	-	Now Eagle Corp. Listed industrial
East Coast Minerals NL	998.5	94.3	-	.22	-	Still listed
Eastern Copper Mines NL	1420.1	220	12.5	$1.20	-	Still listed
Eastmet Minerals NL	1200	368.7	91.1	$6.50	-	Still listed
Endeavour Oil Co NL	3547.5	3654.1	6.8	$1.40	-	Still listed
Endurance Mining Corporation NL	-	1471.5	-	$10.60	-	Suspended 1973
Epoch Minerals Exploration NL	307.5	-	7.5	.46	-	Still listed
Esperance Minerals Ltd	900	-	-	.16	-	Still listed
Euralba Mining Ltd	615	-	-	.15	-	Suspended 1974
Exoil NL	-	819	-	$1.30	-	Now Oilmin. Still listed
Ferrovanadium Corporation NL	3000	-	-	.22	-	Still listed
Flinders Petroleum NL	1390	10.0	2500	.35	-	Still listed
Forsayth Mineral Exploration NL	600.8	-	-	.65	-	Still listed
Gem Exploration & Minerals Ltd	300	750	5.5	$3.30	-	Still listed
Genoa Oil NL	3000	216.4	-	.75	-	Still listed
Geometals NL	497	810	-	.80	-	Still listed
Giltspur Gems & Minerals NL	550.5	136.4	-	.20	-	Suspended 1973

Gippsland Minerals NL	476	960	-	$9.00	-	Still listed
Glendale Explorations NL	500	175	25	.77	-	Now Glendale Land. Suspended 1977
Glomex Mines NL	1570.8	-	35.8	.30	-	Liquidation 1976
Gold Copper Exploration Ltd	2384.7	-	-	.19	1¢	Still listed
Goldrim Mining Australia Ltd	1250	-	-	$1.90	-	Still listed
Great Boulder Gold Mines Ltd	-	3959	-	$20.00	-	T/O by WMC
Great Northern Mining Corp Ltd	-	1192	-	$1.25	-	Still listed
Greenbushes Tin NL	-	1835.2	-	.90	11.75¢	Still listed
Greenvale Mining NL	815.2	-	-	.27	-	Still listed
Hall's Peak Australia Ltd	2225	0.5	375	$1.10	15¢	Now Arrawatta Holdings. Still listed
Hamersley Holdings Ltd	-	31500	-	$20.00	75¢	Still listed
Haoma Gold Mines NL	-	133.8	-	$1.88	-	Still listed
Harbourside Oil NL	2000	11036.7	-	.17	-	Liquidator 1974
Hartog Oil NL	3000	424.1	-	.55	-	Still listed
Hastings Exploration NL	1250	11.2	-	.52	-	Delisted 1975
Hawkestone Minerals Ltd	-	2893.1	638.6	$1.75	-	Now listed industrial
Highland Gold Development NL	600	-	-	.78	-	Now Nema. Listed industrial
Hill 50 Consolidated NL	640.1	-	-	$1.70	-	Now Hill 50 Uranium. Still listed
Hill 50 Gold Mine NL	-	2700	-	$3.00	-	Still listed
Hill Minerals NL	1360.9	-	-	$1.02	-	Still listed
Hopetoun Minerals NL	620	-	-	.10	-	T/O by Bridge Oil
Horizon Explorations Ltd	1625	-	-	.30	-	Now Horizon Pacific. Listed industrial
Hunter Mining & Investment Ltd	700	150	-	.45	-	Delisted 1976
Inland Mining NL	424.9	-	-	.25	-	Still listed
International Mining Corp NL	500	132.2	-	$9.40	-	Still listed

Company	Initial Capital (Raised between July 68 & June71	Placements, Issues, Calls & Option Payments	Vendor Shares inc. takeovers	Highest Price per share (unadjusted)	Divs Paid	What Happened
International Oils Exploration NL	6000	-	667	.96	-	Still listed
Investment Mining & Finance Ltd	1200	51.3	-	.95	-	Still listed
Jensen Mining & Investments Ltd	-	480	-	$2.05	-	Now Jensen Investments. Listed industrial
Jimberlana Minerals NL	2460	-	-	.26	-	Still listed
Jingellic Minerals NL	750	-	7.6	.52	-	Still listed
Jones Mining NL	796.6	-	-	$1.60	-	Still listed
Kambalda Minerals NL	1020.5	29.4	19.3	.73	-	Delisted 1976
Kamilaroi Oil Co Ltd	-	4312.3	40	$3.70	-	Delisted 1976
Karangi Minerals NL	810	-	24.3	.34	-	Delisted 1975
Kathleen Investments (Australia) Ltd	-	5053.6	-	$17.50	25¢	Still listed
Keywest Explorations NL	600	-	-	.19	-	Still listed
Kia Ora Gold Corporation NL	1107	390	-	$10.50	-	Still listed
Kia Ora Investments Ltd	210.1	246	3444	$2.90	3.75¢	Still listed
Kimberley Mineral Holdings Ltd	750.2	-	15	.15	-	Delisted 1972
Kimberley Securities Ltd	501	-	-	.43	-	Still listed
King Mountain Mining NL	723	-	-	.15	-	Delisted 1975
Kratos Uranium NL	1050.5	-	-	.58	-	Still listed
Lake View & Star Ltd	-	603.1	-	$8.40	15¢	T/O by Poseidon 1971
Lamadec Exploration Ltd	801.5	7.4	397	$1.50	-	Delisted 1974
Laverton Nickel NL	1000	-	64.5	$8.00	-	Suspended 1977
Lefroy Minerals Ltd	236	114	-	$4.60	-	Delisted 1974
Leighton Mining NL	3500	-	-	.20	-	Still listed
Lennard Oil NL	1906	97.0	-	.60	-	Still listed

Leonora Nickel NL	660	-	-	.40	-	Suspended 1978
Leopold Minerals NL	1802	35.1	-	$8.80	-	Now Kitchener Mining. Still listed
Lightning Ridge Mining NL	550.3	-	-	.07	-	Still listed
Lone Star Exploration NL	2250	-	-	.17	-	Delisted 1975
Longreach Metals NL	3500	698.3	-	.75	-	Still listed
Longreach Oil Ltd	-	3700	-	$1.40	-	Still listed
Ludbrooks Ltd	-	-	449.2	.35	-	Now Qintex Ltd. Listed industrial
Magellan Petroleum Australia Ltd	-	723.8	-	$5.20	-	Still listed
Magnet Metals Ltd	344.9	147.3	100	.60	-	Still listed
Magnum Exploration NL	1000.3	50	0.5	$1.65	-	Still listed
Mallina Mining & Exploration NL	354	450	122.5	$6.50	-	Still listed
Mareeba Mining NL	1250	-	-	$1.40	-	Still listed
Marks, Percy, Mining Co Ltd	286.1	-	-	.10	-	Now Thomas & Coffey. Listed industrial
Meekatharra Minerals NL	400	1200	-	$6.90	-	Still listed
Metals Exploration NL	-	4525	78	$12.00	-	Still listed
Metals Investment Holdings NL	3751.5	-	40	.31	-	Delisted 1975
Metana Minerals NL	300.3	-	-	.28	-	Still listed
Metramar Minerals Ltd	530	2375	-	$4.24	-	Still listed
Mid-East Minerals NL	1080.4	1322.9	-	$4.10	-	Still listed
Mid-Eastern Oil NL	-	2020	-	$1.95	-	T/O by Woodside-Burmah
Mincorp Ltd	751.3	-	-	.49	-	Liquidated 1976
Minefields Exploration NL	1600	-	-	$1.00	-	Still listed
Mineral Resources Australia Ltd	220	-	-	$3.60	-	Now Vincellow. Listed industrial
Mineral Securities Australia Ltd	-	17314.8	-	$23.00	12.75¢	Liquidated 1971

Company	Initial Capital (Raised between July 68 & June71	Placements, Issues, Calls & Option Payments	Vendor Shares inc. takeovers	Highest Price per share (unadjusted)	Divs Paid	What Happened
Minerals Mining & Metallurgy Ltd	608.7	-	-	.57	12.5¢	Still listed
Minerals Recovery (Australasia) NL	800	-	-	.51	-	Still listed
Minerol Investments Ltd	300	-	-	.26	-	Still listed
Mining & Associated Industries Ltd	750	-	-	$1.30	-	Now Great Western A'asia. Listed industrial
Mining Finance Corporation Ltd	5878	-	5825.1	$5.50	-	T/O by Oakbridge
Mining Houses of Australia Ltd	505	-	-	.60	-	Still listed
Mining Traders Ltd	-	3108.8	654.8	$3.40	16.8¢	Name changed to Patrick Corp. Liquidated 1975
Minvest Ltd	300	-	-	.25	-	Now Loftus Properties
Mogul Mining NL	5400	-	-	$1.20	-	Still listed
Mountain View Gold NL	-	30.2	-	$2.05	-	Now Rosenthal Jewellery. Listed industrial
Mt Arthur Molybdenum NL	1512.9	-	-	$1.15	-	Still listed
Mt Carrington Mines Ltd	3000	600	-	.80	-	Still listed
Mt Coora Mining NL	478.3	-	-	.80	-	Still listed
Mt Hope Minerals NL	1001	-	-	.25	-	T/O by IMC
Mt Lyell Mining & Railway Co Ltd	-	6510	-	$3.65	75¢	Still listed
Mt Margaret Nickel Mines NL	200.1	-	-	.55	5.1¢	Still listed
Munduna Investments Ltd	500	-	-	.78	-	Liquidated 1976
Munga Creek Minerals NL	1252.5	1824.7	-	.46	-	Liquidated 1974
Murchison Central Mines Ltd	250	1625	-	$5.60	-	Suspended 1977
Murphyores Holdings Ltd	-	-	666.7	$4.30	11.25¢	Still listed

Murumba Minerals NL	1944.2	-	-	.35	-	Delisted 1974
Murumba Oil NL	3001	110.6	-	.29	-	Delisted 1974
Naracoopa Rutile Ltd	1409.7	1.3	-	$1.20	-	Receivership 1972
Narla Minerals NL	1200	250	-	.75	-	T/O by Gollin
New Broken Hill Consolidated Ltd	-	23538.8	-	$25.00	191.25¢	Liquidated 1973 in CRA reconstruction
Newmetal Mines Ltd	668	1939.2	313.5	$3.17	-	Still listed
Newmex Exploration Ltd	1412	-	-	$2.01	-	Still listed
N.S.W. Oil & Gas NL	10000	-	-	.48	-	Merged into Planet Resources
Nickelfields of Australia NL	1804	-	-	.67	-	Suspended 1971
Nickelore NL	249.9	-	-	$1.20	-	Still listed
Nicron Ltd	350	-	-	.15	-	Still listed
Nobelex NL	3158	342	2500	$1.80	-	Still listed
Norseman Gold NL	-	3663.6	129	$16.50	-	Now Norseman Mining. Still listed
Norseman Titanium NL	1500	-	-	.30	-	Receiver 1975
North Deborah Mining Co NL	496	2752.5	-	$7.50	-	Delisted 1974
Northern Mining Corporation NL	1000	840.8	33.2	$2.09	-	Still listed
North Flinders Mines NL	1300	72.5	-	$14.00	-	Still listed
North Kalgurli (1912) Ltd	-	3320.7	-	$16.25	-	Still listed
Northland Minerals Ltd	500.3	-	-	.65	-	Still listed
North West Mining NL	-	750	-	$2.50	-	Still listed
Northwest Oil & Minerals Co NL	1815.5	-	125	.91	-	Delisted 1975
Nullarbor Mining & General Investment Ltd	500	-	15	.65	-	Now Nullarbor Holdings. Listed industrial
Nuttall Holdings Ltd	-	729.4	-	$1.10	-	Still listed
Ocean Resources NL	1011.9	-	40	.67	-	Still listed
Offshore Oil NL	3000	343.8	-	.45	-	Still listed
Oil & Minerals Quest NL	-	97.6	-	.95	-	Still listed

Company	Initial Capital (Raised between July 68 & June71	Placements, Issues, Calls & Option Payments	Vendor Shares inc. takeovers	Highest Price per share (unadjusted)	Divs Paid	What Happened
Oil Investments Ltd	-	907.5	-	$2.60	12.75¢	Still listed
Oil Search Ltd	-	7660.2	-	$1.55	-	Still listed
Omega Oil NL	1320	134.1	-	.95	-	Still listed
Opalton Mining NL	500.3	250.2	-	$1.20	-	Suspended 1976
Ora Banda Broad Arrow Minerals Ltd	655	-	-	.15	-	Now Pyrophyllite Corporation Ltd
Otter Exploration NL	1690	-	-	.45	-	Still listed
Overland Mining NL	700	-	26.8	$2.20	-	Delisted 1972
Pacific Copper Explorations Ltd	1250	1152.5	52.1	$5.70	-	Still listed
Pacific Island Mines Ltd	-	4303.1	116.4	$1.45	-	Merged into Planet Resources
Pacific Mining Ltd	-	2963	-	$2.20	-	Still listed
Palgrave Corporation Ltd	555	-	-	$3.10	-	Now Swan Tyre Corp. Still listed industrial
Pancontinental Mining Ltd	-	159.5	-	$21.00	-	Still listed
Papuan Nickel Exploration NL	1124	-	-	.27	-	T/O by Newmetal
Paringa Mining & Exploration Co Ltd	-	258	-	$2.50	-	Still listed
Peerless Mining Ltd	-	361.7	161.8	$1.40	-	Now Aust. Asiatic Corp. Listed industrial
Pegmin Ltd	1087.5	-	-	$1.70	-	Still listed
Peko-Wallsend Ltd	-	11997.2	276.3	$15.10	136.25¢	Still listed
Petrocarb Exploration NL	500	450	-	.41	-	Still listed
Petroleum Securities Australia Ltd	-	375	-	$4.10	17.5¢	Still listed
Petromin NL	2400	-	-	$1.00	-	Still listed
Pexa Oil NL	2158.5	4000	50	$1.22	-	Still listed

Pine Vale Mines Ltd	729.3	1099.4	659.2	$4.60	-	Now Pine Vale Investments. Listed industrial
Pioneer Mines NL	-	786	15.9	$1.65	-	Still listed
Planet Gold Ltd	-	1200	1000	.69	-	Merged into Planet Resources. Liquidated 1976
Planet Metals Ltd	-	-	81.2	$1.90	-	Merged into Planet Resources. Liquidated 1976
Platina Developments NL	600	420	-	$2.25	-	Suspended 1977
Poseidon NL	-	3073.4	335.4	$280.00	-	Delisted 1976
Project Mining Corporation Ltd	477.5	17.4	200	$1.80	-	Still listed
Pursuit Oil NL	1875	0.5	-	.65	-	Merged with Crusader
Queensland Mines Ltd	-	5393.1	-	$46.00	-	Still listed
Quest Mining & Exploration NL	1125	-	-	$3.40	-	Still listed
Ravensthorpe Copper Mines NL	-	1481.3	415	.80	-	Now Hollandia Ravensthorpe. Listed industrial
Reef Oil NL	3000	-	-	.28	-	Still listed
Renison Ltd	-	2200	-	$15.00	220.2¢	Still listed
Reserves & Mining Investment Ltd	500	90.5	-	$1.80	-	T/O by Macdonald Hamilton
Resource Exploration NL	920.2	673	-	.45	-	Now Serex. Listed industrial
Resources of Australia Investment Co Ltd	2000	-	-	$1.02	-	T/O by Pine Vale
Rimibo Resources Ltd	750	-	-	.20	-	Delisted 1975
Robe River Ltd	29200	1161	-	$2.70	3¢	Still listed
Russgar Minerals NL	725	-	-	.28	-	Delisted 1976
Samin Ltd	1755.5	305.8	-	$33.00	-	T/O by Poseidon
Santos Ltd	-	3377.7	-	$3.60	2¢	Still listed
Scamander Mining Corporation NL	304.4	592.5	277.2	$3.45	-	T/O by Zephyr

Company	Initial Capital (Raised between July 68 & June71	Placements, Issues, Calls & Option Payments	Vendor Shares inc. takeovers	Highest Price per share (unadjusted)	Divs Paid	What Happened
Scorpio Investments Ltd	210	-	-	.60	-	Changed to Anglo-Pacific Estates. Delisted 1976
Secmin Ltd	1000.5	100	-	.34	-	Became Falkirk Properties. Suspended 1976
Sedimentary Uranium NL	1000	-	-	.75	-	Still listed
Seeko Nickel NL	700	220	-	$1.80	-	Now Seeko Investments. Still listed
Selcast Exploration Ltd	3400	-	-	$3.30	-	Still listed
Selected Mining Holdings	2000	-	200	.20	-	Became Greyfell. Delisted 1977
Shaw River Alluvials NL	-	187.1	5	$1.65	-	Liquidated 1974
Siberia Nickel NL	655	-	-	.30	-	Still listed
Silver Valley Minerals NL	-	1195.9	1929.3	$4.60	-	Still listed
Sipos Mining & Exploration NL	400	452	-	$5.10	-	Still listed
Sirus Mining & Investment Ltd	301	-	-	.28	-	Became Dow Properties. Suspended 1974
S.A. Barytes Ltd	-	1612.5	-	$1.72	6¢	Receiver 1973
Southern Cross Exploration NL	1701.9	11.8	30	.40	-	Still listed
Southern Pacific Petroleum NL	-	2200	-	$3.25	-	Still listed
Southland Mining Ltd	1250	8778.1	1281.5	$4.56	4¢	Still listed
Spargo's Exploration NL	-	1080.3	12.5	$10.10	-	Still listed
Stellar Mining NL	1375	-	200	$1.95	-	Still listed
Sturts Meadows Prospecting Syndicate NL	-	311.6	10	$6.50	-	Suspended 1974
Sub-Oceanic Minerals NL	350	-	-	$1.10	-	T/O by Scamander

Sundowner Minerals NL	826	-	-	.17	-	Still listed
Sunhill Corporation Ltd	-	105	180	$2.70	-	Delisted 1976
Surveys & Mining Ltd	-	10298	470.2	$3.95	1¢	Liquidated 1971
Talga Exploration Ltd	1005	-	-	.37	-	Delisted 1976
Target Petroleum NL	2000	875.3	-	$1.09	-	Still listed
Tasman Minerals NL	770.9	262.5	53.8	$1.60	-	Liquidated 1975
Tasman Nickel NL	836.4	-	-	.30	-	Liquidated 1975
Tasminex NL	500	1001.7	62.5	$90.00	-	Still listed
Tennant Creek Metals NL	-	452.1	-	$3.70	-	Delisted 1974
Tennyson Minerals Ltd	600	-	-		-	Still listed
Theseus Exploration NL	670	-	4	$2.05	-	Delisted 1977
Timor Oil Ltd	-	3912.3	-	$1.95	-	Still listed
Transcontinental Mining Corporation Ltd	658	-	-	.06	-	Now Transcontinental Projects. Listed industrial
Transoil NL	-	810	-	$1.55	-	Still listed
Trendex Mineral Corporation Ltd	2150	300	-	$1.15	-	Now Lemarne Corp. Listed industrial
Triako Mines NL	6300	-	-	$2.00	-	Still listed
Triton Investment Corporation Ltd	600	-	-	.90	-	T/O by Waters Holdings
Tropic Mineral Holdings NL	2718.4	568.3	-	.40	-	Delisted 1973
United Petroleum Reserves NL	2575.9	-	-	.98	-	Still listed
United Uranium NL	-	4056	-	$3.01	-	Receiver 1973
Uranium & Nickel Exploration NL	-	277.5	87.5	$1.60	-	Still listed
Uranium Consolidated NL	-	424	-	$3.29	-	Still listed
Uranium Development & Prospecting Ltd	-	475	-	.82	-	Now Udap Investments. Listed industrial
Utah Mining Australia Ltd	5000	-	-	$12.10	80.5¢	Still listed

Company	Initial Capital (Raised between July 68 & June71	Placements, Issues, Calls & Option Payments	Vendor Shares inc. takeovers	Highest Price per share (unadjusted)	Divs Paid	What Happened
Valiant Exploration NL	800	-	-	.66	-	Still listed
VAM Ltd	-	15478.8	2331.5	$9.00	-	Receiver 1971
Vamgas NL	3200	1312.3	780	$1.20	-	Still listed
Vultan Minerals Ltd	970.8	-	-	$1.60	-	Still listed
Wattle Gully Gold Mines NL	-	163	129.1	$1.10	0.5¢	Still listed
Weatherly Oil NL	3500	5.2	-	.25	-	Now Olympus Petroleum. Still listed
West Coast Holdings Ltd	-	528.8	274	$3.30	-	Still listed
Western Compass Minerals NL	450	-	-	$1.30	-	Suspended 1975
Western Mining Corporation Ltd	-	9301.1	-	$19.00	52¢	Still listed
Western Resources Ltd	200	-	-	.17	-	Listed industrial
Western Titanium NL	-	20	-	$7.10	114¢	T/O by Associated Minerals
Westmex Ltd	800	45	40	$2.50	-	Still listed
Westmoreland Minerals Ltd	251	1285.1	-	$2.50	-	Delisted 1974
Westralian Nickel Exploration NL	840	2582.3	515.7	$14.00	-	Still listed
Westralian Sands Ltd	-	2501.5	-	$2.01	16.2¢	Still listed
Westside Mines NL	550	-	-.45		-	Still listed
Whim Creek Consolidated NL	2200	-	1.9	$6.13	-	Still listed
Woodside (Lakes Entrance) NL	-	10600	-	$4.08	-	Now Woodside Petroleum. Still listed
Woodsreef Mines Ltd	40000	-	-	.30	-	Still listed
Zephyr Minerals Ltd	380	-	-	$3.25	-	T/O by Winthrop Investments

CHAPTER TWENTY

Where Were the Watchdogs?

As the boom subsided and former glamor companies were collapsing, an inevitable reappraisal began of the roles played by the various parties involved. The promoters were roundly abused, but many of them suffered no worse fate. And at least the promoters had clear motivations. The motivations of the various institutions which purport to protect the public were more obscure. This chapter will conclude the book with a brief examination of the watchdogs.

1. The Financial Press.

In the final analysis the financial press has no clear duty to anyone, except to sell publications on behalf of the various proprietors. There is no universal agreement on what its role ought to be. Company directors and stockbrokers sometimes consider the financial press to be part of the business establishment and take with ill grace any criticism of themselves. However, very few financial journalists in Australia would see themselves as part of such an establishment. Their general attitude would be that the job of the financial press is to keep the investing public promptly and accurately informed, and where necessary to defend the interests of the investing public at large in situations where the public has no other voice. It is not often recognised that this is largely a self-imposed duty. Life would be much easier for working financial journalists if they simply accepted without question what they were told by companies and exchanges.

Taking the financial press on this estimation — that its prime duty is to defend the public interest — its performance in the mining boom was patchy. The worst criticism of any individual pressman was that handed to Cliff Riggs, the finance editor of the Sydney Daily Mirror, by the inspectors into Endurance. The inspectors said: "There can be no doubt that Riggs was more concerned with filling space allotted to him in the Daily Mirror than with accuracy. He consistently quoted as his own considered opinion matters which were given to him by others who, from the material supplied, were not disinterested parties ... Riggs told us in the course of his examination that many of the statements in his published articles were just 'fill'; in that he had space to fill to meet newspaper deadlines and made statements without any regard to their accuracy or authenticity. This is deplorable."

In defence of Riggs it should be pointed out that he only ever held one small parcel of Endurance and that only briefly. He did not gain any tangible benefit, on the evidence, from his bullish articles about the stock. The practice of using 'fill' material to plug holes in a newspaper approaching edition deadline is standard throughout the world. Overseas cables are most frequently used for this purpose, for reasons of both safety and dispensability.

Some journalists did excellent work during this era. The best investigative reporting was done by John Gilmour of the Financial Review. The best commentaries were by Tom Fitzgerald as Finance Editor of the Sydney Morning Herald. John Pierce of the Financial Review and Bryan Frith of The Australian provided the best technical coverage of the mining industry.

The Sydney evening newspapers were later widely criticised as "tip sheets" during the boom. The truth is that there was a wide range of publications available to read during the boom. Some tried to evaluate what was happening soberly and responsibly while others repeated tips and rumors. Newspaper buyers have a perfect freedom of choice to read what they wish. To buy a Sydney evening newspaper to read the finance pages and then complain that the writers are tipping shares is on par with buying Best Bets and complaining that it tips racehorses.

There was certainly a great deal of inaccurate information reported in the press during the boom, but this was by no means entirely the fault of the press. Many statements issued by promoters during the boom deliberately put the best possible gloss on their companies' prospects. The truth was frequently distorted.

Of equal importance, companies frequently did not issue statements promptly to correct previous statements. Reserve estimates and drill assays were allowed to stand uncorrected for long periods. Working under pressure of deadlines, with a mass of mining reports coming in every day, financial journalists often had little option but to simply take a promoter's word for what was happening. Every financial journalist who tried to cover the boom was later amazed by the revelations unearthed in official investigations. What we had been covering was the tip of the iceberg.

Ideally, newspapermen should not write stories about companies whose shares they hold. If this is unavoidable, the interest should be disclosed in the article. Since the Securities Industry Act has been in force all finance journalists must keep registers of their shareholdings and these can be inspected at any time by officers of the Corporate Affairs Commission.

There is also the question of whether newspapermen should take shares in floats. In the discussions about the ethics of participating in floats, it sometimes seems to be forgotten that the shares have to go to *someone* and that there has to be some inducement to take them. My own attitude has always been that if I do not take the shares, they will only go to someone less deserving.

Tom Fitzgerald, discussing the question at the time of the Comalco float, wrote: "Some of those who have a policy of accepting special offers from companies argue that this particular kind of participation sharpens a journalist's sense of the market place and his attention to company affairs. They say there is no serious problem in maintaining impartiality as journalists in the circumstances; that the tendency is sometimes rather to bend over backward in objectivity. The other school of opinion, which may be in the majority among regular commentators, prefers not to be involved in angular decisions as to whether they are bending either forward or backwards." Fitzgerald suggested establishment of journalists' share registers, saying: "Objectivity itself would seem to need an objective criterion." This in effect is what now happens.

2. Geologists.

Geologists work in a technical area where their best calculations are no better than approximations. An unexpected intrusion of barren rock, unsuspected variations in grade, metallurgical difficulties in ore treatment or a host of other factors can make a legitimate and expert calculation of an orebody's worth hopelessly wrong. There is an old saying that nobody knows the reserves of an oilfield until the last barrel is pumped out and the same is true of mines. Some attempt has to be made at calculating tonnage and grade of ore before mining begins or nobody would finance the project. These calculations will necessarily be imperfect and where there is room for error there is much room for fraud.

One disturbing feature of boom-time geological statements is their misleading air of precision. Poseidon's statement of 3.56 percent nickel looked like a fine calculation to one-hundredth of one percent. In fact, Poseidon had no basis on which to make any such calculation at the time and the actual assay of the core turned out to be substantially lower, although still of ore grade.

A statement by Endurance that it has indicated reserves of 500,000 tons or 1.4 percent tungstic oxide or a statement by Queensland Mines that it has 55,000 tons of 540 lb grade uranium oxide has every appearance of precision. On what grounds would a layman or an expert investor be expected to disbelieve these statements? It was generally accepted that there was some margin for error in these statements, as in all reserve estimates, but nobody dreamed they were completely wrong. The effects of the errors in these statements were compounded when the companies did not correct them at the earliest possible date.

One suggestion has been that mining companies should be subject to some form of geological audit as well as financial audit. There is much logic in this idea. At present exploration companies are subject only to a normal financial audit. A typical small company might have little cash or other assets, but a mineral deposit worth anywhere between zero and $10 million. In these circumstances the financial audit is almost meaningless. The investing public values the company upon its mineral prospects, not upon its net tangible asset backing (which is often slight) nor its earnings (probably negative).

It might not be necessary for every company to have a geological audit every year. Such an audit could be ordered by the Corporate Affairs Commission where a company's statement or a movement in its shares warrants investigation. A geological audit need not take long. If a geologist can visit a mine site and has access to all assays and other exploration data he should be able to prepare a report (necessarily approximate) within a few days.

3. Stockbrokers.

Stockbrokers enjoyed palmy days through the mid-1960s before seeing a staggering increase in their business during 1969-70. Broking at that time was dominated by the largest firms in Sydney and Melbourne. The five largest firms in Sydney and the five largest in Melbourne together accounted for 32 percent of the revenue and 28 percent of the profit declared by all 154 firms in Australia's six capital city exchanges.

Melbourne and Sydney Stock Exchanges dominate the broking business of the country, generating about 85 percent of the total revenue. About half the business of Adelaide, Brisbane and Perth would be done on behalf of Sydney or Melbourne orders.

The size of the firms varies immensely. In its survey from 1966 to 1971 the Rae Committee found one large firm at one stage had assets totalling about $60 million, whereas several firms had assets of less than $20,000. In Melbourne in 1969-70 one firm received revenue of $7.5 million whereas another firm's was only $7300.

Superficially, brokers appeared to be doing well during the boom, but beneath the surface lurked hidden perils. A stockbroker, in what might be called his pure form, is an agent who buys and sells shares for clients. In agency business therefore, a broker relies on volume of trading because it is on the value of each transaction that he earns his commissions. During the boom, brokers began greatly expanding the credit advanced to their clients. They also began taking principal positions in the share market and other areas of investment. At the same time their costs were rising as they had to employ extra staff to handle the paperwork. Going more deeply into debt at a time when their costs were rising, they became more vulnerable.

After the boom brokers were given a bad reputation for having mistreated their clients. On the whole this was unfair. For every client caught by a broker's failure or who suffered loss through a broker's incompetence, there were many more who were carried by a broker until they could afford to pay or who skipped town leaving a broker with a bad debt. If a broker executed an order wrongly and the result was a loss, the client could refuse to pay. But if the result was a profit, the client would invariably accept it.

However, the failure of a number of brokers soon after the boom, particularly in Melbourne, exposed a number of slipshod practices in the industry and they were disturbingly widespread. Trust accounts were not properly kept, clients' scrip was pledged as security, speculation was rife among employees and some brokers appeared deficient in basic financial skills. Allan Green, the senior partner in the Adelaide firm of A.J. Green, Burchell & Co, when asked if he could "interpret the trading and profit-and-loss account and the balance sheet (of his own firm) when it was prepared each year", said "No".[1] Worse, the auditors of stockbroking firms repeatedly failed to detect signs of trouble. In Melbourne, W.J. North & Co was given a clean bill of health by the Melbourne Stock Exchange only weeks before it collapsed.

Bad debts expanded as clients defaulted on orders for higher-priced stocks. At first brokers were inhibited from suing by a professional sense of delicacy. As the scale of bad debts mounted in a sliding market, the attitude that it was 'not done' to sue clients changed quickly. Some interesting names began to crop up in the law columns. One was the master of the Bass Strait drillship Glomar III, who left for Honolulu neglecting to pay $4125 to Sydney broker Lionel A.MacFadyen & Co for a losing punt on Pacific Island Mines and Planet Gold.

One way a broker could avoid this problem was to insist that a client put up cash before buying and scrip before selling. Many brokers did not do this, first because it would have taken some extra time and trouble and secondly, because clients may have taken their business elsewhere. Brokers had to offset the volume of business being generated against the increased likelihood of bad debts. This was a deliberate risk because while a broker makes two percent on each transaction, he can lose much more if the client defaults. As the saying goes, it takes a lot of two percents to make 100 percent.

Enough brokers were prepared to take this risk, with the result that debts owing to brokers rose rapidly in the latter half of the sixties. Figures collected by the Rae Committee showed that the aggregate balance sheets of member firms of the six Australian capital city stock exchanges revealed total debtors of just under $30 million, out of total assets of $82 million at June 30, 1966. In 1967 the aggregate debt had risen to $48 million and in June 1968 soared to $190 million. It stayed at this level until June 1969, when the market was relatively quiet, but by June 1970 had almost doubled to $370 million. By mid-1971 the rapid decline in the market had seen the debt fall to $120 million. The $370 million debt at June 1970 had risen to 74 percent. $500 million. Debtors had accounted for 36.6 percent of brokers' assets in 1966 but by 1970 had risen to 74 percent.

Now all this debt would not have been by clients. While agency business is the bread and butter of broking, firms undertake a range of other activities. In some firms, agency business does not even account for half the profit or turnover. Apart from agency business brokers often dealt in shares as principals, thereby increasing both their potential profits and their potential risks. They also lent and borrowed money on the short-term markets. Some became very large dealers in Commonwealth bonds, commercial bills and other fixed interest securities. Many broking firms formed merchant banks and were involved in extensive transactions with them. Brokers also underwrote issues of equities and fixed-interest securities. All these activities created debts, particularly the money-market activities, and these would be included in the $370 million.

To finance this increasing business the broking firms had to expand their own debts, mainly by enlarging their bank overdrafts. Bank overdrafts of member firms of the six capital city stock exchanges mushroomed. The table below shows the overdrafts at the end of the financial years 1965-6 to 1970-1.

BANK OVERDRAFTS OF AUSTRALIAN SHAREBROKERS

June 30, 1966	$13.6 million
June 30, 1967	$17.2 million
June 30, 1968	$57.7 million
June 30, 1969	$44.3 million
June 30, 1970	$55.6 million
June 30, 1971	$25.5 million

These figures, taken from the Rae Report, would certainly understate the position at the peak of the Poseidon boom, which occurred in the middle of the 1969-70 financial year. Ord Minnett alone had a bank overdraft of around $20 million at the height of the boom and Patricks would have been even larger. On the Rae Report figures, Melbourne brokers appear to have generally carried higher overdrafts than Sydney's. This would have been caused by the fact that Melbourne's big three firms — J.B. Were & Son, A.C. Goode & Co and Potter Partners — all carry large fixed-interest portfolios. The peak overdraft figure shown in the table for June 30, 1968, was caused by the growth of Melbourne brokers' overdrafts in that year from $11.4 million at the previous balance date to $36 million. The table should also be treated with reservations because it merely shows the position at June 30 each year. In practice, overdrafts vary greatly from day to day and there would have been some effort to reduce them for balance-sheet purposes at the end of the financial year.

The boom was run mostly on bank credit. Banks provided credit to brokers so that brokers could provide credit to clients. As long as unlimited cheap money was available to fuel the boom, speculation continued until the market broke under its own weight. The Reserve Bank, if it chose, could have curtailed the boom sharply by urging that trading banks stop advances to the broking industry and the associated merchant banks. Clients would then have been forced to trade in cash. Given the chaos in share registries and settlement in those days, speculation would have been greatly dampened.

Of course, the Reserve Bank did no such thing. What did happen was that in April 1970 the bank moved the long bond rate up by a percentage point. In terms of monetary management this was a dramatic move, but attention focused so fixedly on the collapsing market that it took a little time for the rate increase and its implications to be noticed. The movement in the long bond rate was the first signal that money would be dearer throughout the seventies. The Reserve Bank's action was taken for deeper economic purposes than curbing a stock-exchange boom, but it soon had an effect of the market.

Share yields moved up, which meant share prices were correspondingly depressed, and brokers found themselves paying higher interest rates on their uncomfortably large overdrafts. In a rising market where speculators expected to double their money in a matter of days, the cost of borrowed money had been academic. In a falling market it suddenly became less academic.

As brokers tightened credit, the bad debts began to appear. Bad debts reported by Australian brokers had totalled $42,000 in 1966-7. This figure climbed to nearly $400,000 the following year and rose to $630,000 in 1968-9. By June 30, 1970, bad debts hit $5.2 million and in the following year were only slightly lower at $4.7 million.

These figures probably understate the true level of bad debts. This is because it takes some time to recognise which debts are doubtful. Also not all member firms report their bad debt provisions. Further, the back offices of many brokers had become so chaotic by the end of the boom that they could not reconcile their scrip and cash positions. Eventually, all brokers were forced to simply rule off their books because it was uneconomic to spend any more time trying to reconcile them. They bought where they appeared short and sold where they appeared long. Any client subsequently appearing with a proven claim was settled.

Some bad debts were pretty well recognised from the moment they were incurred. This happened where the investment manager of a group or small institution would start playing the market on his own account through a broking firm. One or two of these managers never had any intention of putting up cash. If they won they collected the profits and if they lost the bad debt was the price the broking firm paid for getting their group's business.

The bad debts were worse in Sydney than in Melbourne. Sydney firms wrote off $2.5 million during 1970-1, which represented six percent of total debts by clients. These bad debts included some from failed broking firms. Profits of member firms of the six exchanges had risen from just below $5 million in 1965-6 to $32 million in 1967-8 and in the next two years rose to $26.4 million and $34.7 million. In 1971 the backlog of paperwork carried over from the boom years left brokers with heavy office expenses at a time when they were being hit by falling revenues. Profits fell almost to zero. The 174 member firms in Australia between them aggregated total profits of $387,000 in 1970-1. This miniscule profit would have been turned into a loss if the figures from some failed Melbourne brokers had been included.

One byproduct of the losses was a widespread cutback in research by broking firms. Research is a highly volatile feature of broking. In fat years, the sum spent on research mounts. It carries considerable prestige and impresses institutions. In tough times it is the first department to face the axe. One Brisbane firm increased its expenditure on research from $7000 in 1965-6 to $95,000 in 1969-70 and then in the trauma of the following year slashed research to only $200. You don't get much research for $200.

Despite the bad debts and the back-office problems, agency business must have generally been the most profitable area for brokers throughout the boom and after. It is not possible to be precise on this point but from figures collated from the Rae Report it would appear that the brokers' major losses were in other areas.

Brokers had become heavy share traders on their own accounts during the boom. One medium-sized Melbourne firm turned over $35 million worth of scrip as principal between November 1969 and February 1970. If the firm was doing as much business as this on its own account it is fair to wonder how much time the partners could have been sparing for their clients' affairs. Sydney broking firms lost a total of $1.4 million on share trading as principals in both 1969 and 1970 and this loss soared to $3.6 million in 1971. At least one Sydney firm also lost heavily on bond trading in 1970-1.

These figures have to be treated with some caution. Some of the reported losses as principals would be the result of operating errors which would normally be booked to the house account.

Also, in years when profits were running high the partners might deliberately divert losses from associated companies into the broking firm to minimise tax. And under the double-deductibility laws which prevailed for most of the boom shares in oil companies could be sold at a face loss and still represent a profit after the tax relief was claimed.

In the case of firms that failed, only Donovan and Co came to grief solely through agency business. The failures of others were generally rooted in speculations on their own account, although agency business was sometimes a contributory factor. Michael Ricketson & Co, for example, collapsed in Melbourne with bad debts by several clients, but losses in the house account were also a significant factor.

John T.Martin & Co failed with a deficiency of $1.3 million — after only 13 months' operation. A major contributor to its disaster was a Rushcutters Bay client who had shorted Leopold between 48 cents and $1.06, the last selling order having been completed just before Leopold reported 5.33 percent nickel. Then Leopold shares shot to $8.80 and Martin & Co tried to cover their client's loss (if he defaulted, the firm would have to pay) by going long in the stock. Realising that the firm could suffer a fatal loss if the client defaulted, the principals of John T.Martin decided to go long in Leopold on their own account. They bought heavily until the shares were delisted. Then the stories began circulating about the salted drill core, making it obvious that the shares were worth only a fraction of the price paid at the top by the firm and its client. Leopold and Martin crashed together.

In Adelaide, A.J. Green, Burchell & Co got into trouble through speculation by employees and this was compounded by the cancellation of an underwriting agreement. The Melbourne firm of W.J. North & Co failed as a result of writing options as principal.

While mining shares are notoriously risky investment, broking houses also lost very large sums of money at the opposite end of the investment spectrum in Commonwealth bonds. Losses in this area are a function of changes in the bond rate. Let us say an investor buys a 20-year bond at par of $100 at the current (for 1969) interest rate of 6 percent. The interest payment on the bond will be $6 per year. Then the Commonwealth lifts the long bond rate to 7 percent. All bonds previously issued will have their price adjusted on market to return a yield of 7 percent. As the bond our investor bought brings in only $6 interest each year, its face value will be marked down on market to bring it in line with a 7 percent yield. The calculation is 6 divided by seven multiplied by 100, which comes to $85.71. A one percent mark-up in the interest rate has cost our investor 14.3 percent of the face value of his bond. If he bought $1 million worth of bonds he is losing $142,900.[2] The bigger broking firms were very large dealers in bonds and had to cope with exactly such an interest rate rise in late 1969-70. One large Sydney firm lost $1 million on bond trading in the two years to June 1970. It still showed a profit of about $1.2 million over that period because of large commission earnings from the booming market. Other large firms also suffered substantial bond losses.

4. The Stock Exchanges.

Since 1937 the six capital city stock exchanges together form the Australian Associated Stock Exchanges. The two main functions of the AASE have been to promote the interests of its members and to lay down the listing requirements for public companies. These requirements are enforced by the individual stock exchanges, which have the power to refuse listing to a new company or to suspend or delist an existing company which fails to comply with the listing requirements. Each company that lists in Australia has a 'home' exchange (Melbourne for BHP, Sydney for Lend Lease Corporation, Brisbane for Thiess and so on). That home exchange is responsible for ensuring that each of its companies complies with listing requirements. Companies are required to keep exchanges informed of profits, alterations to capital, board changes, takeover bids and a range of other basic information. If there is a sudden move in a company's share price, the home exchange may ask the directors if they know of any reason for it. If there is no answer, or an unsatisfactory answer, the stock may be suspended from trading.

Apart from policing the companies, each exchange also lays down a set of rules and by-laws governing the conduct of its members. These prescribe the conduct of brokers, including the procedure for share trading, the accounts they must keep and the minimum levels of operating capital they require.

Each exchange is governed by a committee elected by and from the members. The day-to-day running of the exchange is mostly in the hands of an executive staff who are answerable to this committee.

The exchanges take pride in their ability to regulate themselves. "I think we have almost attained perfection in Perth, and I believe the other exchanges are in the same situation," Geoffrey Hynam told the Rae Committee. In hindsight, the boom showed that the exchanges were still short of perfection and that some defects were structural. The most obvious defect is that the exchanges have little real power. The listing officer can query a company but he has no power to force it to reply or to check whether a reply is accurate. The exchanges have the power to suspend trading in a stock, but they are most reluctant to do this as a punishment because the people who suffer the most are the existing small shareholders in the company.

The listing officer, being an employee of the exchange, is under tacit pressure to be polite to companies with whom the bigger brokers have close connections. And while listing officers can bully small companies with some success, large companies can defy them. In 1969-70 all the trading banks and the Herald and Weekly Times Ltd refused to disclose their interim profit movements in percentage terms. This flat defiance went unpunished. The organisations concerned were not only large and an important source of revenue to brokers, in terms of shares traded, but they posed a potent threat to the existing exchanges. With the large scrip holdings of the bank nominee companies and the Australia-wide (apart from Sydney) publishing network of the Herald and Weekly Times, these organisations had two of the fundamental requisites needed to set up an alternative stock market. Eventually, in their own good time, the banks and the Herald complied with AASE requirements.

In many cases the exchanges cannot even enforce their will upon small companies. A notable example during the boom was the affair of Goldrim Mining Australia Ltd.

Goldrim was floated in March 1970 with a prospectus for a public issue of three million fully-paid 25 cent shares. In addition, two million vendor shares were allotted to the Goldfield Corporation of New York (which received 840,000), Chemalloy Minerals of Toronto (560,000) and Goldrim Mining Co of Toronto (600,000). The public float was underwritten by Melbourne broker Richard Parsons & Co and was fully subscribed. The float's purpose was to raise funds to exploit a tantalite deposit at Wodgina in the north of WA. Tantalite is the ore of tantalum, a metal used in the manufacture of electrical components, particularly capacitors for computers.

In November 1968, the chairman of Goldrim of Toronto, Dr W.A. Morgan, had exuded enthusiasm about Wodgina during a visit to Australia. He said it could be one of the world's largest tantalite mines and that a drilling program had been showing good mineralisation. He also said the Goldfields Corporation had agreed to supply funds for further development and, if merited, finance the scheme to the stage of production. "This deposit could be the biggest single tantalite mine in the world," he said.

The founding chairman of Goldrim Australia was Herman Becker-Fluegel, president of the Goldfield Corporation. In the Goldrim Australia prospectus Becker-Fluegel said: "All mining exploration ventures are essentially speculative. However, on the basis of the reports submitted to the company and exploration to date the directors believe that a sustained exploration program is definitely warranted." The prospectus also said production of tantalum from the Wodgina area had begun in 1905. "Tantalum production ceased in 1909 due to a drop in the world market price of tantalum caused by oversupply from Wodgina," the prospectus said. On all these facts and statements, investors must have been encouraged to believe Wodgina was a promising mine and would be backed by the vendors.

Goldrim shares listed in mid-March and showed a tidy profit to stags, ranging between 40 cents and 75 cents in the opening week. After the stags had quit, the shares began to edge upward. One prime reason for buying — apart from remnants of general speculative fever — would have been the exotic allure of tantalum combined with the backing of the American companies. By September, Goldrim shares had reached $1.60. By September, also, rumors began circulating that one of the American vendors was selling its shares.

The rumors were disquieting because the company had only been floated six months before and the investing public had been under the impression that the vendors' aim was to make a thorough investigation of what appeared to be a promising deposit. The vendor-shares were not eligible for listing for another six months. Russell Knowles, then secretary of the Melbourne Stock Exchange, asked the company what had been happening. On September 9 directors of Goldrim Australia told the exchange that none of the vendor shares or options had changed hands.

Knowles knew that several brokers and investment counsellors had been offered the scrip privately. He was disturbed because the fast resale of vendor shares in this fashion implied little confidence in the future of the mine. On October 1 Goldrim announced that Goldfield Corporation of New York had sold the 840,000 shares it had received in the float. Goldfield

Corporation had paid only $73,000 for its interest and had sold privately at 75 cents a share (the market was $1.40 at the time) for a profit of over $500,000.

Also on October 1, the Goldrim directors announced that the company had outlined a tantalite-bearing zone at Wodgina which was "considered large by world standards". Goldrim shares hit $1.90 briefly then began a long downward drift. The tantalite never proved economic.

The episode illustrates several difficulties faced by exchanges. The home exchange knew what was happening but was powerless to do anything except inquire. There was no legal restriction on Goldfield Corporation selling its shares. However, as they had not been admitted for listing on an Australian Stock exchange they could not have been sold in the open market. As a result of this and similar episodes the AASE introduced new restrictions on the sale of vendor scrip.

If the exchange had wished to make a gesture it could have suspended quotation of Goldrim shares, but this would have penalised local investors by denying them a market in their stock and would not necessarily have stopped the Goldfield sale. Even if Goldfield Corporation had breached listing requirements the Melbourne Stock Exchange could have taken no effective action against a New York company.

Closer to home the exchanges attempted, unsuccessfully, to improve the reporting standards of exploration companies. In January 1969, well before the Poseidon boom took off, the Sydney and Melbourne Stock Exchanges adopted broad guidelines on reporting standards expected from mining companies.

Under the guidelines, which are worth quoting at some length in view of what happened later, companies had to:

. Inform the exchanges immediately exploratory drilling located encouraging mineral shows. Such reports had to contain details of the hole location and number.

. Report assays on a weighted average basis for each drill-hole where encouraging shows had been located immediately the hole had been completed.

. Supplement the information on weighted assays by furnishing details of individual assays over limited sections encountered in the hole.

. Furnish reports of development operations at least quarterly, even if no exploration or development had occurred in the period.

The exchanges said mining companies should not report

assay values until a drill-hole had been completed. They also said companies should not wait for a drilling program involving several holes to be completed before informing the market. The exchanges feared this would leave too many opportunities for insider trading.

The guidelines were contradictory. Companies were supposed to report encouraging shows immediately but were not to report on uncompleted holes. It is impossible to define what constitutes an "encouraging" show.

Obviously, the exchanges were trying to ensure that the public was informed promptly of important finds and at the same time trying to ensure that the market would not be misled by assays from a single hole or part of a hole. It is doubtful whether there is a satisfactory solution to this dilemma. A potential mine is normally discovered and proved by a series of steps. It might begin with geological mapping, seismic surveys, magnetometer surveys and geochemical sampling, then go on to costeaning and percussion drilling. Diamond drilling is usually the last stage of exploration. The most scrupulous exploration program can easily record a number of misleading assays or readings as it progresses. A high geiger-counter reading on a radiometric anomaly, for instance, may indicate only thorium or potassium instead of uranium. Assays from outcrop mineralisation may yield high readings for base metals which have no relation to the assays of the orebody (if any) at depth. Exploratory diamond drilling in faulted structures can produce a puzzlingly wide range of readings.

In lay terms, the gradual evolution of knowledge about a mineral prospect can be compared to an unassembled jigsaw puzzle lying face down. As the pieces are turned over and linked together the picture becomes clearer, but early pieces can give a misleading impression. The true size of the mine is only known with complete accuracy when the last shovelful of ore has been extracted, by which time its value will be nil.

One time when an exchange really does have power over a company is when it applies for listing of its securities. If it is an operational company its financial records can be scrutinised. If it is a new float its prospectus can be examined. Many boom-time companies would never have attempted floating if they had not been able to gain quick listing. Exchanges were handling prospectuses at the rate of up to a dozen a week at the height of the boom. Given the weight of other pressures on the exchanges at this time, their executives could not have had the time really to investigate all these prospectuses. As long as the prospectus showed — or purported to show — such basic items as issued capital and names of directors, and its shareholdings had attained the minimum spread ordained by the listing requirements, the company was eligible to be listed subject only to the discretion of the listing committee. Very few companies were denied listing although the information in some of the prospectuses was scant. This particularly applied to the empty prospectuses produced by the investment companies who listed late in the boom. Barewa was refused listing on suspicion, not on facts.

Apart from policing listed companies the exchanges are also responsible for policing their own members. It cannot be said that this area was effectively policed in the boom. There were too many instances where firms broke exchange rules on dealing, trust funds and liquidity requirements. In the case of Ricketson, the Rae Committee found that his trust account had been inadequately managed for some considerable time without the matter being discovered by the regulatory authorities. The Rae Report said: " . . . given only average skill on the part of the investigator in understanding a broker's accounts a short, inexpensive check would have quickly revealed the weaknesses in the management of Mr Ricketson's trust accounts months before the final collapse."

Many of the irregularities went undetected. W.J. North & Co folded within weeks of having been given a clean bill of health by the Melbourne Stock Exchange. Mr Justice Street, in his judgment in the case Bonds & Securities (Trading) Pty Ltd v. Glomex Mines and others, said of John T. Martin & Co: "Its deficiencies in observance of proper and honorable dealing as a broker might well have passed unnoticed, had it not committed the cardinal sin of running out of money."

5. The Companies Offices.

The stock exchanges are policemen with severely restricted powers. The policing function also sits uneasily with the brokers' role as businessmen and entrepreneurs. The Companies Offices of the Commonwealth and the States[3] have considerably more power and no such conflicts of interest. The Companies Offices at the time of the boom had the responsibility of policing the corporate area to detect any breaches of the Companies Act, which lays down penalties for offending directors. Directors can also be tried under the fraud provisions of the Crimes Act.

The Companies Offices were poorly staffed at the time of the boom, particularly in the territories and smaller States. The vast bulk of the work of a Companies Office consisted of handling documents. The staff are occupied filing company returns, registering business names and handling a mass of other paperwork. Investigators account for only a tiny proportion of the total staff.

During the boom Victoria and NSW each had only a small number of full-time investigators. These were mainly trained accountants, backed by a few police seconded from the Fraud Squads.

A thorough investigation of a public company is a labor-intensive business. Minute books, share registers, ledger cards and bank accounts, together with other documents, have to be combed painstakingly, and witnesses have to be examined to piece together an outline of what has happened. The affairs of companies which have crashed or where fraud has been committed are invariably complicated, sometimes deliberately. Sorting out what happened may take months or even years.

With limited manpower, the Victorian and NSW investigators never got around to investigating many cases which might have warranted their attention. They tended to concentrate on what looked like the most important cases and to ignore relatively minor infractions.

The NSW Companies Office tended to try for convictions on indictable offences, which meant the case had to be heard before a magistrate before the accused were committed to a higher court for trial. Where an accused director decided to fight the case in the commital stage, the case could drag out for ages. The Victorian Companies Office tended to concentrate on summary cases heard by magistrates. They lost a few but on balance were more successful. There were a few convictions in Victoria for boom-time offences, including the jailing of Cornelius, whereas in NSW, which was the centre of boom activity, nobody has yet been convicted of a major offence relating to 1969-70

The situation was bad enough in NSW and Victoria, but it was even worse in other states. The Queensland and South Australian Companies Offices appear to have undertaken no significant investigations during the boom. If they did, little resulted from them. In WA the Companies Office used to turn investigations over to the police. The Commonwealth Government undertook no significant investigations, although a number of important companies were based in the Australian Capital Territory. Corporate criminals were aware of these weaknesses and sometimes deliberately based their company in a State with a quiescent Companies Office.

Normal Companies Office investigators have only limited powers, which are often all that is needed for an investigation. However, where a matter has attracted considerable public attention, the Attorney-General may decide it is necessary to appoint an inspector under Part VI of the Companies Act to inquire into some or all aspects of a company's affairs. The inspector in such a case is normally a barrister, although he may be aided by one or more Companies Office investigators. The advantage of a Part VI investigation is that it is very thorough and can probe deeply to discover what happened to a company. The inspector's report can be very long (Barewa's ran to 700 pages) and may form the basis for subsequent criminal proceedings.

The disadvantages of a Part VI investigation are that they take a long while to complete, often three years between the appointment of an inspector and the completion of his report. By this time the directors may have fled and the company may be ruined, with nothing left for shareholders. A successful prosecution may bring belated justice, but it will be cold consolation for those investors who lost money. The costs of the Part VI investigation are normally levied on the company concerned, eroding shareholders' funds still further.

A Part VI inspector's report is, despite its dry legal prose, a highly sophisticated detective story. The report pieces together shreds of evidence from cheques, minute books, assay reports and other documents, the details of how a company foundered or was defrauded. They are extremely valuable from a commercial and social aspect because they provide an in-depth view of how a particular set of directors or officers brought a company to grief. The reports show the nine-tenths of the corporate iceberg that is normally hidden from the public view when a disaster occurs. Often the only punishment of the men who ruined a company is the publicity which follows the tabling of these reports in parliament.

Unfortunately, they are not always made public. Sir Kenneth McCaw, while Attorney-General of NSW, held back the Endurance report while prosecutions dragged on for three years. The Barewa report's size restricted printing to the point where only one copy was available to the public.

Finally, the powers of either an inspector or a normal CAC investigator stop at Australia's shores. They have no powers beyond that of an ordinary citizen if they have to carry an investigation overseas. In several countries, especially tax havens, it is impossible to discover the identity of the people behind a private company.

One example of this occurred when the NSW Corporate Affairs Commission (as it became) was investigating share trading in Queensland Mines Ltd and Kathleen Investments Ltd in 1971. Some trading was on behalf of a Hong Kong company called Bavieca Pty Ltd. All the CAC investigators could learn was that the principals of the company were Lowe, Bingham and Mathews, a firm of Hong Kong solicitors. Later, in evidence before the Rae Committee, Sir Tristan Antico volunteered that he had been trading in the shares through Bavieca. Sir Tristan neither owned nor controlled Bavieca and no ordinary investigation would have uncovered the link.

If it comes to that, there have been difficulties when an investigation crosses State boundaries within Australia. If the Companies Office of one State was investigating a company whose affairs extended into another State during the boom, they had to seek special temporary authority, which was arranged between the Attorney-Generals of the two States.

Penny-pinching State administrations used to cut corners by asking the investigators in another State to inverview a witness or inspect records. But on receiving the replies — often weeks later — there were invariably further questions to be asked. It was a most unsatisfactory way of conducting a detailed investigation. These interstate difficulties have been partly eased by the formation of the Interstate Corporate Affairs Commission, but the international difficulties will always remain.

6. The Law and the Courts.

Company investigators were not only hampered by paucity of staff during the boom; they also struck trouble with the law they were trying to enforce. The most glaring deficiency was the lack of a securities industry Act. Each State and the Commonwealth now has such an Act which seeks to prevent insider trading, to register investment advisers and to make certain dealers and financial journalists keep share registers. There was no effective bar to insider trading during the boom and a number of transactions occurred which would now be illegal. It is still an open question whether the new laws on insider trading can be effectively enforced (in many situations, prosecution would be impossible unless one insider betrayed another), but at least legislation is now on the statute book.

One of the worst legal defeats suffered during the era was by the NSW Companies Office when it prosecuted the principals of Endurance. A total of 13 charges were laid against people connected with Endurance and Attunga. Tom Edgar, a director of A.E.Goodwin Ltd, was fined $1000 on each of two charges relating to offences under Section 124 of the Companies Act (failure to act honestly and diligently as a director). Conviction under this section also carries a five-year bar on the offender being a company director. Edgar's convictions, however were not in relation to the main activities which had been criticised in Endurance. Edgar was also committed for trial on two other offences, but no bill was filed by the Attorney-General, which meant the charges were not pressed in Sydney Quarter Sessions. Paul Murray and Maguire were committed for trial on a conspiracy charge (relating to the Metz prospect, not Attunga) but again no bill was filed. Four other charges against Murray and Maguire were either dismissed or discontinued. Two charges aginst Brudenell-Woods were dismissed. Two charges against Ms Laurie East, of aiding and abetting Edgar, were dismissed and discontinued. Score for the CAC was two out of 13.

The NSW CAC spent considerable time and effort studying whether Roy Hudson should be prosecuted for his statements about Nabarlek. Hudson's statement on September 1 was made before the Securities Industry Act became law in NSW.

Even if the Act had been in force, Hudson's statement of Nabarlek reserves would not have breached the law.

There was a possibility of prosecution under Section 124 of the Companies Act (Companies Ordinance in the ACT) which requires a director to act honestly and diligently in the discharge of his duties. This is a catch-all provision in the Act, designed to secure convictions where a director has acted improperly but has not committed a specific offence.

This raised the problem of jurisdiction. Hudson had made his statement as chairman of Queensland Mines, which was registered in the ACT, but the share dealings by Hudson had been outside the ACT. The NSW Commissioner for Corporate Affairs, Frank Ryan, reported to the NSW Attorney-General in 1975: "The advice was that, as all the relevant events and acts took place outside the ACT there was no likelihood of sustaining a successful prosecution under the ACT Companies Ordinance."

Ryan also held the opinion that Hudson would not have been convicted, even if the Securities Industry Act had been in force. The relevant provision in that Act is Section 73, which provides

"A person shall not, with respect to any securities, make any statement or disseminate any information which at the time it is made or disseminated, he knows or has reasonable grounds for knowing is false or misleading in a material particular." To constitute an offence not only must the statement be false and misleading about something material, but it must also be a statement "with respect to" securities.

Ryan reported to the Attorney-General: "Fowler's Dictionary of Modern English Usage suggests that the phrases 'with respect to' or 'in respect of' may sometimes be replaced by the word 'about'. I am inclined to think that the phrase 'with respect to' in Section 73 could be taken to mean 'about' and therefore a statement such as that by Mr Hudson concerning a company's mineral prospect is not a statement 'with respect to' the company's securities."

If this is the law, it is working in a strange way. A statement that a company's share certificates were colored pink when they were really colored green would presumably be a false and misleading statement with respect to securities, yet an overstatement of the value of the company's main asset — luring countless investors to buy the company's shares at inflated prices — would not be an offence.

Those trying to police the corporate area have also encountered obstacles in the interpretation of law in the courts. The most notable instance of this was the judgment handed down by the Chief Justice of the NSW Supreme Court, Mr Justice Taylor, in the case of the Crown versus McMahon, Nestel and others.

McMahon, Nestel and other directors of Mineral Securities were charged with having published a false statement when they announced a profit of more than $3.5 million for the 1970-1 half year. It will be remembered that this profit statement was withdrawn following advice from Mr Bainton, QC, because it contained a $7 million profit on the Robe River transfer. So 10 days after announcing its profit, Minsec retracted the statement and announced a $3.3 million loss. The Minsec directors were charged with breaches of Section 176 of the Crimes Act (publishing a false statement of profit) and of Section 73 of the Securities Industry Act.

After a long trial, Mr Justice Taylor ordered the jury to find the directors not guilty. One of his main reasons was that the Crown had not established the true profit of the Minsec group for the period. He said: "If the Crown wishes to establish that this profit statement is false then it might be expected, as they have had the whole of the documents, files, records of the group since the inspectors were appointed in February 1971, that they would have produced from an accountant or other qualified person, what they say is a true statement of the profits for this six months' period. But this they have chosen not to do. The suggestion that the jury themselves can find the falsity by examination of the reports and accompanying statements is, in my opinion, quite wrong. It is not open to the Crown to seek to prove its case in this fashion. It follows, therefore, that in my view there is no evidence fit to go to the jury to consider that this profit statement is false."

On Section 73 of the Securities Industry Act, Mr Justice Taylor agreed with Ryan's opinion. He said: "Mr Gyles in relation to the second count took a further point which I think is valid. He submitted that Section 73 of the Securities Industry Act No 35 of 1970 — which provides 'A person shall not with respect to any securities make any statement or disseminate any information which at the time it is made or disseminated'— is confined to statements and information in respect of that specific subject 'securities', which in this instance in the Act, by definition, would mean shares of a company, and that the statement here alleged to be false was not made in respect of that specific subject 'securities', but was made in relation to the trading profits of a company."

The judgment had disturbing implications. His Honor's judgment could be reduced to the proposition that a charge of announcing a false profit cannot succeed unless the prosecution can establish what the true profit was. This may seem a reasonable proposition to a layman, but in practice it is impossible. The profit of a company is the end product of a number of assumptions made when preparing the balance sheet. A variation in any one of a number of items will produce a variation in the profit.

It is important to remember that accountancy is by no means rigid in its interpretation of many items. Stocks of goods, for example, may be valued in a number of different ways — cost, net realisable value, replacement value, first-in-first-out (FIFO), last-in-first-out (LIFO).

Depreciation rates can be varied, assets can be revalued, the estimate of likely bad debts can go up or down, or a host of other assumptions can be made which are perfectly valid in accounting terms but which will produce a variation in the profit. Two scrupulously honest accountants studying the figures of a scrupulously honest company can come to widely varying opinions on its profit or loss for the period. Where there is deliberate attempt to defraud, the assumptions can be stretched a long way.

The judgment may have been correct in law but it is regrettable in practice. In practical terms, Mr Justice Taylor's judgment gives directors immunity when stating the results of their companies, even if their stated profits or losses are wildly wrong. Mr Justice Taylor's reference to Section 73 of the Securities Industry Act, combined with Ryan's opinion, makes this section nearly useless. Its only practical application would appear to be where a company made a false statement about the size of its issued capital.

Mr Justice Taylor's judgment established a precedent which was soon followed by Mr Berman, SM, in a committal hearing in which Alexander and Thomas Barton were charged with publishing a false profit statement for Brins Australia Ltd in the half-year to December 1970. Mr Berman dismissed the charges on the same grounds as laid down by Mr Justice Taylor.

A more subtle enemy than the state of the law is the state of mind sometimes displayed by judges and magistrates. The bench has an occasional tendency to be more deferential to a company director or stockbroker who has landed in the dock, although the charges may relate to defalcations involving large sums. Mr Justice Gillard, sentencing Cornelius after the Barewa case, said some of the witnesses at the trial had been no less culpable in the crime in many ways and Cornelius had otherwise been a man of good character. His Honor described Cornelius as an intelligent, ambitious and successful businessman who had brought financial ruin upon himself. Mr Justice Gillard said he had little sympathy for the victims of the conspiracy, who were speculators in the mining boom. "Most were advised by brokers, who must have seen the emptiness in the prospectus and the grave risk their clients were taking," His Honor said. "The victims were seeking the same benefit as you and equally relying on the greed and stupidity of a speculating public." In the business world Cornelius lived in, his conduct had been no different to that of many other people.

Mr Justice Gillard said he had been tempted to impose a non-custodial sentence upon Cornelius until he had learned that Cornelius and another man had plotted to steal about 400 exhibits — the evidence in the Crown case — from the offices of the Crown Law Department.

[1] Rae Report 6.4

[2] This example of bond pricing is given in terms of simple arithmetic to make it intelligible. In practice, bonds are valued on a yield to redemption. The principle remains true that a rise in bond rates means a fall in bond values.

[3] They were later restyled Corporate Affairs Commissions. As I am here discussing their role during the boom I have used their former names.

EPITAPH

The years 1969 and 1970 have gone, but there is no reason to suppose that mining booms are a thing of the past. Australia has had them before and will surely have them again. In a well-mineralised and affluent country this is only to be expected.

Since 1969-70 the listing requirements of the Australian Associated Stock Exchanges have been expanded, and the number of Corporate Affairs Commission investigators has been increased. There are stirrings of Australia's first National Companies and Securities Commission. These changes may frustrate the tactics of some promoters, but promoters have a knack of finding a way around irksome restrictions.

Investors are warier than in 1969 but they are also richer. The moral climate improved markedly during the recession of the mid-1970s but greed is one of the most powerful human emotions and ethics are a frail defence against the temptations of sudden wealth. The odds against being convicted for any malpractice are high. And the intense excitement of a full-scale boom is irresistable.

The mining boom of 1969-70 developed into a form of mass hysteria. This may be deplorable, but until we can change the fundamentals of the human psyche, people will be prone to mass hysteria occasionally and the stock exchange is probably the most harmless outlet for it.

After all,. it's only money.

GLOSSARY

Anomaly: The Oxford dictionary describes an anomaly as an irregularity or deviation from the natural order, which is precisely what it means in geological terms. A magnetic or radiometric anomaly is one giving readings out of the ordinary.
Arbitraging: Buying and selling the same stock simultaneously on different exchanges to take advantage of a price differential.
Assay: The testing of ores or minerals to determine the percentages of elements therein.
Assets: From the French assez (enough). The total wealth of a company.
Averaging Down: Buying more stock on a falling market to lower the average price per share.
Bear: Pessimist who believes market will fall.
Blue chip: First class stock.
Bull: Optimist who believes market will rise.
Call Option: Paying a premium for the right to buy a certain number of shares at current prices some time in the future.
Calls: Payments of unpaid capital on shares.
Chalcopyrite: A copper-iron sulphide, one of the most common ores of copper.
Claim: A portion of mining land held under Federal or State law.
Concentrate: A product containing the valuable metal after eliminating much of the waste material in the ore.
Core: A cylindrical rock sample recovered by a diamond drill.
Costean: Trench.
Country rock: Waste rock.
Debenture: A loan to a company at a fixed rate of interest in return for a charge over the company's assets.
Diamond drill: Rotary drill with a diamond-tipped cutting edge. Used for deep and precise evaluation of orebody.

Disseminated Ore: Ore carrying small particles of valuable metals spread fairly uniformly through waste material.
Earning Rate: Net profit as a percentage of paid capital. Often expressed in cents per share.
Equities: Shares. If you hold shares in a company you are a part-owner, which means you have equity in the company.
Float: To launch a public company by offering shares.
Flotation: A milling process by which minerals are separated.
Gearing: Ratio of borrowings to shareholders funds.
Gossan: Iron oxide deposit resulting from the oxidation and leaching of a sulphide orebody containing minerals such as pyrite.
Hive-off: Company floated by an existing company. Also called spin-off.
Interim: First half of company's financial year.
Leaching: Natural process by which ground waters dissolve minerals.
Listed stock: Company listed on stock exchange.
Listing requirement: Requirements laid down by Australian Associated Stock Exchanges for listed companies.
Long: When you buy shares you are long in them: when you sell you are short.
Market capitalisation: Total value of a company on share market prices. A company with 1 million shares selling at $5 on market has a capitalisation of $5 million.
Marriage: The matching by a broker of a buying and selling order within his office.
Massive ore: Ore containing very little waste material, as opposed to disseminated ore.
Metamorphic rocks: Rocks which have changed texture or composition through some agency such as heat or pressure.
Net tangible asset backing: Amount standing to credit of shareholders after all liabilities and intangibles have been deducted from total worth of company's assets.
No Liability: A form of company whose shareholders are not liable to pay calls on shares.
Operator: Broker's employee who executes buying and selling orders on exchange floor.
Option: The right to take up shares in a company at a set price at some future date. Also the right to acquire a mining claim.
Orebody: A solid and fairly continuous mass of rock containing valuable mineral and which is separate from the country rock around it.
Ore reserve: Tonnage of ore actually available for extraction.

Outcrop: A mineral deposit exposed on the surface.
Overburden: Worthless material lying above the orebody.
Par: Face value of a stock.
Peg: To mark out a claim or lease. The key peg is called a datum peg.
Percussion drill: Similar to pneumatic drills used on roadworks, these are used for preliminary and relatively shallow drilling of an orebody. They are cheap, but the material recovered is in the form of pulverised rock and they do not give the same accuracy of orebody thicknesses that can be ascertained from a diamond drill core.
Pitch: The inclination or dip of an orebody or vein of rock.
Pitcheblende: A uranium ore containing high percentage of uranium oxide (U308).
Premium: A 50 cent par value share selling at $1.50 on market has a premium of $1.
Price earning ratio: Also called price-earnings multiple. A stock with earnings of five cents per annum which is selling at 50 cents on market has a p/e ratio of 10.
Prospectus: Legal document issued when offering shares to public in company float.
Proxy: Written authority to substitute for a shareholder at a company meeting.
Put Options: Options to sell shares at current prices at some future date. Opposite to call options.
Seismic surveying: Measurement of speed of passage of sound waves through ground.
Stag: Subscriber to company float who sells shares on market soon after listing.
Strike: The direction or bearing of the dipping plane of ore. The strike is at right angles to the dip.
Subsidiary: Company in which more than 50 percent of shares are held by another company.
Takeover: Purchase of one company by another.
Underwriter: Underwriters guarantee to take up the shares in a float if nobody else will. They receive a commission for their risk.
Vendor shares: Shares issued in lieu of cash to somebody selling an asset to a company.
Yield: The return from investment in a security at current market prices. If a share pays a dividend of 8 cents per annum and is currently selling on market at $1, it is yielding 8 per cent.
Zone: An area or region distinct from the surrounding rock.

INDEX